Data-Driven SEO with Python

Solve SEO Challenges with Data Science Using Python

Andreas Voniatis
Foreword by Will Critchlow,
Founder and CEO, SearchPilot

Apress®

Data-Driven SEO with Python: Solve SEO Challenges with Data Science Using Python

Andreas Voniatis
Surrey, UK

ISBN-13 (pbk): 978-1-4842-9174-0
https://doi.org/10.1007/978-1-4842-9175-7

ISBN-13 (electronic): 978-1-4842-9175-7

Managing Director, Apress Media LLC: Welmoed Spahr
Acquisitions Editor: Celestin Suresh John
Development Editor: James Markham
Coordinating Editor: Mark Powers

Cover designed by eStudioCalamar

Cover image by Pawel Czerwinski on Unsplash (www.unsplash.com)

Distributed to the book trade worldwide by Apress Media, LLC, 1 New York Plaza, New York, NY 10004, U.S.A. Phone 1-800-SPRINGER, fax (201) 348-4505, e-mail orders-ny@springer-sbm.com, or visit www.springeronline.com. Apress Media, LLC is a California LLC and the sole member (owner) is Springer Science + Business Media Finance Inc (SSBM Finance Inc). SSBM Finance Inc is a **Delaware** corporation.

For information on translations, please e-mail booktranslations@springernature.com; for reprint, paperback, or audio rights, please e-mail bookpermissions@springernature.com.

Apress titles may be purchased in bulk for academic, corporate, or promotional use. eBook versions and licenses are also available for most titles. For more information, reference our Print and eBook Bulk Sales web page at http://www.apress.com/bulk-sales.

Any source code or other supplementary material referenced by the author in this book is available to readers on GitHub (https://github.com/Apress). For more detailed information, please visit http://www.apress.com/source-code.

Printed on acid-free paper

To Julia.

Table of Contents

About the Author .... **xiii**

About the Contributing Editor ....**xv**

About the Technical Reviewer .... **xvii**

Acknowledgments ....**xix**

Why I Wrote This Book ....**xxi**

Foreword ....**xxv**

Chapter 1: Introduction **1**

The Inexact (Data) Science of SEO 1

   Noisy Feedback Loop 1

   Diminishing Value of the Channel 2

   Making Ads Look More like Organic Listings.................................2

   Lack of Sample Data 2

   Things That Can't Be Measured..3

   High Costs 4

Why You Should Turn to Data Science for SEO..................................4

   SEO Is Data Rich...4

   SEO Is Automatable 5

   Data Science Is Cheap...5

Summary..5

Chapter 2: Keyword Research **7**

Data Sources..7

Google Search Console (GSC) 8

   Import, Clean, and Arrange the Data..9

   Segment by Query Type...11

Round the Position Data into Whole Numbers 12

Calculate the Segment Average and Variation 13

Compare Impression Levels to the Average 15

Explore the Data 15

Export Your High Value Keyword List 18

Activation 18

Google Trends 19

Single Keyword 19

Multiple Keywords 20

Visualizing Google Trends 23

Forecast Future Demand 24

Exploring Your Data 25

Decomposing the Trend 27

Fitting Your SARIMA Model 30

Test the Model 33

Forecast the Future 35

Clustering by Search Intent 38

Starting Point 40

Filter Data for Page 1 41

Convert Ranking URLs to a String 41

Compare SERP Distance 43

SERP Competitor Titles 57

Filter and Clean the Data for Sections Covering Only What You Sell 58

Extract Keywords from the Title Tags 60

Filter Using SERPs Data 61

Summary 62

Chapter 3: Technical **63**

Where Data Science Fits In 64

Modeling Page Authority 64

Filtering in Web Pages 66

Examine the Distribution of Authority Before Optimization 67

Calculating the New Distribution 70

Internal Link Optimization 77

By Site Level 81

By Page Authority 97

Content Type 107

Anchor Texts 111

Anchor Text Relevance 117

Core Web Vitals (CWV) 125

Summary 150

Chapter 4: Content and UX **151**

Content That Best Satisfies the User Query 152

Data Sources 152

Keyword Mapping 152

String Matching 153

Content Gap Analysis 160

Getting the Data 161

Creating the Combinations 168

Finding the Content Intersection 169

Establishing Gap 171

Content Creation: Planning Landing Page Content 174

Getting SERP Data 176

Extracting the Headings 182

Cleaning and Selecting Headings 187

Cluster Headings 191

Reflections 197

Summary 198

Chapter 5: Authority **199**

Some SEO History 199

A Little More History 200

Authority, Links, and Other 200

Examining Your Own Links 201

    Importing and Cleaning the Target Link Data 202

    Targeting Domain Authority 206

    Domain Authority Over Time 208

    Targeting Link Volumes 212

Analyzing Your Competitor's Links 216

    Data Importing and Cleaning 216

    Anatomy of a Good Link 221

    Link Quality 225

    Link Volumes 231

    Link Velocity 234

    Link Capital 235

Finding Power Networks 238

    Taking It Further 243

    Summary 244

Chapter 6: Competitors 245

And Algorithm Recovery Too! 245

Defining the Problem 245

    Outcome Metric 246

    Why Ranking? 246

    Features 246

Data Strategy 246

Data Sources 248

Explore, Clean, and Transform 249

Import Data – Both SERPs and Features 250

Start with the Keywords 252

Focus on the Competitors 254

Join the Data 268

Derive New Features 270

Single-Level Factors (SLFs) 274

Rescale Your Data 277

Near Zero Variance (NZVs) 279

Median Impute 284

One Hot Encoding (OHE) 286

Eliminate NAs 288

Modeling the SERPs 289

Evaluate the SERPs ML Model 292

The Most Predictive Drivers of Rank 293

How Much Rank a Ranking Factor Is Worth 296

The Winning Benchmark for a Ranking Factor 299

Tips to Make Your Model More Robust 299

Activation 299

    Automating This Analysis 299

    Summary 300

Chapter 7: Experiments 301

How Experiments Fit into the SEO Process 301

Generating Hypotheses 302

    Competitor Analysis 302

    Website Articles and Social Media 302

    You/Your Team's Ideas 303

    Recent Website Updates 303

    Conference Events and Industry Peers 303

    Past Experiment Failures 304

Experiment Design 304

    Zero Inflation 308

    Split A/A Analysis 311

    Determining the Sample Size 320

Running Your Experiment 327

    Ending A/B Tests Prematurely 327

    Not Basing Tests on a Hypothesis 328

    Simultaneous Changes to Both Test and Control 328

Non-QA of Test Implementation and Experiment Evaluation......................................329

    Split A/B Exploratory Analysis...332

    Inconclusive Experiment Outcomes 340

Summary...341

Chapter 8: Dashboards **343**

Data Sources...343

    Don't Plug Directly into Google Data Studio 344

    Using Data Warehouses..344

Extract, Transform, and Load (ETL)...344

    Extracting Data 345

    Transforming Data 365

    Loading Data 370

Visualization...373

    Automation 374

Summary...374

Chapter 9: Site Migration Planning **377**

Verifying Traffic and Ranking Changes 377

    Identifying the Parent and Child Nodes 379

    Separating Migration Documents...385

Finding the Closest Matching Category URL 389

    Mapping Current URLs to the New Category URLs...393

    Mapping the Remaining URLs to the Migration URL......................................395

    Importing the URLs 399

Migration Forensics 412

    Traffic Trends 413

    Segmenting URLs 423

    Time Trends and Change Point Analysis 437

    Segmented Time Trends 440

    Analysis Impact 442

Diagnostics 454

Road Map 463

Summary 467

Chapter 10: Google Updates **469**

Algo Updates 470

Dedupe 477

Domains 479

Reach Stratified 485

Rankings 493

WAVG Search Volume 495

Visibility 496

Result Types 504

Cannibalization 512

Keywords 520

Token Length 520

Token Length Deep Dive 525

Target Level 533

Keywords 533

Pages 537

Segments 544

Top Competitors 544

Visibility 550

Snippets 557

Summary 561

Chapter 11: The Future of SEO **563**

Aggregation 563

Distributions 564

String Matching 564

Clustering 565

Machine Learning (ML) Modeling... 565

Set Theory 566

What Computers Can and Can't Do... 566

For the SEO Experts 566

Summary.. 567

Index.. 569

About the Author

 **Andreas Voniatis** is the founder of Artios and a SEO consultant with over 20 year's experience working with ad agencies (PHD, Havas, Universal Mcann, Mindshare and iProspect), and brands (Amazon EU, Lyst, Trivago, GameSys). Andreas founded Artios in 2015 – to apply an advanced mathematical approach and cloud AI/Machine Learning to SEO.

With a background in SEO, data science and cloud engineering, Andreas has helped companies gain an edge through data science and automation. His work has been featured in publications worldwide including The Independent, PR Week, Search Engine Watch, Search Engine Journal and Search Engine Land.

Andreas is a qualified accountant, holds a degree in Economics from Leeds University and has specialised in SEO science for over a decade. Through his firm Artios, Andreas helps grow startups providing ROI guarantees and trains enterprise SEO teams on data driven SEO.

About the Contributing Editor

Simon Dance is the Chief Commercial Officer at Lyst.com, a fashion shopping platform serving over 200M users a year; an angel investor; and an experienced SEO having spent a 15-year career working in senior leadership positions including Head of SEO for Amazon's UK and European marketplaces and senior SEO roles at large-scale marketplaces in the flights and vacation rental space as well as consulting venture-backed companies including Depop, Carwow, and HealthUnlocked. Simon has worn multiple hats over his career from building links, manually auditing vast backlink profiles, carrying our comprehensive bodies of keyword research, and writing technical audit documents spanning hundreds of pages to building, mentoring, and leading teams who have unlocked significant improvements in SEO performance, generating hundreds of millions of dollars of incremental revenue. Simon met Andreas in 2015 when he had just built a rudimentary set of Python scripts designed to vastly increase the scale, speed, and accuracy of carrying out detailed keyword research and classification. They have worked together almost ever since.

About the Technical Reviewer

 **Joos Korstanje** is a data scientist with over five years of industry experience in developing machine learning tools. He has a double MSc in Applied Data Science and in Environmental Science and has extensive experience working with geodata use cases. He has worked at a number of large companies in the Netherlands and France, developing machine learning for a variety of tools.

Acknowledgments

It's my first book and it wouldn't have been possible without the help of a few people. I'd like to thank Simon Dance, my contributing editor, who has asked questions and made suggested edits using his experience as an SEO expert and commercial director. I'd also like to thank all of the people at Springer Nature and Apress for their help and support. Wendy for helping me navigate the commercial seas of getting published. Will Critchlow for providing the foreword to this book. All of my colleagues, clients, and industry peers including SEOs, data scientists, and cloud engineers that I have had the pleasure of working with. Finally, my family, Petra and Julia.

Why I Wrote This Book

Since 2003, when I first got into SEO (by accident), much has changed in the practice of SEO. The ingredients were lesser known even though much of the focus was on getting backlinks, be they reciprocal, one-way links or from private networks (which are still being used in the gaming space). Other ingredients include transitioning to becoming a recognized brand, producing high-quality content which is valuable to users, a delightful user experience, producing and organizing content by search intent, and, for now and tomorrow, optimizing the search journey.

Many of the ingredients are now well known and are more complicated with the advent of mobile, social media, and voice and the increasing sophistication of search engines.

Now more than ever, the devil is in the details. There is more data being generated than ever before from ever more third-party data sources and tools. Spreadsheets alone won't hack it. You need a sharper blade, and data science (combined with your SEO knowledge) is your best friend.

I created this book for you, to make your SEO data driven and therefore the best it can be.

And why now in 2023? Because COVID-19 happened which gave me time to think about how I could add value to the world and in particular the niche world of SEO.

Even more presciently, there are lots of conversations on Twitter and LinkedIn about SEOs and the use of Python in SEO. So we felt the timing is right as the SEO industry has the appetite and we have knowledge to share.

I wish you the very best in your new adventure as a data-driven SEO specialist!

Who This Book Is For

We wrote this book to help you get ahead in your career as an SEO specialist. Whether you work in-house for a brand, an advertising agency, a consultant, or someone else (please write to us and introduce yourself!), this book will help you see SEO from a different angle and probably in a whole new way. Our goals for you are as follows:

- *A data science mindset to solving SEO challenges*: You'll start thinking about the outcome metrics, the data sources, the data structures to feed data into the model, and the models required to help you solve the problem or at the very least remove some of the disinformation surrounding the SEO challenge, all of which will take you several steps nearer to producing great SEO recommendations and ideas for split testing.

- *A greater insight into search engines*: You'll also have a greater appreciation for search engines like Google and a more contextual understanding of how they are likely to rank websites. After all, search engines are computer algorithms, not people, and so building your own algorithms and models to solve SEO challenges will give you some insight into how a search engine may reward or not reward certain features of a website and its content.

- *Code to get going*: The best way to learn naturally is by doing. While there are many courses in SEO, the most committed students of SEO will build their own websites and test SEO ideas and practices. Data science for SEO is no different if you want to make your SEO data driven. So, you'll be provided with starter scripts in Python to try your own hand in clustering pages and content, analyzing ranking factors. There will be code for most things but not for everything, as not everything has been coded for (yet). The code is there to get you started and can always be improved upon.

- *Familiarity with Python*: Python is the mainstay of data science in industry, even though R is still widely used. Python is free (open source) and is highly popular with the SEO community, data scientists, and the academic community alike. In fact, R and Python are quite similar in syntax and structure. Python is easy to use, read, and learn. To be clear, in no way do we suggest or advocate one language is better than the other, it's purely down to user preference and convenience.

Beyond the Scope

While this book promises and delivers on making your SEO data driven, there are a number of things that are better covered by other books out there, such as

- *How to become an SEO specialist*: What this book won't cover is how to become an SEO expert although you'll certainly come away with a lot of knowledge on how to be a better SEO specialist. There are some fundamentals that are beyond the scope of this book.

For example, we don't get into how a search engine works, what a content management system is, how it works, and how to read and code HTML and CSS. We also don't expose all of the ranking factors that a search engine might use to rank websites or how to perform a site relaunch or site migration.

This book assumes you have a rudimentary knowledge of how SEO works and what SEO is. We will give a data-driven view of the many aspects of SEO, and that is to reframe the SEO challenge from a data science perspective so that you have a useful construct to begin with.

- *How to become a data scientist*: This book will certainly expose the data science techniques to solve SEO challenges. What it won't do is teach you to become a data scientist or teach you how to program in the Python computing language.

To become a data science professional requires a knowledge of maths (linear algebra, probability, and statistics) in addition to programming. A true data scientist not only knows the theory and underpinnings of the maths and the software engineering to obtain and transform the data, they also know how and when to deploy certain models, the pros and cons of each (say Random Forest vs. AdaBoost), and how to rebuild each model from scratch. While we won't teach you how to become a fully fledged data scientist, you'll understand the intuition behind the models and how a data scientist would approach an SEO challenge.

There is no one answer of course; however, the answers we provide are based on experience and will be the best answer we believe at the time of writing. So you'll certainly be a data-driven SEO specialist, and if you take the trouble to learn data science properly, then you're well on your way to becoming an SEO scientist.

How This Book Works

Each chapter covers major work streams of SEO which will be familiar to you:

1. Keyword research

2. Technical

3. Content and UX

4. Authority

5. Competitor analysis

6. Experiments

7. Dashboards

8. Migration planning and postmortems

9. Google updates

10. Future of SEO

Under each chapter, we will define as appropriate

- SEO challenge(s) from a data perspective

- Data sources

- Data structures

- Models

- Model output evaluation

- Activation suggestions

I've tried to apply data science to as many SEO processes as possible in the areas identified earlier. Naturally, there will be some areas that could be applied that have not. However, technology is changing, and Google is already releasing updates to combat AI-written content. So I'd imagine in the very near future, more and more areas of SEO will be subject to data science.

Foreword

The data we have access to as SEOs has changed a lot during my 17 years in the industry. Although we lost analytics-level keyword data, and Yahoo! Site Explorer, we gained a wealth of opportunity in big data, proprietary metrics, and even some from the horse's mouth in Google Search Console.

You don't have to be able to code to be an effective SEO. But there is a certain kind of approach and a certain kind of mindset that benefits from wrangling data in all its forms. If that's how you prefer to work, you will very quickly hit the limits of spreadsheets and text editors. When you do, you'll do well to turn to more powerful tools to help you scale what you're capable of, get things done that you wouldn't even have been able to do without a computer helping, and speed up every step of the process.

There are a lot of programming languages, and a lot of ways of learning them. Some people will tell you there is only one right way. I'm not one of those people, but my personal first choice has been Python for years now. I liked it initially for its relative simplicity and ease of getting started, and very quickly fell for the magic of being able to import phenomenal power written by others with a single line of code. As I got to know the language more deeply and began to get some sense of the "pythonic" way of doing things, I came to appreciate the brevity and clarity of the language. I am no expert, and I'm certainly not a professional software engineer, but I hope that makes me a powerful advocate for the approach outlined in this book - because I have been the target market.

When I was at university, I studied neural networks among many other things. At the time, they were fairly abstract concepts in operations research. At that point in the late 90s, there wasn't the readily available computing power plus huge data sets needed to realise the machine learning capabilities hidden in those nodes, edges, and statistical relationships. I've remained fascinated by what is possible and with the help of magical import statements and remarkably mature frameworks, I have even been able to build and train my own neural networks in Python. As a stats geek, I love that it's all stats under the hood, but at the same time, I appreciate the beauty in a computer being able to do something a person can't.

A couple of years after university, I founded the SEO agency Distilled with my co-founder Duncan Morris, and one of the things that we encouraged among our SEO consultants was taking advantage of the data and tools at their disposal. This led to fun innovation - both decentralised, in individual consultants building scripts and notebooks to help them scale their work, do it faster, or be more effective, and centrally in our R&D team.

That R&D team would be the group who built the platform that would become SearchPilot and launched the latest stage of my career where we are very much leading the charge for data aware decisions in SEO. We are building the enterprise SEO A/B testing platform to help the world's largest websites prove the value of their on-site SEO initiatives. All of this uses similar techniques to those outlined in the pages that follow to decide how to implement tests, to consume data from a variety of APIs, and to analyse their results with neural networks.

I believe that as Google implements more and more of their own machine learning into their ranking algorithms, that SEO becomes fundamentally harder as the system becomes harder to predict, and has a greater variance across sites, keywords, and topics. It's for this reason that I am investing so much time, energy, and the next phase of my career into our corner of data driven SEO. I hope that this book can set a whole new cohort of SEOs on a similar path.

I first met Andreas over a decade ago in London. I've seen some of the things he has been able to build over the years, and I'm sure he is going to be an incredible guide through the intricacies of wrangling data to your benefit in the world of SEO. Happy coding!

Will Critchlow, CEO, SearchPilot

September 2022

CHAPTER 1

Introduction

Before the Google Search Essentials (formerly Webmaster Guidelines), there was an unspoken contract between SEOs and search engines which promised traffic in return for helping search engines extract and index website content. This chapter introduces you to the challenges of applying data science to SEO and why you should use data.

The Inexact (Data) Science of SEO

There are many trends that motivate the application of data science to SEO; however, before we get into that, why isn't there a rush of data scientists to the industry door of SEO? Why are they going into areas of paid search, programmatic advertising, and audience planning instead?

Here's why:

- Noisy feedback loop
- Diminishing value of the channel
- Making ads look more like organic listings
- Lack of sample data
- Things that can't be measured
- High costs

Noisy Feedback Loop

Unlike paid search campaigns where changes can be live after 15 mins, the changes that affect SEO, be it on a website or indeed offsite, can take anywhere between an hour and many weeks for Google and other search engines to take note of and process the change

© Andreas Voniatis 2023
A. Voniatis, *Data-Driven SEO with Python*, https://doi.org/10.1007/978-1-4842-9175-7_1

within their systems before it gets reflected in the search engine results (which may or may not result in a change of ranking position).

Because of this variable and unpredictable time lag, this makes it rather difficult to undertake cause and effect analysis to learn from SEO experiments.

Diminishing Value of the Channel

The diminishing value of the channel will probably put off any decision by a data scientist to move into SEO when weighing up the options between computational advertising, financial securities, and other industries. SEO is likely to fall by the wayside as Google and others do as much as possible to reduce the value of organic traffic in favor of paid advertising.

Making Ads Look More like Organic Listings

Google is increasing the amount of ads shown before displaying the organic results, which diminishes the return of SEO (and therefore the appeal) to businesses. Google is also monetizing organic results such as Google Jobs, Flights, Credit Cards, and Shopping, which displaces the organic search results away from the top.

Lack of Sample Data

It's the lack of data points that makes data-driven SEO analysis more challenging. How many times has an SEO run a technical audit and taken this as a reflection of the SEO reality? How do we know this website didn't have an off moment during that particular audit?

Thank goodness, the industry-leading rank measurement tools are recording rankings on a daily basis. So why aren't SEO teams auditing on a more regular basis?

Many SEO teams are not set up to take multiple measurements because most do not have the infrastructure to do so, be it because they

- Don't understand the value of multiple measurements for data science

- Don't have the resources or don't have the infrastructure

- Rely on knowing when the website changes before having to run another audit (albeit tools like ContentKing have automated the process)

To have a dataset that has a true representation of the SEO reality, it must have multiple audit measurements which allow for statistics such as average and standard deviations per day of

- Server status codes
- Duplicate content
- Missing titles

With this type of data, data scientists are able to do meaningful SEO science work and track these to rankings and UX outcomes.

Things That Can't Be Measured

Even with the best will to collect the data, not everything worth measuring can be measured. Although this is likely to be true of all marketing channels, not just SEO, it's not the greatest reason for data scientists not to move into SEO. If anything, I'd argue the opposite in the sense that many things in SEO are measurable and that SEO is data rich.

There are things we would like to measure such as

- *Search query*: Google, for some time, has been hiding the search query detail of organic traffic, of which the keyword detail in Google Analytics is shown as "Not Provided." Naturally, this would be useful as there are many keywords to one URL relationship, so getting the breakdown would be crucial for attribution modeling outcomes, such as leads, orders, and revenue.

- *Search volume*: Google Ads does not fully disclose search volume per search query. The search volume data for long tail phrases provided by Ads is reallocated to broader matches because it's profitable for Google to encourage users to bid on these terms as there are more bidders in the auction. Google Search Console (GSC) is a good substitute, but is first-party data and is highly dependent on your site's presence for your hypothesis keyword.

- *Segment*: This would tell us who is searching, not just the keyword, which of course would in most cases vastly affect the outcomes of any machine-learned SEO analysis because a millionaire searching for "mens jeans" would expect different results to another user of more modest means. After all, Google is serving personalized results. Not knowing the segment simply adds noise to any SERPs model or otherwise.

High Costs

Can you imagine running a large enterprise crawling technology like Botify daily? Most brands run a crawl once a month because it's cost prohibitive, and not just on your site. To get a complete dataset, you'd need to run it on your competitors, and that's only one type of SEO data.

Cost won't matter as much to the ad agency data scientist, but it will affect whether they will get access to the data because the agency may decide the budget isn't worthwhile.

Why You Should Turn to Data Science for SEO

There are many reasons to turn to data science to make your SEO campaigns and operations data driven.

SEO Is Data Rich

We don't have the data to measure everything, including Google's user response data to the websites listed in the Search Engine Results Pages (SERPs), which would be the ultimate outcome data. What we do have is first-party (your/your company's data like Google/Adobe Analytics) and third-party (think rank checking tools, cloud auditing software) export data.

We also have the open source data science tools which are free to make sense of this data. There are also many free highly credible sources online that are willing to teach you how to use these tools to make sense of the ever-increasing deluge of SEO data.

SEO Is Automatable

At least in certain aspects. We're not saying that robots will take over your career. And yet, we believe there is a case that some aspects of your job as an SEO a computer can do instead. After all, computers are extremely good at doing repetitive tasks, they don't get tired nor bored, can "see" beyond three dimensions, and only live on electricity.

Andreas has taken over teams where certain members spent time constantly copying and pasting information from one document to another (the agency and individual will remain unnamed to spare their blushes).

Doing repetitive work that can be easily done by a computer is not value adding, emotionally engaging, nor good for your mental health. The point is we as humans are at our best when we're thinking and synthesizing information about a client's SEO; that's when our best work gets done.

Data Science Is Cheap

We also have the open source data science tools (R, Python) which are free to make sense of this data. There are also many free highly credible sources online that are willing to teach you how to use these tools to make sense of the ever-increasing deluge of SEO data.

Also, if there is too much data, cloud computing services such as Amazon Web Services (AWS) and Google Cloud Platform (GCP) are also rentable by the hour.

Summary

This brief introductory chapter has covered the following:

- The inexact science of SEO

- Why you should turn to data science for SEO

CHAPTER 2

Keyword Research

Behind every query a user enters within a search engine is a word or series of words. For instance, a user may be looking for a "hotel" or perhaps a "hotel in New York City." In search engine optimization (SEO), keywords are invariably the target, providing a helpful way of understanding demand for said queries and helping to more effectively understand various ways that users search for products, services, organizations, and, ultimately, answers.

As well as SEO starting from keywords, it also tends to end with the keyword as an SEO campaign may be evaluated on the value of the keyword's contribution. Even if this information is hidden from us by Google, attempts have been made by a number of SEO tools to infer the keyword used by users to reach a website.

In this chapter, we will give you data-driven methods for finding valuable keywords for your website (to enable you to have a much richer understanding of user demand).

It's also worth noting that given keyword rank tracking comes at a cost (usually charged per keyword tracked or capped at a total number of keywords), it makes sense to know which keywords are worth the tracking cost.

Data Sources

There are a number of data sources when it comes to keyword research, which we'll list as follows:

- **Google Search Console**
- **Competitor Analytics**
- **SERPs**
- **Google Trends**
- Google Ads
- Google Suggest

© Andreas Voniatis 2023
A. Voniatis, *Data-Driven SEO with Python*, https://doi.org/10.1007/978-1-4842-9175-7_2

We'll cover the ones highlighted in bold as they are not only the more informative of the data sources, they also scale as data science methods go. Google Ads data would only be so appealing if it were based on actual impression data.

We will also show you how to make forecasts of keyword data both in terms of the amount of impressions you get if you achieve a ranking on page 1 (within positions 1 to 10) and what this impact would be over a six-month horizon.

Armed with a detailed understanding of *how* customers search, you're in a much stronger position to benchmark where you index vs. this demand (in order to understand the available opportunity you can lean into), as well as be much more customer focused when orienting your website and SEO activity to target that demand.

Let's get started.

Google Search Console (GSC)

Google Search Console (GSC) is a (free) first-party data source, which is rich in market intelligence. It's no wonder Google does everything possible to make it difficult to parse, let alone obfuscate, the data when attempting to query the API at date and keyword levels.

GSC data is my first port of call when it comes to keyword research because the numbers are consistent, and unlike third-party numbers, you'll get data which isn't based on a generic click through a rate mapped to ranking.[1]

The overall strategy is to look for search queries that have impressions that are significantly above the average for their ranking position. Why impressions? Because impressions are more plentiful and they represent the opportunity, whereas clicks tend to come "after the fact," that is, they are the outcome of the opportunity.

What is significant? This could be any search query with impression levels more than two standard deviations (sigmas) above the mean (average), for example.

---

[1] In 2006, AOL shared click-through rate data based upon over 35 million search queries, and since then it has inspired numerous models to try and estimate the click-through rate (CTR) by search engine ranking position. That is, for every 100 people searching for "hotels in New York," 30% (for example) click on the position 1 ranking, with just 16% clicking on position 2 (hence the importance of achieving the top ranked position, in order to, effectively, double your traffic (for that keyword))

There is no hard and fast rule. Two sigmas simply mean that there's a less than 5% chance that the search query is actually less like the average search query, so a lower significance threshold like one sigma could easily suffice.

Import, Clean, and Arrange the Data

```
import pandas as pd
import numpy as np
import glob
import os
```

The data are several exports from Google Search Console (GSC) of the top 1000 rows based on a number of filters. The API could be used, and some code is provided in Chapter 10 showing how to do so.

For now, we're reading multiple GSC export files stored in a local folder.

Set the path to read the files:

```
data_dir = os.path.join('data', 'csvs')
gsc_csvs = glob.glob(data_dir + "/*.csv")
```

Initialize an empty list that will store the data being read in:

```
gsc_li = []
```

The for loop iterates through each export file and takes the filename as the modifier used to filter the results and then appends it to the preceding list:

```
for cf in gsc_csvs:
    df = pd.read_csv(cf, index_col=None, header=0)
    df['modifier'] = os.path.basename(cf)
    df.modifier = df.modifier.str.replace('_queries.csv', '')
    gsc_li.append(df)
```

Once the list is populated with the export data, it's combined into a single dataframe:

```
gsc_raw_df = pd.DataFrame()
gsc_raw_df = pd.concat(gsc_li, axis=0, ignore_index=True)
```

The columns are formatted to be more data-friendly:

```
gsc_raw_df.columns = gsc_raw_df.columns.str.strip().str.lower().str.
replace(' ', '_').str.replace('(', '').str.replace(')', '')
```

```
gsc_raw_df.head()
```

This produces the following:

	top_queries	clicks	impressions	ctr	position	modifier
0	ps4 cd keys	17	206	8.25%	13.40	cdkeys
1	cheap ps4 cd keys	13	40	32.5%	9.38	cdkeys
2	ps4 cd key	12	34	35.29%	19.44	cdkeys
3	cheap cd keys ps4	11	21	52.38%	8.71	cdkeys
4	xbox cd keys	8	89	8.99%	13.46	cdkeys

With the data imported, we'll want to format the column values to be capable of being summarized. For example, we'll remove the percent signs in the ctr column and convert it to a numeric format:

```
gsc_clean_ctr_df['ctr'] = gsc_clean_ctr_df['ctr'].str.replace('%', '')
gsc_clean_ctr_df['ctr'] = pd.to_numeric(gsc_clean_ctr_df['ctr'])
```

GSC data contains a funny character "<" in the impressions and clicks columns for values less than 10; our job is to clean this up by removing them and then arranging impressions in descending order. In Python, this would look like

```
gsc_clean_ctr_df['impressions'] = gsc_clean_ctr_df.impressions.str.
replace('<', '')
pd.to_numeric(gsc_import_df.impressions)
```

We'll also deduplicate the top_queries column:

```
gsc_dedupe_df = gsc_clean_ctr_df.drop_duplicates(subset='top_queries',
keep="first")
```

Segment by Query Type

The next step is to segment the queries by type. The reason for this is that we want to compare the impression volumes within a segment as opposed to the overall website.

This makes numbers more meaningful in terms of highlighting opportunities within segments. Otherwise, if we compared impressions to the website average, then we may miss out on valuable search query opportunities.

The approach we're using in Python is to categorize based on modifier strings found in the query column:

```
retail_vex = ['cdkeys', 'argos', 'smyth', 'amazon', 'cyberpunk', 'GAME']
platform_vex = ['ps5', 'xbox', 'playstation', 'switch', 'ps4', 'nintendo']
title_vex = ['blackops', 'pokemon', 'minecraft', 'mario',
'outriders','fifa', 'animalcrossing', 'resident', 'spiderman',
'newhorizons', 'callofduty']
network_vex = ['ee', 'o2', 'vodafone','carphone']

gsc_segment_strdetect = gsc_dedupe_df[['query', 'clicks', 'impressions',
'ctr', 'position']]
```

Create a list of our conditions:

```
query_conds = [
    gsc_segment_strdetect['query'].str.contains('|'.join(retail_vex)),
    gsc_segment_strdetect['query'].str.contains('|'.join(platform_vex)),
    gsc_segment_strdetect['query'].str.contains('|'.join(title_vex)),
    gsc_segment_strdetect['query'].str.contains('|'.join(network_vex))
]
```

Create a list of the values we want to assign for each condition:

```
segment_values = ['Retailer', 'Console', 'Title', 'Network'] #, 'Title',
'Accessories', 'Network', 'Top1000', 'Broadband']
```

Create a new column and use np.select to assign values to it using our lists as arguments:

```
gsc_segment_strdetect['segment'] = np.select(query_conds, segment_values)

gsc_segment_strdetect
```

Here is the output:

	query	clicks	impressions	ctr	position	segment
0	ps4 cd keys	17	206	8.25	13.40	Console
1	cheap ps4 cd keys	13	40	32.5	9.38	Console
2	ps4 cd key	12	34	35.29	19.44	Console
3	cheap cd keys ps4	11	21	52.38	8.71	Console
4	xbox cd keys	8	89	8.99	13.46	Console
...	...	...	...	...	...	...
18567	nintendo switch lite on credit	48	94	51.06	2.73	Console
18568	nintendo switch limited edition	47	3430	1.37	13.27	Console
18569	the cheapest nintendo switch	47	1182	3.98	5.16	Console
18570	nintendo switch lite grey bundle	47	721	6.52	8.03	Console
18571	where to buy a nintendo switch uk	47	620	7.58	11.03	Console

15867 rows × 6 columns

Round the Position Data into Whole Numbers

Given the position column is a floating number (i.e., contains decimals), the reason we'd like to do this is because we'll be calculating the impression statistics per rounded ranking position. This will give us 100 statistics. Now imagine if we didn't round it, we could have impression statistics for 10,000 ranking positions and not all of them are useful.

```
gsc_segment_strdetect['rank_bracket'] = gsc_segment_strdetect.position.
round(0)
gsc_segment_strdetect
```

This results in the following:

	query	clicks	impressions	ctr	position	segment	rank_bracket
0	ps4 cd keys	17	206	8.25	13.40	Console	13.0
1	cheap ps4 cd keys	13	40	32.5	9.38	Console	9.0
2	ps4 cd key	12	34	35.29	19.44	Console	19.0
3	cheap cd keys ps4	11	21	52.38	8.71	Console	9.0
4	xbox cd keys	8	89	8.99	13.46	Console	13.0
...	...	...	...	...	...	...	...
18567	nintendo switch lite on credit	48	94	51.06	2.73	Console	3.0
18568	nintendo switch limited edition	47	3430	1.37	13.27	Console	13.0
18569	the cheapest nintendo switch	47	1182	3.98	5.16	Console	5.0
18570	nintendo switch lite grey bundle	47	721	6.52	8.03	Console	8.0
18571	where to buy a nintendo switch uk	47	620	7.58	11.03	Console	11.0

Calculate the Segment Average and Variation

Now the data is segmented, we compute the average impressions and the lower and upper percentiles of impressions for the ranking position. The aim is to identify queries that have impressions two standard deviations or more above the ranking position. This means the query is likely to be a great opportunity for SEO and well worth monitoring.

The reason we're doing it this way, as opposed to just selecting high impression keywords per se, is because many keyword queries have high impressions just by virtue of being in the top 20 in the first place. This would make the number of queries to track rather large and expensive.

```
queries_rank_imps = gsc_segment_strdetect[['rank_bracket', 'impressions']]
group_by_rank_bracket = queries_rank_imps.groupby(['rank_bracket'], as_
index=False)

def imp_aggregator(col):
    d = {}
    d['avg_imps'] = col['impressions'].mean()
    d['imps_median'] = col['impressions'].quantile(0.5)
    d['imps_lq'] = col['impressions'].quantile(0.25)
    d['imps_uq'] = col['impressions'].quantile(0.95)
    d['n_count'] = col['impressions'].count()
```

```
return pd.Series(d, index=['avg_imps', 'imps_median', 'imps_lq', 'imps_
uq', 'n_count'])
```

```
overall_rankimps_agg = group_by_rank_bracket.apply(imp_aggregator)
overall_rankimps_agg
```

This results in the following:

	rank_bracket	avg_imps	imps_median	imps_lq	imps_uq	n_count
0	1.0	784.795848	132.0	43.25	2970.35	578.0
1	2.0	991.002639	153.0	24.00	2930.20	1137.0
2	3.0	1816.848187	159.5	35.00	6628.15	1324.0
3	4.0	2234.595041	151.0	22.00	8387.55	1452.0
4	5.0	2529.486692	153.0	22.00	9174.60	1315.0
...	...	...	...	...	...	...
97	98.0	36.000000	36.0	20.50	63.90	2.0
98	99.0	5.666667	6.0	5.00	6.90	3.0
99	105.0	1.000000	1.0	1.00	1.00	1.0
100	108.0	1.000000	1.0	1.00	1.00	1.0
101	110.0	4.000000	4.0	4.00	4.00	1.0

In this case, we went with the 25th and 95th percentiles. The lower percentile number doesn't matter as much as we're far more interested in finding queries with averages beyond the 95th percentile. If we can do that, we have a juicy keyword. Quick note, in data science, a percentile is known as a "quantile."

Could we make a table for each and every segment? For example, show the statistics for impressions by ranking position by section. Yes, of course, you could, and in theory, it would provide a more contextual analysis of queries performed vs. their segment average. The deciding factor on whether to do so or not depends on how many data points (i.e., ranked queries) you have for each rank bracket to make it worthwhile (i.e., statistically robust). You'd want at least 30 data points in each to go that far.

Compare Impression Levels to the Average

Okay, now let's left join (think vlookup or index match) the table from the previous set and then join it to the segmented data. Then we have a dataframe that shows the query data vs. the expected average and upper quantile.

Join accessories_rankimps_agg onto accessory_queries by rank_bracket:

```
query_quantile_stats = gsc_segment_strdetect.merge(overall_rankimps_agg, on
=['rank_bracket'], how='left')
query_quantile_stats
```

This results in the following:

	query	clicks	impressions	ctr	position	segment	rank_bracket	avg_imps	imps_median	imps_lq	imps_uq	n_count
0	ps4 cd keys	17	206	8.25	13.40	Retailer	13.0	5619.412587	50.0	6.00	12414.50	286.0
1	cheap ps4 cd keys	13	40	32.5	9.38	Retailer	9.0	2823.633851	72.0	8.25	12480.15	1158.0
2	ps4 cd key	12	34	35.29	19.44	Retailer	19.0	1385.688889	9.0	1.00	8170.05	90.0
3	cheap cd keys ps4	11	21	52.38	8.71	Retailer	9.0	2823.633851	72.0	8.25	12480.15	1158.0
4	xbox cd keys	8	89	8.99	13.46	Retailer	13.0	5619.412587	50.0	6.00	12414.50	286.0
...	...	...	...	...	...	...	...	...	...	...	...	...
15862	nintendo switch lite on credit	48	94	51.06	2.73	Console	3.0	1816.848187	159.5	35.00	6628.15	1324.0
15863	nintendo switch limited edition	47	3430	1.37	13.27	Console	13.0	5619.412587	50.0	6.00	12414.50	286.0
15864	the cheapest nintendo switch	47	1182	3.98	5.16	Console	5.0	2529.486692	153.0	22.00	9174.60	1315.0
15865	nintendo switch lite grey bundle	47	721	6.52	8.03	Console	8.0	2164.032761	59.0	8.00	10308.25	1282.0
15866	where to buy a nintendo switch uk	47	620	7.58	11.03	Console	11.0	4645.429894	22.0	5.00	13906.00	756.0

Explore the Data

Now you might be wondering, how many keywords are punching above and below their weight (i.e., above and below their quantile limits relative to their ranking position) and what are those keywords?

Get the number of keywords with high volumes of impressions:

```
query_stats_uq = query_quantile_stats.loc[query_quantile_stats.impressions
> query_quantile_stats.imps_uq]
query_stats_uq['query'].count()
```

This results in the following:

8390

Get the number of keywords with impressions and ranking beyond page 1:

```
query_stats_uq_p2b = query_quantile_stats.loc[(query_quantile_stats.
impressions > query_quantile_stats.imps_uq) & (query_quantile_stats.rank_
bracket > 10)]
query_stats_uq_p2b['query'].count()
```

This results in the following:

```
2510
```

Depending on your resources, you may wish to track all 8390 keywords or just the 2510. Let's see how the distribution of impressions looks visually across the range of ranking positions:

```
import seaborn as sns
import matplotlib.pyplot as plt
from pylab import savefig
```

Set the plot size:

```
sns.set(rc={'figure.figsize':(15, 6)})
```

Plot impressions vs. rank_bracket:

```
imprank_plt = sns.relplot(x = "rank_bracket", y = "impressions",
                hue = "quantiled", style = "quantiled",
                kind = "line", data = overall_rankimps_agg_long)
```

Save Figure 2-1 to a file for your PowerPoint deck or others:

```
imprank_plt.savefig("images/imprank_plt.png")
```

What's interesting is the upper quantile impression keywords are not all in the top 10, but many are on pages 2, 4, and 6 of the SERP results (Figure 2-1). This indicates that the site is either targeting the high-volume keywords but not doing a good job of achieving a high ranking position or not targeting these high-volume phrases.

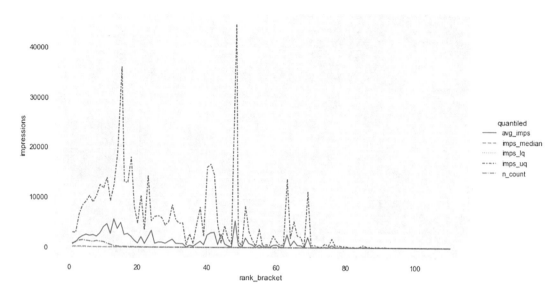

Figure 2-1. *Line chart showing GSC impressions per ranking position bracket for each distribution quantile*

Let's break this segment down.

Plot impressions vs. rank_bracket by segment:

```
imprank_seg = sns.relplot(x="rank_bracket", y="impressions",
            hue="quantiled", col="segment",
            kind="line", data = overall_rankimps_agg_long, facet_
            kws=dict(sharex=False))
```

Export the file:

```
imprank_seg.savefig("images/imprank_seg.png")
```

Most of the high impression keywords are in Accessories, Console, and of course Top 1000 (Figure 2-2).

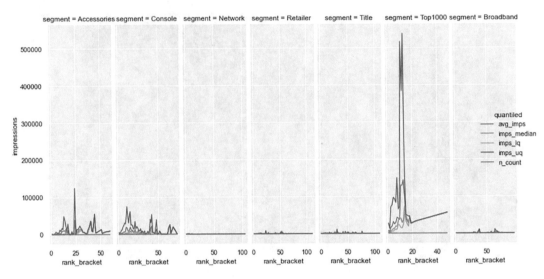

Figure 2-2. *Line chart showing GSC impressions per ranking position bracket for each distribution quantile faceted by segment*

Export Your High Value Keyword List

Now that you have your keywords, simply filter and export to CSV.

Export the dataframe to CSV:

```
query_stats_uq_p2b.to_csv('exports/query_stats_uq_p2b_TOTRACK.csv')
```

Activation

Now that you've identified high impression value keywords, you can

- Replace or add those keywords to the ones you're currently tracking and campaigning

- Research the content experience required to rank on the first page

- Think about how to integrate these new targets into your strategy

- Explore levels of on-page optimization for these keywords, including where there are low-hanging fruit opportunities to more effectively interlink landing pages targeting these keywords (such as through blog posts or content pages)

- Consider whether increasing external link popularity (through content marketing and PR) across these new landing pages is appropriate

Obviously, the preceding list is reductionist, and yet as a minimum, you have better nonbrand targets to better serve your SEO campaign.

Google Trends

Google Trends is another (free) third-party data source, which shows time series data (data points over time) up to the last five years for any search phrase that has demand. Google Trends can also help you compare whether a search is on the rise (or decline) while comparing it to other search phrases. It can be highly useful for forecasting.

Although no Google Trends API exists, there are packages in Python (i.e., pytrends) that can automate the extraction of this data as we'll see as follows:

```
import pandas as pd
from pytrends.request import TrendReq
import time
```

Single Keyword

Now that you've identified high impression value keywords, you can see how they've trended over the last five years:

```
kw_list = ["Blockchain"]
pytrends.build_payload(kw_list, cat=0, timeframe='today 5-y', geo='GB',
gprop='')
pytrends.interest_over_time()
```

This results in the following:

date	Blockchain	isPartial
2016-06-19	8	False
2016-06-26	6	False
2016-07-03	7	False
2016-07-10	6	False
2016-07-17	7	False
...	...	...
2021-05-09	42	False
2021-05-16	55	False
2021-05-23	43	False
2021-05-30	30	False
2021-06-06	34	True

260 rows × 2 columns

Multiple Keywords

As you can see earlier, you get a dataframe with the date, the keyword, and the number of hits (scaled from 0 to 100), which is great, and what if you had 10,000 keywords that you wanted trends for?

In that case, you'd want a for loop to query the search phrases one by one and stick them all into a dataframe like so:

Read in your target keyword data:

```
csv_raw = pd.read_csv('data/your_keyword_file.csv')
keywords_df = csv_raw[['query']]
keywords_list = keywords_df['query'].values.tolist()
keywords_list
```

Here's the output of what keywords_list looks like:

```
['nintendo switch',
 'ps4',
 'xbox one controller',
 'xbox one',
 'xbox controller',
 'ps4 vr',
 'Ps5' ...]
```

Let's now get Google Trends data for all of your keywords in one dataframe:

```
dataset = []
exceptions = []

for q in keywords_list:
    q_lst = [q]
    try:
        pytrends.build_payload(kw_list=q_lst, timeframe='today 5-y',
        geo='GB', gprop='')
        data = pytrends.interest_over_time()
        data = data.drop(labels=['isPartial'],axis='columns')
        dataset.append(data)
        time.sleep(3)
    except:
        exceptions.append(q_lst)

gtrends_long = pd.concat(dataset, axis=1)
```

This results in the following:

date	nintendo switch	ps4	xbox one controller	xbox one	xbox controller	ps4 vr	ps5	ps5 console	ps5 pre order	xbox series x
2016-06-19	0	30	30	37	29	9	0	0	0	0
2016-06-26	0	31	26	34	31	6	0	0	0	0
2016-07-03	0	31	26	36	26	7	0	0	0	0
2016-07-10	0	26	23	31	26	6	0	0	0	0
2016-07-17	0	27	17	29	19	4	0	0	0	0
...	...	...	...	...	...	...	...	...	...	...
2021-05-09	20	24	18	14	28	12	13	14	1	8
2021-05-16	18	22	14	15	29	9	12	10	0	8
2021-05-23	19	22	17	14	30	7	10	10	1	7
2021-05-30	20	23	13	15	33	9	10	9	1	6
2021-06-06	17	20	12	14	27	7	11	10	1	7

260 rows × 10 columns

Let's convert to long format:

```
gtrends_long = gtrends_raw.melt(id_vars=['date'], var_name = 'query',
value_name = 'hits')
gtrends_long
```

This results in the following:

	date	query	hits
0	2016-06-19	nintendo switch	0
1	2016-06-26	nintendo switch	0
2	2016-07-03	nintendo switch	0
3	2016-07-10	nintendo switch	0
4	2016-07-17	nintendo switch	0
...	...	...	...
3115	2021-05-09	id	hits
3116	2021-05-16	id	hits
3117	2021-05-23	id	hits
3118	2021-05-30	id	hits
3119	2021-06-06	id	hits

Looking at Google Trends raw, we now have data in long format showing

- Date

- Keyword

- Hits

Let's visualize some of these over time. We start by subsetting the dataframe:

```
k_list = ['ps5', 'xbox one', 'ps4', 'xbox series x', 'nintendo switch']
keyword_gtrends = gtrends_long.loc[gtrends_long['query'].isin(k_list)]
keyword_gtrends
```

This results in the following:

	date	query	hits
0	2016-06-19	nintendo switch	0
1	2016-06-26	nintendo switch	0
2	2016-07-03	nintendo switch	0
3	2016-07-10	nintendo switch	0
4	2016-07-17	nintendo switch	0
...	...	...	...
2595	2021-05-09	xbox series x	8
2596	2021-05-16	xbox series x	7
2597	2021-05-23	xbox series x	6
2598	2021-05-30	xbox series x	6
2599	2021-06-06	xbox series x	7

Visualizing Google Trends

Okay, so we're now ready to plot the time series data as a chart, starting with the library import:

```
import seaborn as sns
```

Set the plot size:

```
sns.set(rc={'figure.figsize':(15, 6)})
```

Build and plot the chart:

```
keyword_gtrends_plt = sns.lineplot(data = keyword_gtrends, x = 'date', y =
'hits', hue = 'query')
```

Save the image to a file for your PowerPoint deck or others:

```
keyword_gtrends_plt.figure.savefig("images/keyword_gtrends.png")
keyword_gtrends_plt
```

Here, we can see that the "ps5" and "xbox series x" show a near identical trend which ramp up significantly, while other models are fairly stable and seasonal until the arrival of the new models.

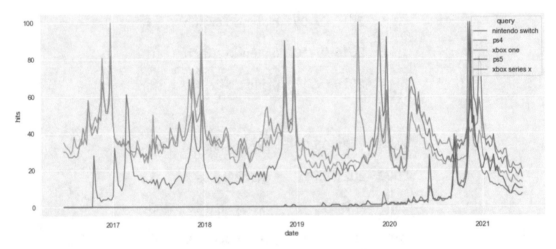

Figure 2-3. *Time series plot of Google Trends keywords*

Forecast Future Demand

While it's great to see what's happened in the last five years, it's also great to see what might happen in the future. Thankfully, Python provides the tools to do so. The most obvious use cases for forecasts are client pitches and reporting.

Exploring Your Data

```
import pandas as pd
from statsmodels.tsa.statespace.sarimax import SARIMAX
from statsmodels.graphics.tsaplots import plot_acf,plot_pacf
from statsmodels.tsa.seasonal import seasonal_decompose
from sklearn.metrics import mean_squared_error
from statsmodels.tools.eval_measures import rmse
import warnings
warnings.filterwarnings("ignore")
from pmdarima import auto_arima
```

Import Google Trends data:

```
df = pd.read_csv("exports/keyword_gtrends_df.csv", index_col=0)
df.head()
```

This results in the following:

	date	query	hits
1815	2021-05-09	ps5	12
1816	2021-05-16	ps5	11
1817	2021-05-23	ps5	10
1818	2021-05-30	ps5	10
1819	2021-06-06	ps5	10

As we'd expect, the data from Google Trends is a very simple time series with date, query, and hits spanning a five-year period. Time to format the dataframe to go from long to wide:

```
df_unstacked = ps_trends.set_index(["date", "query"]).unstack(level=-1)
df_unstacked.columns.set_names(['hits', 'query'], inplace=True)
ps_unstacked = df_unstacked.droplevel('hits', axis=1)
ps_unstacked.columns = [c.replace(' ', '_') for c in ps_unstacked.columns]
ps_unstacked = ps_unstacked.reset_index()
ps_unstacked.head()
```

This results in the following:

	date	ps4	ps5
0	2016-06-19	30	0
1	2016-06-26	30	0
2	2016-07-03	29	0
3	2016-07-10	27	0
4	2016-07-17	27	0

We no longer have a hits column as these are the values of the queries in their respective columns. This format is not only useful for SARIMA[2] (which we will be exploring here) but also neural networks such as long short-term memory (LSTM). Let's plot the data:

```
ps_unstacked.plot(figsize=(10,5))
```

From the plot (Figure 2-4), you'll note that the profiles of both "PS4" and "PS5" are different.

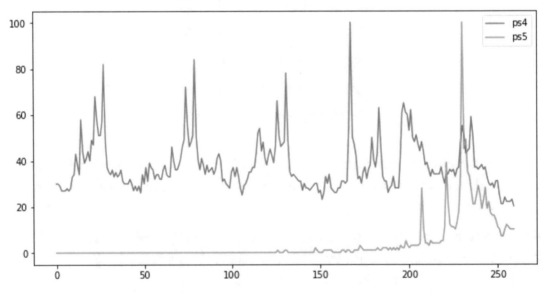

Figure 2-4. *Time series plot of both ps4 and ps5*

---

[2] Seasonal Autoregressive Integrated Moving Average

For the nongamers among you, "PS4" is the fourth generation of the Sony PlayStation console, and "PS5" the fifth. "PS4" searches are highly seasonal and have a regular pattern apart from the end when the "PS5" emerged. The "PS5" didn't exist five years ago, which would explain the absence of trend in the first four years of the preceding plot.

Decomposing the Trend

Let's now decompose the seasonal (or nonseasonal) characteristics of each trend:

```
ps_unstacked.set_index("date", inplace=True)
ps_unstacked.index = pd.to_datetime(ps_unstacked.index)

query_col = 'ps5'
a = seasonal_decompose(ps_unstacked[query_col], model = "add")
a.plot();
```

Figure 2-5 shows the time series data and the overall smoothed trend showing it rises from 2020.

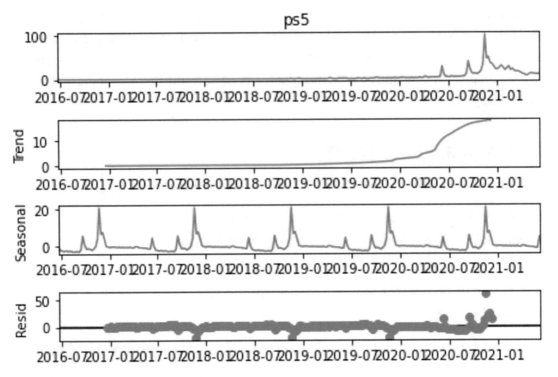

Figure 2-5. *Decomposition of the ps5 time series*

The seasonal trend box shows repeated peaks which indicates that there is seasonality from 2016, although it doesn't seem particularly reliable given how flat the time series is from 2016 until 2020. Also suspicious is the lack of noise as the seasonal plot shows a virtually uniform pattern repeating periodically.

The Resid (which stands for "Residual") shows any pattern of what's left of the time series data after accounting for seasonality and trend, which in effect is nothing until 2020 as it's at zero most of the time.

For "ps4," see Figure 2-6.

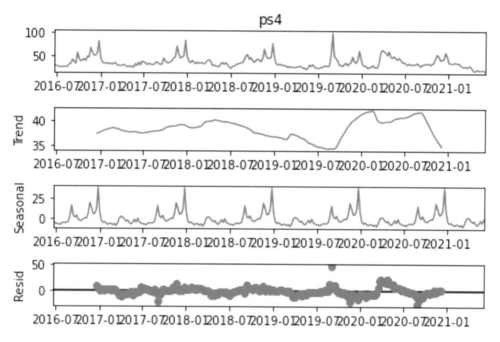

Figure 2-6. *Decomposition of the ps4 time series*

We can see fluctuation over the short term (Seasonality) and long term (Trend), with some noise (Resid). The next step is to use the augmented Dickey-Fuller method (ADF) to statistically test whether a given time series is stationary or not:

```
from pmdarima.arima import ADFTest
adf_test = ADFTest(alpha=0.05)
adf_test.should_diff(ps_unstacked[query_col])
```

```
PS4: (0.09760939899434763, True)
PS5: (0.01, False)
```

We can see that the p-value of "PS5" shown earlier is more than 0.05, which means that the time series data is not stationary and therefore needs differencing. "PS4" on the other hand is less than 0.05 at 0.01, meaning it's stationery and doesn't require differencing.

The point of all this is to understand the parameters that would be used if we were manually building a model to forecast Google searches.

Fitting Your SARIMA Model

Since we'll be using automated methods to estimate the best fit model parameters (later), we're not going to estimate the number of parameters for our SARIMA model.

To estimate the parameters for our SARIMA model, note that we set m to 52 as there are 52 weeks in a year which is how the periods are spaced in Google Trends. We also set all of the parameters to start at 0 so that we can let the auto_arima do the heavy lifting and search for the values that best fit the data for forecasting:

```
ps5_s = auto_arima(ps_unstacked['ps4'],
            trace=True,
            m=52, #there are 52 period per season (weekly data)
            start_p=0,
            start_d=0,
            start_q=0,
            seasonal=False)
```

This results in the following:

```
Performing stepwise search to minimize aic

ARIMA(3,0,3)(0,0,0)[0]             : AIC=1842.301, Time=0.26 sec
ARIMA(0,0,0)(0,0,0)[0]             : AIC=2651.089, Time=0.01 sec
ARIMA(1,0,0)(0,0,0)[0]             : AIC=1865.936, Time=0.02 sec
ARIMA(0,0,1)(0,0,0)[0]             : AIC=2370.569, Time=0.05 sec
ARIMA(2,0,3)(0,0,0)[0]             : AIC=1845.911, Time=0.12 sec
ARIMA(3,0,2)(0,0,0)[0]             : AIC=1845.959, Time=0.16 sec
ARIMA(4,0,3)(0,0,0)[0]             : AIC=1838.349, Time=0.34 sec
ARIMA(4,0,2)(0,0,0)[0]             : AIC=1846.701, Time=0.22 sec
ARIMA(5,0,3)(0,0,0)[0]             : AIC=1843.754, Time=0.25 sec
ARIMA(4,0,4)(0,0,0)[0]             : AIC=1842.801, Time=0.27 sec
ARIMA(3,0,4)(0,0,0)[0]             : AIC=1841.447, Time=0.36 sec
ARIMA(5,0,2)(0,0,0)[0]             : AIC=1841.893, Time=0.24 sec
ARIMA(5,0,4)(0,0,0)[0]             : AIC=1845.734, Time=0.29 sec
ARIMA(4,0,3)(0,0,0)[0] intercept   : AIC=1824.187, Time=0.82 sec
ARIMA(3,0,3)(0,0,0)[0] intercept   : AIC=1824.769, Time=0.34 sec
ARIMA(4,0,2)(0,0,0)[0] intercept   : AIC=1826.970, Time=0.34 sec
ARIMA(5,0,3)(0,0,0)[0] intercept   : AIC=1826.789, Time=0.44 sec
```

```
ARIMA(4,0,4)(0,0,0)[0] intercept   : AIC=1827.114, Time=0.43 sec
ARIMA(3,0,2)(0,0,0)[0] intercept   : AIC=1831.587, Time=0.32 sec
ARIMA(3,0,4)(0,0,0)[0] intercept   : AIC=1825.359, Time=0.42 sec
ARIMA(5,0,2)(0,0,0)[0] intercept   : AIC=1827.292, Time=0.40 sec
ARIMA(5,0,4)(0,0,0)[0] intercept   : AIC=1829.109, Time=0.51 sec

Best model:  ARIMA(4,0,3)(0,0,0)[0] intercept
Total fit time: 6.601 seconds
```

The preceding printout shows that the parameters that get the best results are

```
PS4: ARIMA(4,0,3)(0,0,0)
PS5: ARIMA(3,1,3)(0,0,0)
```

The PS5 estimate is further detailed when printing out the model summary:

```
ps5_s.summary()
```

This results in the following:

SARIMAX Results

Dep. Variable:	y	No. Observations:	260
Model:	SARIMAX(4, 0, 3)	Log Likelihood	-903.094
Date:	Fri, 30 Jul 2021	AIC	1824.187
Time:	13:02:02	BIC	1856.233
Sample:	0	HQIC	1837.070
	- 260		
Covariance Type:	opg		

	coef	std err	z	P>|z|	[0.025	0.975]
intercept	8.7756	2.661	3.298	0.001	3.561	13.990
ar.L1	1.2577	0.325	3.872	0.000	0.621	1.894
ar.L2	-0.1521	0.701	-0.217	0.828	-1.527	1.223
ar.L3	-0.8086	0.652	-1.241	0.215	-2.086	0.469
ar.L4	0.4698	0.240	1.961	0.050	0.000	0.939
ma.L1	-0.6256	0.316	-1.979	0.048	-1.245	-0.006
ma.L2	-0.2806	0.455	-0.617	0.537	-1.172	0.611
ma.L3	0.7784	0.294	2.648	0.008	0.202	1.355
sigma2	60.0816	3.039	19.771	0.000	54.125	66.038

Ljung-Box (L1) (Q):	0.02	Jarque-Bera (JB):	1449.21
Prob(Q):	0.90	Prob(JB):	0.00
Heteroskedasticity (H):	0.54	Skew:	2.24
Prob(H) (two-sided):	0.00	Kurtosis:	13.66

What's happening is the function is looking to minimize the probability of error measured by both the Akaike information criterion (AIC) and Bayesian information criterion:

```
AIC = -2Log(L) + 2(p + q + k + 1)
```

such that L is the likelihood of the data, k = 1 if c ≠ 0, and k = 0 if c = 0.

```
BIC = AIC + [log(T) - 2] + (p + q + k + 1)
```

By minimizing AIC and BIC, we get the best estimated parameters for p and q.

Test the Model

Now that we have the parameters, we can now start making forecasts for both products:

```
ps4_order = ps4_s.get_params()['order']
ps4_seasorder = ps4_s.get_params()['seasonal_order']

ps5_order = ps5_s.get_params()['order']
ps5_seasorder = ps5_s.get_params()['seasonal_order']

params = {
    "ps4": {"order": ps4_order, "seasonal_order": ps4_seasorder},
    "ps5": {"order": ps5_order, "seasonal_order": ps5_seasorder}
}
```

Create an empty list to store the forecast results:

```
results = []
fig, axs = plt.subplots(len(X.columns), 1, figsize=(24, 12))
```

Iterate through the columns to fit the best SARIMA model:

```
for i, col in enumerate(X.columns):
    arima_model = SARIMAX(train_data[col],
                        order = params[col]["order"],
                        seasonal_order = params[col]["seasonal_order"])
    arima_result = arima_model.fit()
```

Make forecasts:

```
    arima_pred = arima_result.predict(start = len(train_data),
                                    end = len(X)-1, typ="levels")\
                        .rename("ARIMA Predictions")
```

Plot predictions:

```
test_data[col].plot(figsize = (8,4), legend=True, ax=axs[i])
arima_pred.plot(legend = True, ax=axs[i])

arima_rmse_error = rmse(test_data[col], arima_pred)
mean_value = X[col].mean()

results.append((col, arima_pred, arima_rmse_error, mean_value))
print(f'Column: {col} --> RMSE Error: {arima_rmse_error} - Mean: {mean_
value}\n')
```

This results in the following:

```
Column: ps4 --> RMSE Error: 8.626764032898576 - Mean: 37.83461538461538
Column: ps5 --> RMSE Error: 27.552818032476257 - Mean: 3.973076923076923
```

For ps4, the forecasts are pretty accurate from the beginning until March when the search values start to diverge (Figure 2-7), while the ps5 forecasts don't appear to be very good at all, which is unsurprising.

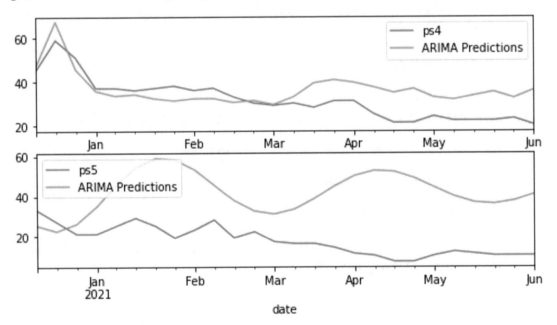

Figure 2-7. *Time series line plots comparing forecasts and actual data for both ps4 and ps5*

The forecasts show the models are good when there is enough history until they suddenly change like they have for PS4 from March onward. For PS5, the models are hopeless virtually from the get-go. We know this because the Root Mean Squared Error (RMSE) is 8.62 for PS4 which is more than a third of the PS5 RMSE of 27.5, which, given Google Trends varies from 0 to 100, is a 27% margin of error.

Forecast the Future

At this point, we'll now make the foolhardy attempt to forecast the future based on the data we have to date:

```
oos_train_data = ps_unstacked
oos_train_data.tail()
```

This results in the following:

date	ps4	ps5
2021-05-09	22	12
2021-05-16	22	11
2021-05-23	22	10
2021-05-30	23	10
2021-06-06	20	10

As you can see from the preceding table extract, we're now using all available data. Now we shall predict the next six months (defined as 26 weeks) in the following code:

```
oos_results = []
weeks_to_predict = 26
fig, axs = plt.subplots(len(ps_unstacked.columns), 1, figsize=(24, 12))
```

Again, iterate through the columns to fit the best model each time:

```
for i, col in enumerate(ps_unstacked.columns):
    s = auto_arima(oos_train_data[col], trace=True)
```

```
oos_arima_model = SARIMAX(oos_train_data[col],
                          order = s.get_params()['order'],
                          seasonal_order = s.get_params()['seasonal_
                          order'])
oos_arima_result = oos_arima_model.fit()
```

Make forecasts:

```
oos_arima_pred = oos_arima_result.predict(start = len(oos_train_data),
                                          end = len(oos_train_data) + weeks_to_
                                          predict, typ="levels").rename("ARIMA
                                          Predictions")
```

Plot predictions:

```
oos_arima_pred.plot(legend = True, ax=axs[i])
axs[i].legend([col]);
mean_value = ps_unstacked[col].mean()

oos_results.append((col, oos_arima_pred, mean_value))
print(f'Column: {col} - Mean: {mean_value}\n')
```

Here's the output:

```
Performing stepwise search to minimize aic
 ARIMA(2,0,2)(0,0,0)[0] intercept   : AIC=1829.734, Time=0.21 sec
 ARIMA(0,0,0)(0,0,0)[0] intercept   : AIC=1999.661, Time=0.01 sec
 ARIMA(1,0,0)(0,0,0)[0] intercept   : AIC=1827.518, Time=0.03 sec
 ARIMA(0,0,1)(0,0,0)[0] intercept   : AIC=1882.388, Time=0.05 sec
 ARIMA(0,0,0)(0,0,0)[0]             : AIC=2651.089, Time=0.01 sec
 ARIMA(2,0,0)(0,0,0)[0] intercept   : AIC=1829.254, Time=0.04 sec
 ARIMA(1,0,1)(0,0,0)[0] intercept   : AIC=1829.136, Time=0.09 sec
 ARIMA(2,0,1)(0,0,0)[0] intercept   : AIC=1829.381, Time=0.26 sec
 ARIMA(1,0,0)(0,0,0)[0]             : AIC=1865.936, Time=0.02 sec

Best model:  ARIMA(1,0,0)(0,0,0)[0] intercept
Total fit time: 0.722 seconds
Column: ps4 - Mean: 37.83461538461538
```

```
Performing stepwise search to minimize aic
 ARIMA(2,1,2)(0,0,0)[0] intercept   : AIC=1657.990, Time=0.19 sec
 ARIMA(0,1,0)(0,0,0)[0] intercept   : AIC=1696.958, Time=0.01 sec
 ARIMA(1,1,0)(0,0,0)[0] intercept   : AIC=1673.340, Time=0.04 sec
 ARIMA(0,1,1)(0,0,0)[0] intercept   : AIC=1666.878, Time=0.05 sec
 ARIMA(0,1,0)(0,0,0)[0]             : AIC=1694.967, Time=0.01 sec
 ARIMA(1,1,2)(0,0,0)[0] intercept   : AIC=1656.899, Time=0.14 sec
 ARIMA(0,1,2)(0,0,0)[0] intercept   : AIC=1663.729, Time=0.04 sec
 ARIMA(1,1,1)(0,0,0)[0] intercept   : AIC=1656.787, Time=0.07 sec
 ARIMA(2,1,1)(0,0,0)[0] intercept   : AIC=1656.351, Time=0.16 sec
 ARIMA(2,1,0)(0,0,0)[0] intercept   : AIC=1672.668, Time=0.04 sec
 ARIMA(3,1,1)(0,0,0)[0] intercept   : AIC=1657.661, Time=0.11 sec
 ARIMA(3,1,0)(0,0,0)[0] intercept   : AIC=1670.698, Time=0.05 sec
 ARIMA(3,1,2)(0,0,0)[0] intercept   : AIC=1653.392, Time=0.33 sec
 ARIMA(4,1,2)(0,0,0)[0] intercept   : AIC=inf, Time=0.40 sec
 ARIMA(3,1,3)(0,0,0)[0] intercept   : AIC=1643.872, Time=0.45 sec
 ARIMA(2,1,3)(0,0,0)[0] intercept   : AIC=1659.698, Time=0.23 sec
 ARIMA(4,1,3)(0,0,0)[0] intercept   : AIC=inf, Time=0.48 sec
 ARIMA(3,1,4)(0,0,0)[0] intercept   : AIC=inf, Time=0.47 sec
 ARIMA(2,1,4)(0,0,0)[0] intercept   : AIC=1645.994, Time=0.52 sec
 ARIMA(4,1,4)(0,0,0)[0] intercept   : AIC=1647.585, Time=0.56 sec
 ARIMA(3,1,3)(0,0,0)[0]             : AIC=1641.790, Time=0.37 sec
 ARIMA(2,1,3)(0,0,0)[0]             : AIC=1648.325, Time=0.38 sec
 ARIMA(3,1,2)(0,0,0)[0]             : AIC=1651.416, Time=0.24 sec
 ARIMA(4,1,3)(0,0,0)[0]             : AIC=1650.077, Time=0.59 sec
 ARIMA(3,1,4)(0,0,0)[0]             : AIC=inf, Time=0.58 sec
 ARIMA(2,1,2)(0,0,0)[0]             : AIC=1656.290, Time=0.10 sec
 ARIMA(2,1,4)(0,0,0)[0]             : AIC=1644.099, Time=0.38 sec
 ARIMA(4,1,2)(0,0,0)[0]             : AIC=inf, Time=0.38 sec
 ARIMA(4,1,4)(0,0,0)[0]             : AIC=1645.756, Time=0.56 sec

Best model:  ARIMA(3,1,3)(0,0,0)[0]
Total fit time: 7.954 seconds
Column: ps5 - Mean: 3.973076923076923
```

This time, we automated the finding of the best-fitting parameters and fed that directly into the model.

The forecasts don't look great (Figure 2-8) because there's been a lot of change in the last few weeks of the data; however, that's in the case of those two keywords.

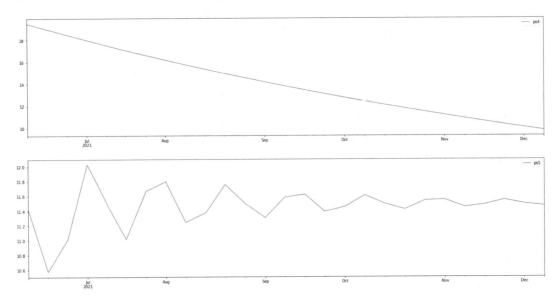

Figure 2-8. *Out-of-sample forecasts of Google Trends for ps4 and ps5*

The forecast quality will be dependent on how stable the historic patterns are and will obviously not account for unforeseeable events like COVID-19.

Export your forecasts:

```
df_pred = pd.concat([pd.Series(res[1]) for res in oos_results], axis=1)
df_pred.columns = [x + str('_preds') for x in ps_unstacked.columns]
df_pred.to_csv('your_forecast_data.csv')
```

What we learn here is where forecasting using statistical models are useful or are likely to add value for forecasting, particularly in automated systems like dashboards, that is, when there's historical data and not when there is a sudden spike like PS5.

Clustering by Search Intent

Search intent is the meaning behind the search queries that users of Google type in when searching online. So you may have the following queries:

"Trench coats"

"Ladies trench coats"

"Life insurance"

"Trench coats" will share the same search intent as "Ladies trench coats" but won't share the same intent as "Life insurance." To work this out, a simple comparison of the top 10 ranking sites for both search phrases in Google will offer a strong suggestion of what Google thinks of the search intent between the two phrases.

It's not a perfect method, but it works well because you're using the ranking results which are a distillation of everything Google has learned to date on what content satisfies the search intent of the search query (based upon the trillions of global searches per year). Therefore, it's reasonable to surmise that if two search queries have similar enough SERPs, then the search intent is shared between keywords.

This is useful for a number of reasons:

- *Rank tracking costs*: If your budget is limited, then knowing the search intent means you can avoid incurring further expense by not tracking keywords with the same intent as those you're tracking. This comes with a risk as consumers change and the keyword not tracked may not share the same intent anymore.

- *Core updates*: With changing consumer search patterns come changing intents, which means you can see if keywords change clusters or not by comparing the search intent clusters of keywords before and after the update, which will help inform your response.

- *Keyword content mapping*: Knowing the intent means you can successfully map keywords to landing pages. This is especially useful in ensuring your site architecture consists of landing pages which map to user search demand.

- *Paid search ads*: Good keyword content mappings also mean you can improve the account structure and resulting quality score of your paid search activity.

Starting Point

Okay, time to cluster. We'll assume you already have the top 100 SERPs[3] results for each of your keywords stored as a Python dataframe "serps_input." The data is easily obtained from a rank tracking tool, especially if they have an API:

serps_input

This results in the following:

	keyword	rank	url	se_results_count
0	xbox one x controller	1	https://www.xbox.com/en-GB/accessories	144000000
1	xbox one x controller	2	None	144000000
2	xbox one x controller	3	https://www.xbox.com/en-GB/accessories/control...	144000000
3	xbox one x controller	4	https://www.argos.co.uk/browse/technology/vide...	144000000
4	xbox one x controller	5	https://www.game.co.uk/en/accessories/xbox-one...	144000000
5	xbox one x controller	6	https://www.currys.co.uk/gbuk/xbox-one-control...	144000000
6	xbox one x controller	7	https://www.amazon.co.uk/xbox-one-controller/s...	144000000
7	xbox one x controller	8	None	144000000
8	xbox one x controller	9	https://www.ebay.co.uk/b/Microsoft-Xbox-One-Co...	144000000
9	xbox one x controller	10	https://www.amazon.com/Xbox-Wireless-Controlle...	144000000
10	xbox one x controller	11	https://www.powera.com/product_platform/xbox-one/	144000000
11	xbox one x controller	12	https://www.pricerunner.com/sp/xbox-one-x-cont...	144000000
12	xbox one x controller	13	https://en.wikipedia.org/wiki/Xbox_Wireless_Co...	144000000
13	xbox one x controller	14	https://scufgaming.com/uk/xbox	144000000
14	xbox one x controller	15	https://www.digitaltrends.com/gaming/how-to-sy...	144000000

Here, we're using DataForSEO's SERP API,[4] and we have renamed the column from "rank_absolute" to "rank."

---

[3] Search Engine Results Pages (SERP)

[4] Available at https://dataforseo.com/apis/serp-api/

Filter Data for Page 1

Because DataForSEO's numbers to individual results are contained within carousels, People Also Ask, etc., we'll want to compare the top 20 results of each SERP to each other to get the approximate results for page 1. We'll also filter out URLs that have the value "None." The programming approach we'll take is "Split-Apply-Combine." What is Split-Apply-Combine?

- Split the dataframe into keyword groups

- Apply the filtering formula to each group

- Combine the keywords of each group

Here it goes:

Split:

```
serps_grpby_keyword = serps_input.groupby("keyword")
```

Apply the function, before combining:

```
def filter_twenty_urls(group_df):
    filtered_df = group_df.loc[group_df['url'].notnull()]
    filtered_df = filtered_df.loc[filtered_df['rank'] <= 20]
    return filtered_df
filtered_serps = serps_grpby_keyword.apply(filter_twenty_urls)
```

Combine and add prefix to column names:

```
normed = normed.add_prefix('normed_')
```

Concatenate with an initial dataframe:

```
filtered_serps_df = pd.concat([filtered_serps],axis=0)
```

Convert Ranking URLs to a String

To compare the SERPs for each keyword, we need to convert the SERPs URL into a string. That's because there's a one (keyword) to many (SERP URLs) relationship. The way we achieve that is by simply concatenating the URL strings for each keyword, using the Split-Apply-Combine approach (again). Convert results to strings using SAC:

```
filtserps_grpby_keyword = filtered_serps_df.groupby("keyword")

def string_serps(df):
    df['serp_string'] = ''.join(df['url'])
    return df

    Combine
strung_serps = filtserps_grpby_keyword.apply(string_serps)
```

Concatenate with an initial dataframe and clean:

```
strung_serps = pd.concat([strung_serps],axis=0)
strung_serps = strung_serps[['keyword', 'serp_string']]#.head(30)
strung_serps = strung_serps.drop_duplicates()
strung_serps
```

This results in the following:

	keyword	serp_string
0	fifa 19 ps4	https://www.amazon.co.uk/Electronic-Arts-22154...
18	gaming broadband	https://www.bt.com/products/broadband/gaminght...
37	playstation vr	https://www.playstation.com/en-gb/ps-vr/https:...
54	ps4	https://www.playstation.com/en-gb/ps4/https://...
72	ps4 console	https://www.game.co.uk/en/hardware/playstation...
91	ps4 controller	https://www.playstation.com/en-gb/ps4/ps4-acce...
109	ps4 controllers	https://www.playstation.com/en-gb/ps4/ps4-acce...
127	ps4 vr	https://www.playstation.com/en-gb/ps-vr/https:...
146	ps5	https://direct.playstation.com/en-us/ps5https:...
162	ps5 console	https://direct.playstation.com/en-us/ps5https:...

Now we have a table showing the keyword and their SERP string, we're ready to compare SERPs. Here's an example of the SERP string for "fifa 19 ps4":

```
strung_serps.loc[1, 'serp_string']
```

This results in the following:

'https://www.amazon.co.uk/Electronic-Arts-221545-FIFA-PS4/dp/
B07DLXBGN8https://www.amazon.co.uk/FIFA-19-GAMES/dp/B07DL2SY2Bhttps://
www.game.co.uk/en/fifa-19-2380636https://www.ebay.co.uk/b/FIFA-19-Sony-
PlayStation-4-Video-Games/139973/bn_7115134270https://www.pricerunner.com/
pl/1422-4602670/PlayStation-4-Games/FIFA-19-Compare-Priceshttps://pricespy.
co.uk/games-consoles/computer-video-games/ps4/fifa-19-ps4--p4766432https://
store.playstation.com/en-gb/search/fifa%2019https://www.amazon.com/FIFA-19-
Standard-PlayStation-4/dp/B07DL2SY2Bhttps://www.tesco.com/groceries/
en-GB/products/301926084https://groceries.asda.com/product/ps-4-games/
ps-4-fifa-19/1000076097883https://uk.webuy.com/product-detail/?id=503094
5121916&categoryName=playstation4-software&superCatName=gaming&title=fi
fa-19https://www.pushsquare.com/reviews/ps4/fifa_19https://en.wikipedia.
org/wiki/FIFA_19https://www.amazon.in/Electronic-Arts-Fifa19SEPS4-Fifa-
PS4/dp/B07DVWWF44https://www.vgchartz.com/game/222165/fifa-19/https://www.
metacritic.com/game/playstation-4/fifa-19https://www.johnlewis.com/fifa-19-
ps4/p3755803https://www.ebay.com/p/22045274968'

Compare SERP Distance

The SERPs comparison will use string distance techniques which allow us to see how similar or dissimilar one keyword's SERPs are. This technique is similar to how geneticists would compare one DNA sequence to another.

Naturally, we need to get the SERPs into a format ready for Python to compare SERPs. To do this, we need to convert each SERP to a string and then put them side by side. Group the table by keyword:

```
filtserps_grpby_keyword = filtered_serps_df.groupby("keyword")
def string_serps(df):
    df['serp_string'] = ' '.join(df['url'])
    return df
```

Combine using the preceding function:

```
strung_serps = filtserps_grpby_keyword.apply(string_serps)
```

Concatenate with an initial dataframe and clean:

```
strung_serps = pd.concat([strung_serps],axis=0)
```

```
strung_serps = strung_serps[['keyword', 'serp_string']]#.head(30)
strung_serps = strung_serps.drop_duplicates()
#strung_serps['serp_string'] = strung_serps.serp_string.str.
replace("https://www\.", "")
strung_serps.head(15)
```

This results in the following:

	keyword	serp_string
0	beige trench coats	https://www.zalando.co.uk/womens-clothing-coat...
9	blue trench coats	https://www.johnlewis.com/browse/women/womens-...
19	buy ps4	https://www.playstation.com/en-gb/ps4/buy-ps4/...
24	ladies trench coats	https://www.johnlewis.com/browse/women/womens-...
34	mens trench coats	https://uk.burberry.com/mens-trench-coats/ htt...
43	ps4	https://www.playstation.com/en-gb/ps4/ https:/...
51	ps4 console	https://www.game.co.uk/en/hardware/playstation...
60	ps4 vr	https://www.playstation.com/en-gb/ps-vr/ https...
69	ps5	https://direct.playstation.com/en-us/ps5 https...
78	ps5 console	https://www.game.co.uk/playstation-5 https://d...
86	ps5 news	https://www.pushsquare.com/ps5 https://www.pla...
95	ps5 pre order	https://www.playstation.com/en-gb/ps5/buy-now/...
104	trench coats	https://uk.burberry.com/womens-trench-coats/ h...
112	xbox controller	https://www.xbox.com/en-GB/accessories/control...
120	xbox one	https://www.xbox.com/ https://www.xbox.com/en-...

Here, we now have the keywords and their respective SERPs all converted into a string which fits into a single cell. For example, the search result for "beige trench coats" is

```
'https://www.zalando.co.uk/womens-clothing-coats-trench-coats/_beige/
https://www.asos.com/women/coats-jackets/trench-coats/cat/?cid=15143
https://uk.burberry.com/womens-trench-coats/beige/ https://www2.hm.com/
```

44

en_gb/productpage.0751992002.html https://www.hobbs.com/clothing/
coats-jackets/trench/beige/ https://www.zara.com/uk/en/woman-outerwear-
trench-l1202.html https://www.ebay.co.uk/b/Beige-Trench-Coats-for-
Women/63862/bn_7028370345 https://www.johnlewis.com/browse/women/womens-
coats-jackets/trench-coats/_/N-flvZ1z0rnyl https://www.elle.com/uk/fashion/
what-to-wear/articles/g30975/best-trench-coats-beige-navy-black/'

Time to put these side by side. What we're effectively doing here is taking a product of the column to itself, that is, squaring it, so that we get all the SERPs combinations possible to put the SERPs side by side.

Add a function to align SERPs:

```
def serps_align(k, df):
    prime_df = df.loc[df.keyword == k]
    prime_df = prime_df.rename(columns = {"serp_string" : "serp_string_a",
    'keyword': 'keyword_a'})
    comp_df = df.loc[df.keyword != k].reset_index(drop=True)
    prime_df = prime_df.loc[prime_df.index.repeat(len(comp_df.index))].
    reset_index(drop=True)
    prime_df = pd.concat([prime_df, comp_df], axis=1)
    prime_df = prime_df.rename(columns = {"serp_string" : "serp_string_b",
    'keyword': 'keyword_b', "serp_string_a" : "serp_string", 'keyword_a':
    'keyword'})
    return prime_df
```

Test the function on a single keyword:

```
serps_align('ps4', strung_serps)
```

Set up desired dataframe columns:

```
columns = ['keyword', 'serp_string', 'keyword_b', 'serp_string_b']
matched_serps = pd.DataFrame(columns=columns)
matched_serps = matched_serps.fillna(0)
```

Call the function for each keyword:

```
for q in queries:
    temp_df = serps_align(q, strung_serps)
    matched_serps = matched_serps.append(temp_df)
```

This results in the following:

	keyword	serp_string	keyword_b	serp_string_b
0	ps4	https://www.playstation.com/en-gb/ps4/ https:/...	beige trench coats	https://www.zalando.co.uk/womens-clothing-coat...
1	ps4	https://www.playstation.com/en-gb/ps4/ https:/...	blue trench coats	https://www.johnlewis.com/browse/women/womens-...
2	ps4	https://www.playstation.com/en-gb/ps4/ https:/...	buy ps4	https://www.playstation.com/en-gb/ps4/buy-ps4/...
3	ps4	https://www.playstation.com/en-gb/ps4/ https:/...	ladies trench coats	https://www.johnlewis.com/browse/women/womens-...
4	ps4	https://www.playstation.com/en-gb/ps4/ https:/...	mens trench coats	https://uk.burberry.com/mens-trench-coats/ htt...
...	...	...	...	...
267	blue trench coats	https://www.johnlewis.com/browse/women/womens-...	trench coats	https://uk.burberry.com/womens-trench-coats/ h...
268	blue trench coats	https://www.johnlewis.com/browse/women/womens-...	xbox controller	https://www.xbox.com/en-GB/accessories/control...
269	blue trench coats	https://www.johnlewis.com/browse/women/womens-...	xbox one	https://www.xbox.com/ https://www.xbox.com/en-...
270	blue trench coats	https://www.johnlewis.com/browse/women/womens-...	xbox one controller	https://www.xbox.com/en-GB/accessories https:/...
271	blue trench coats	https://www.johnlewis.com/browse/women/womens-...	xbox series x	https://www.xbox.com/en-GB/consoles/xbox-serie...

The preceding result shows all of the keywords with SERPs compared side by side with other keywords and their SERPs. Next, we'll infer keyword intent similarity by comparing serp_strings, but first here's a note on the methods like Levenshtein, Jaccard, etc.

Levenshtein distance is edit based, meaning the number of edits required to transform one string (in our case, serp_string) into the other string (serps_string_b). This doesn't work very well because the websites within the SERP strings are individual tokens, that is, not a single continuous string.

Sorensen-Dice is better because it is token based, that is, it treats the individual websites as individual items or tokens. Using set similarity methods, the logic is to find the common tokens and divide them by the total number of tokens present by combining both sets. It doesn't take the order into account, so we must go one better.

M Measure which looks at both the token overlap and the order of the tokens, that is, weighting the order tokens earlier (i.e., the higher ranking sites/tokens) more than the later tokens. There is no API for this unfortunately, so we wrote the function for you here:

```
import py_stringmatching as sm
ws_tok = sm.WhitespaceTokenizer()
```

Only compare the top k_urls results:

```
def serps_similarity(serps_str1, serps_str2, k=15):
    denom = k+1
    norm = sum([2*(1/i - 1.0/(denom)) for i in range(1, denom)])
    #use to tokenize the URLs
```

```
ws_tok = sm.WhitespaceTokenizer()
#keep only first k URLs
serps_1 = ws_tok.tokenize(serps_str1)[:k]
serps_2 = ws_tok.tokenize(serps_str2)[:k]
#get positions of matches
match = lambda a, b: [b.index(x)+1 if x in b else None for x in a]
#positions intersections of form [(pos_1, pos_2), ...]
pos_intersections = [(i+1,j) for i,j in enumerate(match(serps_1,
serps_2)) if j is not None]
pos_in1_not_in2 = [i+1 for i,j in enumerate(match(serps_1, serps_2)) if
j is None]
pos_in2_not_in1 = [i+1 for i,j in enumerate(match(serps_2, serps_1)) if
j is None]

a_sum = sum([abs(1/i -1/j) for i,j in pos_intersections])
b_sum = sum([abs(1/i -1/denom) for i in pos_in1_not_in2])
c_sum = sum([abs(1/i -1/denom) for i in pos_in2_not_in1])

intent_prime = a_sum + b_sum + c_sum
intent_dist = 1 - (intent_prime/norm)
return intent_dist
```

Apply the function:

```
matched_serps['si_simi'] = matched_serps.apply(lambda x: serps_
similarity(x.serp_string, x.serp_string_b), axis=1)
matched_serps[["keyword", "keyword_b", "si_simi"]]
```

This is the resulting dataframe:

	keyword	keyword_b	si_simi
0	ps4	beige trench coats	0.058203
1	ps4	blue trench coats	0.050328
2	ps4	buy ps4	0.314561
3	ps4	ladies trench coats	0.050328
4	ps4	mens trench coats	0.058203
...	...	...	...
267	blue trench coats	trench coats	0.096999
268	blue trench coats	xbox controller	0.050328
269	blue trench coats	xbox one	0.050328
270	blue trench coats	xbox one controller	0.040118
271	blue trench coats	xbox series x	0.063454

272 rows × 3 columns

Before sorting the keywords into topic groups, let's add search volumes for each. This could be an imported table like the following one called "keysv_df":

keysv_df

This results in the following:

	keyword	search_volume
0	best isa rates	40500
1	isa	49500
2	savings account	60500
3	cash isa	14800
4	isa account	9900
5	child savings account	14800
6	fixed rate bonds	12100
7	isa rates	8100
8	savings account interest rate	9900
9	fixed rate isa	5400
10	isa interest rates	5400
11	savings accounts uk	6600
12	cash isa rates	4400
13	easy access savings account	3600
14	savings rates	5400
15	easy access savings	3600
16	fixed rate savings	4400
17	isa savings	3600
18	kids savings account	4400
19	online savings account	2400

Let's now join the data. What we're doing here is giving Python the ability to group keywords according to SERP similarity and name the topic groups according to the keyword with the highest search volume.

Group keywords by search intent according to a similarity limit. In this case, keyword search results must be 40% or more similar. This is a number based on trial and error of which the right number can vary by the search space, language, or other factors.

```
simi_lim = 0.4
```

Append topic vols:

```
keywords_crossed_vols = serps_compared.merge(keysv_df, on = 'keyword', how
= 'left')
keywords_crossed_vols = keywords_crossed_vols.rename(columns = {'keyword':
'topic', 'keyword_b': 'keyword', 'search_volume': 'topic_volume'})
```

Append keyword vols:

```
keywords_crossed_vols = keywords_crossed_vols.merge(keysv_df, on =
'keyword', how = 'left')
```

Simulate si_simi:

```
#keywords_crossed_vols['si_simi'] = np.random.rand(len(keywords_crossed_
vols.index))
keywords_crossed_vols.sort_values('topic_volume', ascending = False)
```

Strip the dataframe of NAN:

```
keywords_filtered_nonnan = keywords_crossed_vols.dropna()
```

We now have the potential topic name, keyword SERP similarity, and search volumes of each. You'll note the keyword and keyword_b have been renamed to topic and keyword, respectively. Now we're going to iterate over the columns in the dataframe using list comprehensions.

List comprehension is a technique for looping over lists. We applied it to the Pandas dataframe because it's much quicker than the .iterrows() function. Here it goes.

Add a dictionary comprehension to create numbered topic groups from keywords_filtered_nonnan:

```
# {1: [k1, k2, ..., kn], 2: [k1, k2, ..., kn], ..., n: [k1, k2, ..., kn]}
```

Convert the top names into a list:

```
queries_in_df = list(set(keywords_filtered_nonnan.topic.to_list()))
```

Set empty lists and dictionaries:

```
topic_groups_numbered = {}
topics_added = []
```

Define a function to find the topic number:

```
def latest_index(dicto):
    if topic_groups_numbered == {}:
        i = 0
    else:
        i = list(topic_groups_numbered)[-1]
    return i
```

Define a function to allocate keyword to topic:

```
def find_topics(si, keyw, topc):
    i = latest_index(topic_groups_numbered)
    if (si >= simi_lim) and (not keyw in topics_added) and (not topc in
    topics_added):
        #print(si, ', kw=' , keyw,', tpc=', topc,', ', i,', ', topic_
        groups_numbered)
        i += 1
        topics_added.extend([keyw, topc])
        topic_groups_numbered[i] = [keyw, topc]
    elif si >= simi_lim and (keyw in topics_added) and (not topc in
    topics_added):
        #print(si, ', kw=' , keyw,', tpc=', topc,', ', i,', ', topic_
        groups_numbered)
        j = [key for key, value in topic_groups_numbered.items() if keyw
        in value]
        topics_added.extend(topc)
        topic_groups_numbered[j[0]].append(topc)
    elif si >= simi_lim and (not keyw in topics_added) and (not topc in
    topics_added):
        #print(si, ', kw=' , keyw,', tpc=', topc,', ', i,', ', topic_
        groups_numbered)
        j = list(mydict.keys())[list(mydict.values()).index(keyw)]
        topic_groups_numbered[j[0]].append(topc)
```

The list comprehension will now apply the function to group keywords into clusters:

```
[find_topics(x, y, z) for x, y, z in zip(keywords_filtered_nonnan.si_simi,
keywords_filtered_nonnan.keyword, keywords_filtered_nonnan.topic)]
topic_groups_numbered
```

This results in the following:

```
{1: ['easy access savings',
  'savings account',
  'savings accounts uk',
  'savings rates',
  'online savings account',
  'online savings account',
  'online savings account'],
 2: ['isa account', 'isa', 'isa savings', 'isa savings'],
 3: ['kids savings account', 'child savings account'],
 4: ['best isa rates',
  'cash isa',
  'fixed rate isa',
  'fixed rate isa',
  'isa rates',
  'isa rates',
  'isa rates'],
 5: ['savings account interest rate',
  'savings accounts uk',
  'online savings account'],
 6: ['easy access savings account', 'savings rates', 'online savings
    account'],
 7: ['cash isa rates', 'fixed rate isa', 'isa rates'],
 8: ['isa interest rates', 'isa rates'],
 9: ['fixed rate savings', 'fixed rate bonds', 'online savings account']}
```

The preceding results are statements printing out what keywords are in which topic group. We do this to make sure we don't have duplicates or errors, which is crucial for the next step to perform properly. Now we're going to convert the dictionary into a dataframe so you can see all of your keywords grouped by search intent:

```
topic_groups_lst = []
for k, l in topic_groups_numbered.items():
    for v in l:
        topic_groups_lst.append([k, v])

topic_groups_dictdf = pd.DataFrame(topic_groups_lst, columns=['topic_group_
no', 'keyword'])
topic_groups_dictdf
```

This results in the following:

	topic_group_no	keyword
0	1	easy access savings
1	1	savings account
2	1	savings accounts uk
3	1	savings rates
4	1	online savings account
5	1	online savings account
6	1	online savings account
7	2	isa account
8	2	isa
9	2	isa savings
10	2	isa savings
11	3	kids savings account
12	3	child savings account
13	4	best isa rates
14	4	cash isa
15	4	fixed rate isa
16	4	fixed rate isa
17	4	isa rates

As you can see, the keywords are grouped intelligently, much like a human SEO analyst would group these, except these have been done at scale using the wisdom of Google which is distilled from its vast number of users. Name the clusters:

```
topic_groups_vols = topic_groups_dictdf.merge(keysv_df, on = 'keyword', how
= 'left')

def highest_demand(df):
```

```
    df = df.sort_values('search_volume', ascending = False)
    del df['topic_group_no']
    max_sv = df.search_volume.max()
    df = df.loc[df.search_volume == max_sv]
    return df

topic_groups_vols_keywgrp = topic_groups_vols.groupby('topic_group_no')
topic_groups_vols_keywgrp.get_group(1)
```

Apply and combine:

```
high_demand_topics = topic_groups_vols_keywgrp.apply(highest_demand).
reset_index()
del high_demand_topics['level_1']
high_demand_topics = high_demand_topics.rename(columns = {'keyword':
'topic'})

def shortest_name(df):
    df['k_len'] = df.topic.str.len()
    min_kl = df.k_len.min()
    df = df.loc[df.k_len == min_kl]
    del df['topic_group_no']
    del df['k_len']
    del df['search_volume']
    return df

high_demand_topics_spl = high_demand_topics.groupby('topic_group_no')
```

Apply and combine:

```
named_topics = high_demand_topics_spl.apply(shortest_name).reset_index()
del named_topics['level_1']
```

Name topic numbered keywords:

```
topic_keyw_map = pd.merge(named_topics, topic_groups_dictdf, on = 'topic_
group_no', how = 'left')
topic_keyw_map
```

The resulting table shows that we now have keywords clustered by topic:

	topic_group_no	topic	keyword
0	1	savings account	savings accounts uk
1	1	savings account	savings account
2	1	savings account	savings account interest rate
3	1	savings account	easy access savings
4	1	savings account	savings rates
5	1	savings account	fixed rate savings
6	1	savings account	fixed rate bonds
7	1	savings account	online savings account
8	1	savings account	easy access savings account
9	2	isa	isa
10	2	isa	isa account
11	2	isa	isa savings
12	3	child savings account	kids savings account
13	3	child savings account	child savings account
14	4	best isa rates	cash isa
15	4	best isa rates	best isa rates

Let's add keyword search volumes:

```
topic_keyw_vol_map = pd.merge(topic_keyw_map, keysv_df, on = 'keyword', how
= 'left')
topic_keyw_vol_map
```

This results in the following:

	topic_group_no	topic	keyword	search_volume
0	1	savings account	savings accounts uk	6600
1	1	savings account	savings account	60500
2	1	savings account	savings account interest rate	9900
3	1	savings account	easy access savings	3600
4	1	savings account	savings rates	5400
5	1	savings account	fixed rate savings	4400
6	1	savings account	fixed rate bonds	12100
7	1	savings account	online savings account	2400
8	1	savings account	easy access savings account	3600
9	2	isa	isa	49500
10	2	isa	isa account	9900
11	2	isa	isa savings	3600
12	3	child savings account	kids savings account	4400
13	3	child savings account	child savings account	14800
14	4	best isa rates	cash isa	14800
15	4	best isa rates	best isa rates	40500

This is really starting to take shape, and you can quickly see opportunities emerging.

SERP Competitor Titles

If you don't have much Google Search Console data or Google Ads data to mine, then you may need to resort to your competitors. You may or may not want to use third-party keyword research tools such as SEMRush. And you don't have to.

Tools like SEMRush, Keyword.io, etc., certainly have a place in the SEO industry. In the absence of any other data, they are a decent ready source of intelligence on what search queries generate relevant traffic.

However, some work will need to be done in order to weed out the noise and extract high value phrases – assuming a competitive market. Otherwise, if your website (or niche) is so new in terms of what it offers that there's insufficient demand (that has yet to be created by advertising and PR to generate nonbrand searches), then these external tools won't be as valuable. So, our approach will be to

1. Crawl your own website

2. Filter and clean the data for sections covering only what you sell

3. Extract keywords from your site's title tags

4. Filter using SERPs data (next section)

Filter and Clean the Data for Sections Covering Only What You Sell

The required data for this exercise is to literally take a site auditor[5] and crawl your website. Let's assume you've exported the crawl data with just the columns: URL and title tag; we'll import and clean:

```
import pandas as pd
import numpy as np

crawl_import_df = pd.read_csv('data/crawler-filename.csv')
crawl_import_df
```

This results in the following:

	deeprank	page_title	url	redirected_to_url	http_status_code	indexable
0	0.71	Growing blueberries in pots - Saga	https://www.saga.co.uk/magazine/home-garden/ga...	NaN	200	True
1	0.66	How to grow succulents - Saga	https://www.saga.co.uk/magazine/home-garden/ga...	NaN	200	True
2	0.49	Creating a wildlife garden: courtyards & small...	https://www.saga.co.uk/magazine/home-garden/ga...	NaN	200	True
3	0.55	Plants for clay soil - Saga	https://www.saga.co.uk/magazine/home-garden/ga...	NaN	200	True
4	0.28	The brambling: diet, identifcation & location ...	https://www.saga.co.uk/magazine/home-garden/ga...	NaN	200	True
...	...	...	...	...	...	...
7026	0.49	The best plants to complement roses - Saga	https://www.saga.co.uk/magazine/home-garden/ga...	NaN	200	True
7027	0.77	How to buy outdoor security lights for your ga...	https://www.saga.co.uk/magazine/home-garden/ga...	NaN	200	True
7028	0.39	Wildlife watch: lesser spotted woodpecker - Saga	https://www.saga.co.uk/magazine/home-garden/ga...	NaN	200	True
7029	0.25	Has my border collie's coat been damaged by cl...	https://www.saga.co.uk/magazine/home-garden/pe...	NaN	200	True
7030	0.32	The best daffodil varieties that bloom all spr...	https://www.saga.co.uk/magazine/home-garden/ga...	NaN	200	False

---

[5] Like Screaming Frog, OnCrawl, or Botify, for instance

The preceding result shows the dataframe of the crawl data we've just imported. We're most interested in live indexable[6] URLs, so let's filter and select the page_title and URL columns:

```
titles_urls_df = crawl_import_df.loc[crawl_import_df.indexable == True]
titles_urls_df = titles_urls_df[['page_title', 'url']]
titles_urls_df
```

This results in the following:

	page_title	url
0	Growing blueberries in pots - Saga	https://www.saga.co.uk/magazine/home-garden/ga...
1	How to grow succulents - Saga	https://www.saga.co.uk/magazine/home-garden/ga...
2	Creating a wildlife garden: courtyards & small...	https://www.saga.co.uk/magazine/home-garden/ga...
3	Plants for clay soil - Saga	https://www.saga.co.uk/magazine/home-garden/ga...
4	The brambling: diet, identifcation & location ...	https://www.saga.co.uk/magazine/home-garden/ga...
...	...	...
7025	Avoid these online tolls and road charges scam...	https://www.saga.co.uk/magazine/motoring/cars/...
7026	The best plants to complement roses - Saga	https://www.saga.co.uk/magazine/home-garden/ga...
7027	How to buy outdoor security lights for your ga...	https://www.saga.co.uk/magazine/home-garden/ga...
7028	Wildlife watch: lesser spotted woodpecker - Saga	https://www.saga.co.uk/magazine/home-garden/ga...
7029	Has my border collie's coat been damaged by cl...	https://www.saga.co.uk/magazine/home-garden/pe...

Now we're going to clean the title tags to make these nonbranded, that is, remove the site name and the magazine section.

```
titles_urls_df['page_title'] = titles_urls_df.page_title.str.replace(' -
Saga', '')
titles_urls_df = titles_urls_df.loc[~titles_urls_df.url.str.contains('/
magazine/')]
titles_urls_df
```

This results in the following:

---

[6] That is, pages with a 200 HTTP response that do block search indexing with "noindex"

	page_title	url
9	Travel money | Saga | Safe and quick order today	https://www.saga.co.uk/insurance/travel-money
21	MySaga	https://www.saga.co.uk/mysaga/manage-policy/html/
31	Direct Choice Contact Us	https://www.saga.co.uk/insurance/car-insurance...
48	Personal SOS Alarms Service | Saga Healthcare	https://www.saga.co.uk/sos-personal-alarm
49	Tailor Your Cover With Optional Extras | Saga ...	https://www.saga.co.uk/insurance/car-insurance...
...	...	...
6991	At home with John Sergeant	https://www.saga.co.uk/membership/articles/at-...
6994	Policy Documents | Policy Books | Saga Home In...	https://www.saga.co.uk/insurance/home-insuranc...
6995	Health and Beauty Offers From Saga	https://www.saga.co.uk/membership/categories/h...
6996	Single Trip Travel Insurance for Over 50s | Sa...	https://www.saga.co.uk/insurance/travel-insura...
7008	Oleanna	https://www.saga.co.uk/membership/tickets/lond...

We now have 349 rows, so we will query some of the keywords to illustrate the process.

Extract Keywords from the Title Tags

We now desire to extract keywords from the page title in the preceding dataframe. A typical data science approach would be to break down the titles into all kinds of combinations and then do a frequency count, maybe weighted by ranking.

Having tried it, we wouldn't recommend this approach; it's overkill and there is probably not enough data to make it worthwhile. A more effective and simpler approach is to break down the titles by punctuation marks. Why? Because humans (or probably some AI nowadays) wrote those titles, so these are likely to be natural breakpoints for target search phrases.

Let's try it; break the titles into n grams:

```
pd.set_option('display.max_rows', 1000)
serps_ngrammed = filtered_serps_df.set_index(["keyword", "rank_absolute"])\
                .apply(lambda x: x.str.split('[-,|?()&:;\[\]=]').
                explode())\
                .dropna()\
                .reset_index()
serps_ngrammed.head(10)
```

This results in the following:

	keyword	rank_absolute	title
0	Care Funding Advice Service	1	care funding advice service
1	Care Funding Advice Service	1	saga
2	Care Funding Advice Service	2	how does the saga care funding advice service ...
3	Care Funding Advice Service	2	
4	Care Funding Advice Service	3	paying for care
5	Care Funding Advice Service	3	money advice service
6	Care Funding Advice Service	4	care funding advice
7	Care Funding Advice Service	4	hub financial solutions
8	Care Funding Advice Service	5	care advice service
9	Care Funding Advice Service	5	paying for care fees in sussex and ...

Courtesy of the explode function, the dataframe has been unnested such that we can see the keyword rows expanded for the different text previously within the same title and conjoined by the punctuation mark.

Filter Using SERPs Data

Now all we have to do is perform a frequency count of the top three titles and then filter for any that appear three times or more:

```
serps_ngrammed_grp = serps_ngrammed.groupby(['keyword', 'title'])
keyword_ideas_df = serps_ngrammed_grp.size().reset_index(name='freq').sort_
values(['keyword', 'freq'], ascending = False)
keyword_ideas_df = keyword_ideas_df[keyword_ideas_df.freq > 2]
keyword_ideas_df = keyword_ideas_df[keyword_ideas_df.title.str.
contains('[a-z]')]
keyword_ideas_df = keyword_ideas_df.rename(columns = {'title':
'keyword_idea'})
keyword_ideas_df
```

This results in the following:

	keyword	rank_absolute	title
0	Care Funding Advice Service	1	care funding advice service
1	Care Funding Advice Service	1	saga
2	Care Funding Advice Service	2	how does the saga care funding advice service ...
3	Care Funding Advice Service	2	
4	Care Funding Advice Service	3	paying for care
5	Care Funding Advice Service	3	money advice service
6	Care Funding Advice Service	4	care funding advice
7	Care Funding Advice Service	4	hub financial solutions
8	Care Funding Advice Service	5	care advice service
9	Care Funding Advice Service	5	paying for care fees in sussex and ...

Eh voila, the preceding result shows a dataframe of keywords obtained from the SERPs. Most of it makes sense and can now be added to your list of keywords for serious consideration and tracking.

Summary

This chapter has covered data-driven keyword research, enabling you to

- Find standout keywords from GSC data

- Obtain trend data from Google Trends

- Forecast future organic traffic using time series techniques

- Cluster keywords by search intent

- Find keywords from your competitors using SERPs data

In the next chapter, we will cover the mapping of those keywords to URLs.

CHAPTER 3

Technical

Technical SEO mainly concerns the interaction of search engines and websites such that

- Website content is made discoverable by search engines.

- The priority of content is made apparent to search engines implied by its proximity to the home page.

- Search engine resources are conserved to access content (known as crawling) intended for search result inclusion.

- Extract the content meaning from those URLs again for search result inclusion (known as indexing).

In this chapter, we'll look at how data-driven approach can be taken toward improving technical SEO in the following manner:

- *Modeling page authority*: This is useful for helping fellow SEO and non-SEOs understand the impact of technical SEO changes.

- *Internal link optimization*: To improve the use of internal links used to make content more discoverable and help signal to search engines the priority of content.

- *Core Web Vitals (CWV)*: While the benefits to the UX are often lauded, there are ranking boost benefits to an improved CWV because of the conserved search engine resources used to extract content from a web page.

By no means will we claim that this is the final word on data-driven SEO from a technical perspective. What we will do is expose data-driven ways of solving technical SEO issues using some data science such as distribution analysis.

© Andreas Voniatis 2023
A. Voniatis, *Data-Driven SEO with Python*, https://doi.org/10.1007/978-1-4842-9175-7_3

Where Data Science Fits In

An obvious challenge of SEO is deciding which pages should be made accessible to the search engines and users and which ones should not. While many crawling tools provide visuals of the distributions of pages by site depth, etc., it never hurts to use data science, which we will go into more detail and complexity, which will help you

- Optimize internal links

- Allocate keywords to pages based on the copy

- Allocate parent nodes to the orphaned URLs

Ultimately, the preceding list will help you build better cases for getting technical recommendations implemented.

Modeling Page Authority

Technical optimization involves recommending changes that often make URLs nonindexable or canonicalized (for a number of reasons such as duplicate content). These changes are recommended with the aim of consolidating page authority onto URLs which will remain eligible for indexing.

The following section aims to help data-driven SEO quantify the beneficial extra page authority. The approach will be to

1. Filter in web pages

2. Examine the distribution of authority before optimization

3. Calculate the new distribution (to quantify the incremental page authority following a decision on which URLs will no longer be made indexable, making their authority available for reallocation)

First, we need to load the necessary packages:

```
import re
import time
import random
import pandas as pd
import numpy as np
import datetime
```

```
import requests
import json
from datetime import timedelta
from glob import glob
import os
from client import RestClient # If using the Data For SEO API
from textdistance import sorensen_dice
from plotnine import *
import matplotlib.pyplot as plt
from pandas.api.types import is_string_dtype
from pandas.api.types import is_numeric_dtype
import uritools
from urllib.parse import urlparse
import tldextract

pd.set_option('display.max_colwidth', None)
%matplotlib inline
```

Set variables:

```
root_domain = 'boundlesshq.com'
hostdomain = 'www.boundlesshq.com'
hostname = 'boundlesshq'
full_domain = 'https://www.boundlesshq.com'
client_name = 'Boundless'
audit_monthyear = 'jul_2022'
```

Import the crawl data from the Sitebulb desktop crawler. Screaming Frog or any other site crawling software can be used; however, the column names may differ:

```
crawl_csv = pd.read_csv('data/boundlesshq_com_all_urls__excluding_
uncrawled__filtered.csv')
```

Clean the column names using a list comprehension:

```
crawl_csv.columns = [col.lower().replace('.','').replace('(','').
replace(')','').replace(' ','_')
                    for col in crawl_csv.columns]

crawl_csv
```

Here is the result of crawl_csv:

	url	crawl_depth	craw
0	https://boundlesshq.com/	0	
1	https://boundlesshq.com/wp-content/uploads/oxygen/css/2920.css	1	
2	https://bat.bing.com/action/0?ti=149000342&tm=gtm002&Ver=2&mid=40c49240-b5a7-4d30-9cc3-29f89fc1e165&sid=d7008f20086b11edbbe6afef94e29e7c&vid=e23f2650fc5911ec911285f6dcda096a&vids=0&gtm_tag_source=1&pi=2083220816&lg=en-US&sw=1512&sh=982&sc=30&tl=Blog%20-%20Boundless&p=https%3A%2F%2Fboundlesshq.com%2Fblog%2F&r=&lt=1347&evt=pageLoad&msclkid=N&sv=1&rn=541225	1	
3	https://boundlesshq.com/wp-content/uploads/2022/06/touch-handy.png	1	
4	https://boundlesshq.com/wp-content/uploads/oxygen/css/5983.css	1	
...	...	...	
5417	https://www.convertcalculator.com/_next/static/chunks/462-a4fa6446d46110f5.js	1	
5418	https://static.cloudflareinsights.com/beacon.min.js	1	
5419	https://fonts.gstatic.com/s/inter/v12/UcC73FwrK3iLTeHuS_fvQtMwCp50KnMa1ZL7W0Q5nw.woff2	1	
5420	https://www.convertcalculator.com/_next/static/4t_ps-wEsto6ZaMDrAxlf/_ssgManifest.js	1	
5421	https://www.convertcalculator.com/_next/static/chunks/webpack-ab4bf6124310705c.js	1	

5422 rows × 25 columns

The dataframe is loaded into a Pandas dataframe. The most important fields are as follows:

- *url*: To detect patterns for noindexing and canonicalizing

- *ur*: URL Rating, Sitebulb's in-house metric for measuring internal page authority

- *content_type*: For filtering

- *passes_pagerank*: So we know which pages have authority

- *indexable*: Eligible for search engine index inclusion

Filtering in Web Pages

The next step is to filter in actual web pages that belong to the site and are capable of passing authority:

```
crawl_html = crawl_csv.copy()
crawl_html = crawl_html.loc[crawl_html['content_type'] == 'HTML']
crawl_html = crawl_html.loc[crawl_html['host'] == root_domain]
crawl_html = crawl_html.loc[crawl_html['passes_pagerank'] == 'Yes']

crawl_html
```

	url	crawl_depth	crawl_status	host	is_subdomain	scheme	crawl_source	
0	https://boundlesshq.com/	0	Success	boundlesshq.com	No	https	Crawler	
6	https://boundlesshq.com/blog/	1	Success	boundlesshq.com	No	https	Crawler	https:/.
7	https://boundlesshq.com/summer-sale/	1	Success	boundlesshq.com	No	https	Crawler	https:/.
28	https://boundlesshq.com/blog/employment/what-is-an-employer-of-record/	1	Success	boundlesshq.com	No	https	Crawler	https:/.
47	https://boundlesshq.com/how-it-works/	1	Success	boundlesshq.com	No	https	Crawler	https:/.
...	...	...	...	...	...	...	...	
5354	https://boundlesshq.com/category/blog/page/3/	4	Success	boundlesshq.com	No	https	Crawler	https://boundlesshq.c
5356	https://boundlesshq.com/category/blog/page/2/	4	Success	boundlesshq.com	No	https	Crawler	https://boundlesshq.c
5358	https://boundlesshq.com/category/blog/page/4/	4	Success	boundlesshq.com	No	https	Crawler	https://boundlesshq.c
5370	https://boundlesshq.com/category/blog/page/1/	5	Redirect	boundlesshq.com	No	https	Crawler	https://boundlesshq.com/cat
5385	https://boundlesshq.com/uk-custom-report/	Not Set	Success	boundlesshq.com	No	https	XML Sitemap	

309 rows × 25 columns

The dataframe has been reduced to 309 rows. For ease of data handling, we'll select some columns:

```
crawl_select = crawl_html[['url', 'ur', 'crawl_depth', 'crawl_source',
'http_status_code', 'indexable',
                'indexable_status', 'passes_pagerank', 'total_impressions',
'first_parent_url', 'meta_robots_response']].copy()
```

Examine the Distribution of Authority Before Optimization

It is useful for groupby aggregation and counting:

```
crawl_select['project'] = client_name
crawl_select['count'] = 1
```

Let's get some quick stats:

```
print(crawl_select['ur'].sum(), crawl_select['ur'].sum()/crawl_select.
shape[0])
```

```
10993 35.57605177993528
```

URLs on this site have an average page authority level (measured as UR). Let's look at some further stats, indexable and nonindexable pages. We'll dimension on (I) indexable and (II) passes pagerank to sum the number of URLs and UR (URL Rating):

```
overall_pagerank_agg = crawl_select.groupby(['indexable',
```

```
                                              'passes_pagerank']).agg
                                              ({'count': 'sum',

                                                                  'ur':
                                                                  'sum'}).
                                                                  reset_
                                                                  index()
```

Then we derive the page authority per URL by dividing the total UR by the total number of URLs:

```
overall_pagerank_agg['PA'] = overall_pagerank_agg['ur'] / overall_pagerank_
agg['count']
overall_pagerank_agg
```

This results in the following:

	indexable	passes_pagerank	count	ur	PA
0	No	Yes	32	929	29.03125
1	Yes	Yes	277	10064	36.33213

We see that there are 32 nonindexable URLs with a total authority of 929 that could be consolidated to the indexable URLs.

There are some more stats, this time analyzed by site level purely out of curiosity:

```
site_pagerank_agg = crawl_select.groupby(['indexable',
                                          'crawl_depth']).
                                          agg({'count': 'sum',

                                                            'ur':
                                                            'sum'}).
                                                            reset_
                                                            index()
site_pagerank_agg['PA'] = site_pagerank_agg['ur'] / site_pagerank_
agg['count']

site_pagerank_agg
```

This results in the following:

	indexable	crawl_depth	count	ur	PA
0	No	1	2	186	93.000000
1	No	2	1	20	20.000000
2	No	3	19	640	33.684211
3	No	4	9	73	8.111111
4	No	5	1	10	10.000000
5	Yes	0	1	95	95.000000
6	Yes	1	13	1127	86.692308
7	Yes	2	46	2052	44.608696
8	Yes	3	196	6470	33.010204
9	Yes	4	20	320	16.000000
10	Yes	Not Set	1	0	0.000000

Most of the URLs that have the authority for reallocation are four clicks away from the home page.

Let's visualize the distribution of the authority preoptimization, using the geom_histogram function:

```
pageauth_dist_plt = (
    ggplot(crawl_select, aes(x = 'ur')) +
    geom_histogram(alpha = 0.7, fill = 'blue', bins = 20) +
    labs(x = 'Page Authority', y = 'URL Count') +
    theme(legend_position = 'none', axis_text_x=element_text(rotation=0,
    hjust=1, size = 12))
)

pageauth_dist_plt.save(filename = 'images/1_pageauth_dist_plt.png',
                        height=5, width=8, units = 'in', dpi=1000)
pageauth_dist_plt
```

As we'd expect from looking at the stats computed previously, most of the pages have between 25 and 50 UR, with the rest spread out (Figure 3-1).

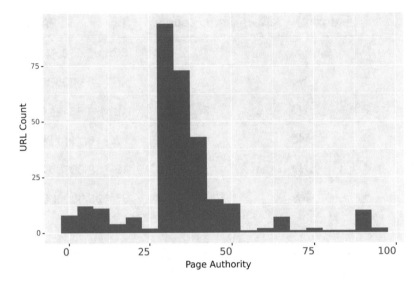

Figure 3-1. *Histogram plot showing URL count of URL Page Authority scores*

Calculating the New Distribution

With the current distribution examined, we'll now go about quantifying the new page authority distribution following optimization.

We'll start by getting a table of URLs by the first parent URL and the URL's UR values which will be our mapping for how much extra authority is available:

```
parent_pa_map = crawl_select[['first_parent_url', 'ur']].copy()
parent_pa_map = parent_pa_map.rename(columns = {'first_parent_url': 'url' ,
'ur': 'extra_ur'})

parent_pa_map
```

This results in the following:

	url	extra_ur
0	None	95
6	https://boundlesshq.com/	80
7	https://boundlesshq.com/	70
28	https://boundlesshq.com/	88
47	https://boundlesshq.com/	91
...	...	...
5354	https://boundlesshq.com/category/blog/	10
5356	https://boundlesshq.com/category/blog/	10
5358	https://boundlesshq.com/category/blog/	10
5370	https://boundlesshq.com/category/blog/page/2/	10
5385	None	0

309 rows × 2 columns

The table shows all the parent URLs and their mapping.

The next step is to mark pages that will be noindexed, so we can reallocate their authority:

```
crawl_optimised = crawl_select.copy()
```

Create a list of URL patterns for noindex:

```
reallocate_conds = [
    crawl_optimised['url'].str.contains('/page/[0-9]/'),
    crawl_optimised['url'].str.contains('/country/')
]
```

Values if the URL pattern conditions are met.

```
reallocate_vals = [1, 1]
```

The reallocate column uses the np.select function to mark URLs for noindex. Any URLs not for noindex are marked as "0," using the default parameter:

```
crawl_optimised['reallocate'] = np.select(reallocate_conds, reallocate_
vals, default = 0)
```

`crawl_optimised`

This results in the following:

first_parent_url	meta_robots_response	project	count	reallocate
None	index, follow, max-image-preview:large, max-snippet:-1, max-video-preview:-1	Boundless	1	0
https://boundlesshq.com/	index, follow, max-image-preview:large, max-snippet:-1, max-video-preview:-1	Boundless	1	0
https://boundlesshq.com/	index, follow, max-image-preview:large, max-snippet:-1, max-video-preview:-1	Boundless	1	0
https://boundlesshq.com/	index, follow, max-image-preview:large, max-snippet:-1, max-video-preview:-1	Boundless	1	0
https://boundlesshq.com/	index, follow, max-image-preview:large, max-snippet:-1, max-video-preview:-1	Boundless	1	0
...	...	...	...	...

The reallocate column is added so we can start seeing the effect of the reallocation, that is, the potential upside of technical optimization.

As usual, a groupby operation by reallocate and the average PA are calculated:

```
reallocate_agg = crawl_optimised.groupby('reallocate').agg({'count': sum,
'ur': sum}).reset_index()
reallocate_agg['PA'] = reallocate_agg['ur'] / reallocate_agg['count']
reallocate_agg
```

This results in the following:

	reallocate	count	ur	PA
0	0	285	10312	36.182456
1	1	24	681	28.375000

So we'll be actually reallocating 681 UR from the noindex URLs to the 285 indexable URLs. These noindex URLs have an average UR of 28.

We filter the URLs just for the ones that will be noindexed to help us in determining what the extra page authority will be:

```
no_indexed = crawl_optimised.loc[crawl_optimised['reallocate'] == 1]
```

We aggregate by the first parent URL (the parent node) for the total URLs within and their URL, because the UR is likely to be reallocated to the remaining indexable URLs that share the same parent node:

```
no_indexed_map = no_indexed.groupby('first_parent_url').agg({'count':
'sum', 'ur': sum}).reset_index()
```

add_ur is a new column created representing the additional authority as a result of the optimization. This is the total UR divided by the number of URLs:

```
no_indexed_map['add_ur'] = (no_indexed_map['ur'] / no_indexed_
map['count']).round(0)
```

Drop columns not required for joining later:

```
no_indexed_map.drop(['ur', 'count'], inplace = True,  axis = 1)
no_indexed_map
```

This results in the following:

	first_parent_url	add_ur
0	https://boundlesshq.com/category/blog/	10.0
1	https://boundlesshq.com/category/blog/employment/	2.0
2	https://boundlesshq.com/category/blog/employment/page/2/	2.0
3	https://boundlesshq.com/category/blog/page/2/	10.0
4	https://boundlesshq.com/guides/australia/	35.0
5	https://boundlesshq.com/guides/brazil/	38.0
6	https://boundlesshq.com/guides/canada/	35.0
7	https://boundlesshq.com/guides/chile/end-of-employment/	31.0

The preceding table will be merged into the indexable URLs by the first parent URL.

Filter the URLs just for the indexable and add more authority as a result of the noindexing reallocate URLs:

```
crawl_new = crawl_optimised.copy()
crawl_new = crawl_new.loc[crawl_new['reallocate'] == 0]
```

Join the no_indexed_map to get the amount of authority to be added:

```
crawl_new = crawl_new.merge(no_indexed_map, on = 'first_parent_url', how
= 'left')
```

Often, when joining data, there will be null values for first parent URLs not in the mapping. np.where() is used to replace those null values with zeros. This enables further data manipulation to take place as you'll see shortly.

```
crawl_new['add_ur'] = np.where(crawl_new['add_ur'].isnull(), 0, crawl_
new['add_ur'])
```

New_ur is the new authority score calculated by adding ur to add_ur:

```
crawl_new['new_ur'] = crawl_new['ur'] + crawl_new['add_ur']
```

```
crawl_new
```

This results in the following:

first_parent_url	meta_robots_response	project	count	reallocate	add_ur	new_ur
None	index, follow, max-image-preview:large, max-snippet:-1, max-video-preview:-1	Boundless	1	0	0.0	95.0
https://boundlesshq.com/	index, follow, max-image-preview:large, max-snippet:-1, max-video-preview:-1	Boundless	1	0	0.0	80.0
https://boundlesshq.com/	index, follow, max-image-preview:large, max-snippet:-1, max-video-preview:-1	Boundless	1	0	0.0	70.0
https://boundlesshq.com/	index, follow, max-image-preview:large, max-snippet:-1, max-video-preview:-1	Boundless	1	0	0.0	88.0
https://boundlesshq.com/	index, follow, max-image-preview:large, max-snippet:-1, max-video-preview:-1	Boundless	1	0	0.0	91.0

The indexable URLs now have their authority scores post optimization, which we'll visualize as follows:

```
pageauth_newdist_plt = (
    ggplot(crawl_new, aes(x = 'new_ur')) +
    geom_histogram(alpha = 0.7, fill = 'lightgreen', bins = 20) +
    labs(x = 'Page Authority', y = 'URL Count') +
    theme(legend_position = 'none', axis_text_x=element_text(rotation=0,
    hjust=1, size = 12))
)

pageauth_newdist_plt.save(filename = 'images/2_pageauth_newdist_plt.png',
                    height=5, width=8, units = 'in', dpi=1000)
pageauth_newdist_plt
```

The pageauth_newdist_plt in Figure 3-2 shows the distribution of page-level authority (page authority).

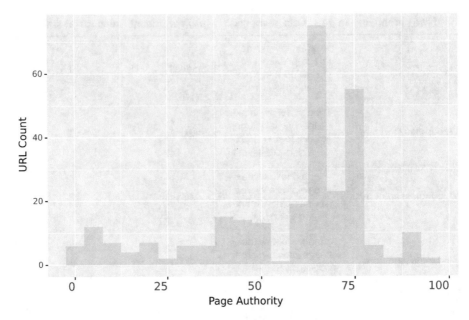

Figure 3-2. *Histogram of the distribution of page-level authority (page authority)*

The impact is noticeable, as we see most pages are above 60 UR post optimization, should the implementation move forward.

There are some quick stats to confirm:

```
new_pagerank_agg = crawl_new.groupby(['reallocate']).agg({'count': 'sum',
                                                           'ur': 'sum',
                                                           'new_ur':
                                                           'sum'}).
                                                           reset_ex()
```

```
new_pagerank_agg['PA'] = new_pagerank_agg['new_ur'] / new_pagerank_
agg['count']
```

```
print(new_pagerank_agg)
```

```
   reallocate  count     ur    new_ur    PA
0            0    285  10312   16209.0  57.0
```

The average page authority is now 57 vs. 36, which is a significant improvement. While this method is not an exact science, it could help you to build a case for getting your change requests for technical SEO fixes implemented.

Internal Link Optimization

Search engines are highly dependent on links in order to help determine the relative importance of pages within a website. That's because search engines work on the basis of assigning probability that content will be found by users at random based on the random surfer concept. That is, a content is more likely to be discovered if there are more links to the content.

If the content has more inbound links, then search engines also assume the content has more value, having earned more links.

Search engines also rely on the anchor text to signal what the hyperlinked URL's content will be about and therefore its relevance to keywords.

Thus, for SEO, internal links play a key role in website optimization, helping search engines decide which pages on the site are important and their associated keywords.

Here, we shall provide methods to optimize internal links using some data science, which will cover

1. Distributing authority by site level

2. Distributing authority by external page authority accrued from external sites

3. Anchor text

```
import pandas as pd
import numpy as np
from textdistance import sorensen_dice
from plotnine import *
import matplotlib.pyplot as plt
from mizani.formatters import import comma_format

target_name = 'ON24'
target_filename = 'on24'
website = 'www.on24.com'
```

The link data is sourced from the Sitebulb auditing software which is being imported along with making the column names easier to work with:

```
link_raw = pd.read_csv('data/'+ client_filename + '_links.csv')
link_data = link_raw.copy()
```

```
link_data.drop('Unnamed: 13', axis = 1, inplace = True)
```

```
link_data.columns = [col.lower().replace('.','').replace('(','').
replace(')','').replace(' ','_')
                    for col in link_data.columns]
```

```
link_data
```

	referring_url	referring_url_rank_ur	target_url	target_url_rank_ur	anchor_text	location	crawl_status	no
0	https://www.on24.com/	96	https://www.on24.com/resources/asset/webinar-b...	92	Register Now	Content	Success	
1	https://www.on24.com/	96	https://www.on24.com/contact-us/	96	Contact Us	Header	Success	
2	https://www.on24.com/	96	https://www.on24.com/login/	100	Login	Header	Success	
3	https://www.on24.com/	96	https://www.on24.com/resources/	96	Resources	Header	Success	
4	https://www.on24.com/	96	https://www.on24.com/live-demo/	96	Live Demo	Header	Success	
...	...	...	...	...	...	...	...	...
406022	https://www.on24.com/zapier/	0	https://www.on24.com/about-us/careers/	96	Careers	Footer	Success	
406023	https://www.on24.com/zapier/	0	https://www.on24.com/newsroom/	96	In The News	Footer	Success	
406024	https://www.on24.com/zapier/	0	https://www.on24.com/press-releases/	96	Press Releases	Footer	Success	
406025	https://www.on24.com/zapier/	0	https://www.on24.com/contact-us/	96	Contact Sales	Footer	Success	
406026	https://www.on24.com/zapier/	0	https://www.on24.co.jp/	0	Japan Website	Footer	Success	

406027 rows × 13 columns

The link dataframe shows us a list of links in terms of

- *Referring URL*: Where they are found

- *Target URL*: Where they point to

- *Referring URL Rank UR*: The page authority of the referring page

- *Target URL Rank UR*: The page authority of the target page

- *Anchor text*: The words used in the hyperlink

- *Location*: Where the link can be found

Let's import the crawl data, also sourced from Sitebulb:

```
crawl_data = pd.read_csv('data/'+ client_filename + '_crawl.csv')
```

```
crawl_data.drop('Unnamed: 103', axis = 1, inplace = True)
```

```
crawl_data.columns = [col.lower().replace('.','').replace('(','').
replace(')','').replace(' ','_')
                        for col in crawl_data.columns]
```

```
crawl_data
```

This results in the following:

	url	base_url	crawl_depth	crawl_status	host	http_protocol	is_subdomain	no_query_string_keys	
0	https://www.on24.com/	No Data	0	Success	www.on24.com	http/1.1	No	0	
1	https://www.on24.com/customer-stories/align-te...	No Data	1	Not Found	www.on24.com	http/1.1	No	0	
2	https://www.on24.com/contact-us/	No Data	1	Success	www.on24.com	http/1.1	No	0	
3	https://www.on24.com/resources/	No Data	1	Success	www.on24.com	http/1.1	No	0	
4	https://www.on24.com/blog/how-juniper-networks...	No Data	1	Success	www.on24.com	http/1.1	No	0	
...	...	...	...	...	...	...	...	...	
8606	https://on24.influitive.com/saml/initialize	No Data	Not Set	Redirect	on24.influitive.com	http/1.1	No	0	
8607	https://www.contenttechsummit.com/	No Data	Not Set	Redirect	www.contenttechsummit.com	http/1.1	No	0	
8608	https://newsnetwork.mayoclinic.org/	No Data	Not Set	Success	newsnetwork.mayoclinic.org	http/1.1	No	0	
8609	https://www.eetimes.com/	No Data	Not Set	Success	www.eetimes.com	http/1.1	No	0	
8610	http://offers.hubspot.com/how-to-promote-a-wor...	No Data	Not Set	Redirect	offers.hubspot.com	http/1.1	No	0	

8611 rows × 104 columns

So we have the usual list of URLs and how they were found (crawl source) with other features spanning over 100 columns.

As you'd expect, the number of rows in the link data far exceeds the crawl dataframe as there are many more links than pages!

Import the external inbound link data:

```
ahrefs_raw = pd.read_csv('data/'+ client_filename + '_ahrefs.csv')
```

```
ahrefs_raw.columns = [col.lower().replace('.','').replace('(','').
replace(')','').replace(' ','_')
                        for col in ahrefs_raw.columns]
```

```
ahrefs_raw
```

This results in the following:

	#	url_rating_desc	page_url	page_title	referring_domains	dofollow	nofollow	redirects	first_seen	size	co(
0	1	81	https://www.on24.com/	Webinar Software & Virtual Event Platform | ON24	3215	64144	18915	48	20/12/2018 17:02	32 kB	2(
1	2	54	http://www.on24.com/	NaN	856	4666	1017	173	08/08/2013 16:21	162 B	3(
2	3	52	https://vshow.on24.com/view/vts/error.html?cod...	Error	643	3455	299	0	12/12/2014 12:07	3 kB	2(
3	4	43	https://on24.com/	NaN	668	46591	4628	0	17/12/2018 03:07	0	3(
4	5	41	https://www.on24.com/live-webcast-elite/	Deliver Better Webcasts on ON24 Webcast Elite ...	208	173	363	0	04/01/2019 01:57	33 kB	2(
...	...	...	...	...	...	...	...	...	...	...	
210399	210400	0	https://www.on24.com/ww19internal/	NaN	1	0	10	0	NaN	0	
210400	210401	0	https://www.on24.com/youre-in/	NaN	1	0	10	0	NaN	0	
210401	210402	0	https://www-staging.on24.com/	NaN	1	2	0	0	NaN	0	
210402	210403	0	http://acueductoportachuelo.xyzevent.on24.com/...	NaN	1	5	0	0	NaN	0	
210403	210404	0	https://golearnership.co.zaevent.on24.com/even...	NaN	1	22	0	0	NaN	0	

210404 rows × 13 columns

There are over 210,000 URLs with backlinks, which is very nice! There's quite a bit of data, so let's simplify a little by removing columns and renaming some columns so we can join the data later:

```
ahrefs_df = ahrefs_raw[['page_url', 'url_rating_desc', 'referring_
domains']]
ahrefs_df = ahrefs_df.rename(columns = {'url_rating_desc': 'page_
authority', 'page_url': 'url'})
ahrefs_df
```

This results in the following:

	url	page_authority	referring_domains
0	https://www.on24.com/	81	3215
1	http://www.on24.com/	54	856
2	https://vshow.on24.com/view/vts/error.html?cod...	52	643
3	https://on24.com/	43	668
4	https://www.on24.com/live-webcast-elite/	41	208
...	...	...	...
210399	https://www.on24.com/ww19internal/	0	1
210400	https://www.on24.com/youre-in/	0	1
210401	https://www-staging.on24.com/	0	1
210402	http://acueductoportachuelo.xyzevent.on24.com/...	0	1
210403	https://golearnership.co.zaevent.on24.com/even...	0	1

210404 rows × 3 columns

Now we have the data in its simplified form which is important because we're not interested in the detail of the links but rather the estimated page-level authority that they import into the target website.

By Site Level

With the data imported and cleaned, the analysis can now commence.

We're always curious to see how many URLs we have at different site levels. We'll achieve this with a quick groupby aggregation function:

```
redir_live_urls.groupby(['crawl_depth']).size()
```

This results in the following:

```
crawl_depth
0              1
1             70
10             5
11             1
```

```
12                   1
13                   2
14                   1
2                  303
3                  378
4                  347
5                  253
6                  194
7                   96
8                   33
9                   19
Not Set          2351
dtype: int64
```

We can see how Python is treating the crawl depth as a string character rather than a numbered category, which we can fix shortly.

Most of the site URLs can be found in the site depths of 2 to 6. There are over 2351 orphaned URLs, which means these won't inherit any authority unless they have backlinks.

We'll now filter for redirected and live links:

```
redir_live_urls = crawl_data[['url', 'crawl_depth', 'http_status_code',
'indexable', 'no_internal_links_to_url', 'host', 'title']]
```

The dataframe is filtered to include URLs that are indexable:

```
redir_live_urls = redir_live_urls = redir_live_urls.loc[redir_live_
urls['indexable'] == 'Yes']
```

Crawl depth is set as a category and ordered so that Python treats the column variable as a number as opposed to a string character type:

```
redir_live_urls['crawl_depth'])
redir_live_urls['crawl_depth'] = redir_live_urls['crawl_depth'].
astype('category')
redir_live_urls['crawl_depth'] = redir_live_urls['crawl_depth'].cat.
reorder_categories(['0', '1', '2', '3', '4', '5', '6', '7', '8', '9', '10',
'Not Set'
```

```
            ])
redir_live_urls = redir_live_urls.loc[redir_live_urls.host == website]
redir_live_urls.drop('host', axis = 1, inplace = True)

redir_live_urls
```

This results in the following:

	url	crawl_depth	http_status_code	indexable	indexable_status	no_internal_links_to_url	title
0	https://www.on24.com/	0	200	Yes	Indexable	4139	Webinar Software & Virtual Event Platform | ON24
2	https://www.on24.com/contact-us/	1	200	Yes	Indexable	11189	Contact Us | Global Office Locations | ON24 Te...
3	https://www.on24.com/resources/	1	200	Yes	Indexable	11155	Webinar, White Paper, and Video Resources | ON24
4	https://www.on24.com/blog/how-juniper-networks...	1	200	Yes	Indexable	17	How Juniper Networks Set Up a Global Virtual S...
6	https://www.on24.com/solutions/manufacturing/	1	200	Yes	Indexable	7414	Platform for Manufacturing Industry | ON24
...	...	...	...	...	...	...	...
6949	https://www.on24.com/customer-stories/shell-ex...	Not Set	200	Yes	Indexable	1	Shell expands global brand awareness and drive...
6950	https://www.on24.com/blog/category/feature-fri...	Not Set	200	Yes	Indexable	0	Feature Friday Archives | ON24
6951	https://www.on24.com/blog/use_cases/certificat...	Not Set	200	Yes	Indexable	0	Certification Archives | ON24
6955	https://www.on24.com/customer-stories/agilent-...	Not Set	200	Yes	Indexable	1	Agilent Optimizes its Global Digital Marketing...
6956	https://www.on24.com/blog/types/beginner/	Not Set	200	Yes	Indexable	0	Beginner Archives | ON24

3483 rows × 7 columns

Let's look at the number of URLs by site level.

```
redir_live_urls.groupby(['crawl_depth']).size()
```

```
crawl_depth
0            1
1           66
2          169
3          280
4          253
5          201
6          122
7           64
```

```
8                17
9                 6
10                1
Not Set     2303
dtype: int64
```

Note how the size has dropped slightly to 2303 URLs. The 48 nonindexable URLs were probably paginated pages.

Let's visualize the distribution:

```
from plotnine import *
import matplotlib.pyplot as plt
pd.set_option('display.max_colwidth', None)
%matplotlib inline

# Distribution of internal links to URL by site level
ove_intlink_dist_plt = (ggplot(redir_live_urls, aes(x = 'no_internal_links_
to_url')) +
                    geom_histogram(fill = 'blue', alpha = 0.6, bins = 7) +
                    labs(y = '# Internal Links to URL') +
                    theme_classic() +
                    theme(legend_position = 'none')
                    )

ove_intlink_dist_plt.save(filename = 'images/1_overall_intlink_dist_
plt.png',
                    height=5, width=5, units = 'in', dpi=1000)
ove_intlink_dist_plt
```

The plot ove_intlink_dist_plt in Figure 3-3 is a histogram of the number of internal links to a URL.

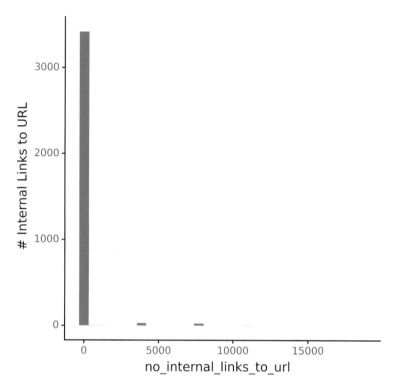

Figure 3-3. *Histogram of the number of internal links to a URL*

The distribution is negatively skewed such that most pages have close to zero links. This would be of some concern to an SEO manager.

While the overall distribution gives one view, it would be good to deep dive into the distribution of internal links by crawl depth:

```
redir_live_urls.groupby('crawl_depth').agg({'no_internal_links_to_url':
['describe']}).sort_values('crawl_depth')
```

This results in the following:

no_internal_links_to_url

describe

crawl_depth	count	mean	std	min	25%	50%	75%	max
0	1.0	4139.000000	NaN	4139.0	4139.00	4139.0	4139.0	4139.0
1	66.0	4808.257576	3507.312178	1.0	3713.00	3732.0	7442.5	18625.0
2	169.0	141.792899	640.547774	0.0	2.00	5.0	8.0	3720.0
3	280.0	3.928571	7.953397	0.0	1.00	2.0	4.0	82.0
4	253.0	2.754941	1.934243	0.0	2.00	2.0	4.0	15.0
5	201.0	2.477612	1.389513	0.0	2.00	2.0	3.0	10.0
6	122.0	2.319672	0.911509	0.0	2.00	2.0	3.0	5.0
7	64.0	1.984375	0.745190	0.0	2.00	2.0	2.0	4.0
8	17.0	1.882353	0.696631	0.0	2.00	2.0	2.0	3.0
9	6.0	1.500000	0.836660	0.0	1.25	2.0	2.0	2.0
10	1.0	2.000000	NaN	2.0	2.00	2.0	2.0	2.0
Not Set	2303.0	0.067304	0.489049	0.0	0.00	0.0	0.0	11.0

The table describes the distribution of internal links by crawl depth or site level. Any URL that is 3+ clicks away from the home page can expect two internal links on average. This is probably the blog content as the marketing team produces a lot of it.

To visualize it graphically

```
from plotnine import *
import matplotlib.pyplot as plt
pd.set_option('display.max_colwidth', None)
%matplotlib inline

intlink_dist_plt = (ggplot(redir_live_urls, aes(x = 'crawl_depth', y = 'no_
internal_links_to_url')) +
                geom_boxplot(fill = 'blue', alpha = 0.8) +
                labs(y = '# Internal Links to URL', x = 'Site Level') +
                theme_classic() +
```

```
                    theme(legend_position = 'none')
                )

intlink_dist_plt.save(filename = 'images/1_intlink_dist_plt.png', height=5,
width=5, units = 'in', dpi=1000)
intlink_dist_plt
```

The plot intlink_dist_plt in Figure 3-4 is a histogram of the number of internal links to a URL by site level.

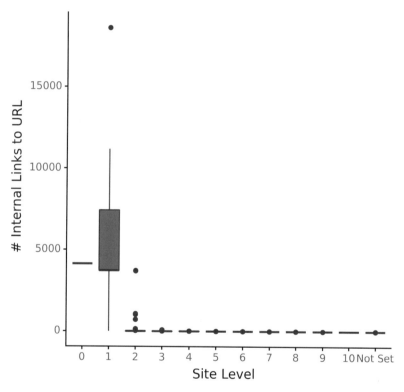

Figure 3-4. *Box plot distributions of the number of internal links to a URL by site level*

As suspected, the most variation is in the first level directly below the home page, with very little variation beyond.

However, we can compare the variation between site levels for content in level 2 and beyond. For a quick peek, we'll use a logarithmic scale for the number of internal links to a URL:

```
from mizani.formatters import comma_format
```

```
intlink_dist_plt = (ggplot(redir_live_urls, aes(x = 'crawl_depth', y = 'no_
internal_links_to_url')) +
                    geom_boxplot(fill = 'blue', alpha = 0.8) +
                    labs(y = '# Internal Links to URL', x = 'Site Level') +
                    scale_y_log10(labels = comma_format()) +
                    theme_classic() +
                    theme(legend_position = 'none')
                    )

intlink_dist_plt.save(filename = 'images/1_log_intlink_dist_plt.png',
height=5, width=5, units = 'in', dpi=1000)
intlink_dist_plt
```

The picture is clearer and more insightful, as we can see how much better and varied the distribution of the lower site levels compared to each other (Figure 3-5).

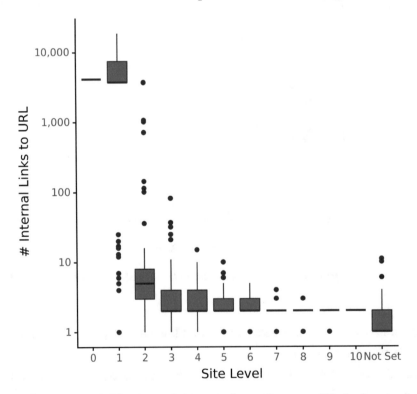

Figure 3-5. *Box plot distribution of the number of internal links by site level with logarized vertical axis*

For example, it's much more obvious that the median number of inbound internal links for pages on site level 2 is much higher than the lower levels.

It's also very obvious that the variation in internal inbound links for pages in site levels 3 and 4 is higher than those in levels 5 and 6.

Remember though the preceding example was achieved using a log scale of the same input variable.

What we've learned here is that having a new variable which is taking a log of the internal links would yield a more helpful picture to compare levels from 2 to 10.

We'll achieve this by creating a new column variable "log_intlinks" which is a log of the internal link count. To avoid negative infinity values from taking a log of zero, we'll add 0.01 to the calculation:

```
redir_live_urls['log_intlinks'] = np.log2(redir_live_urls['no_internal_
links_to_url'] + .01)
```

Now we'll plot using the new logarized variable:

```
intlink_dist_plt = (ggplot(redir_live_urls, aes(x = 'crawl_depth', y =
'log_intlinks')) +
                    geom_boxplot(fill = 'blue', alpha = 0.8) +
                    labs(y = '# Log Internal Links to URL', x = 'Site
                    Level') +
                    theme_classic() +
                    theme(legend_position = 'none')
                   )
```

```
intlink_dist_plt.save(filename = 'images/1c_loglinks_dist_plt.png',
height=5, width=5, units = 'in', dpi=1000)
intlink_dist_plt
```

The intlink_dist_plt plot (Figure 3-6) is quite similar to the logarized scale, only this time the numbers are easier to read because we're using normal scales for the vertical axis. The comparative averages and variations are easier to compare.

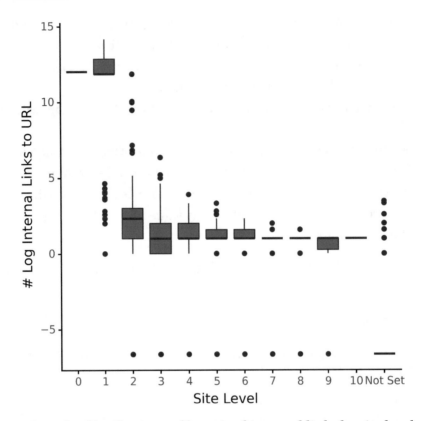

Figure 3-6. *Box plot distributions of logarized internal links by site level*

Site-Level URLs That Are Underlinked

Now that we know the lay of the land in terms of what the distributions look like at the site depth level, we're ready to start digging deeper and see how many URLs are underlinked per site level.

For example, if the 35th percentile number of internal links to a URL is 10 for URLs at a given site level, how many URLs are below that percentile?

That's what we aim to find out. Why 35th and not 25th? It doesn't really matter, a low cutoff point just needs to be picked as the cutoff is arbitrary.

The first step is to calculate the averages of internal links for both nonlog and log versions, which will be joined onto the main dataframe later:

```
intlink_dist = redir_live_urls.groupby('crawl_depth').agg({'no_internal_
links_to_url': ['mean'],

                                                    'log_intlinks':
                                                    ['mean']
```

```
                                                }).reset_index()
intlink_dist.columns = ['_'.join(col) for col in intlink_dist.
columns.values]
intlink_dist = intlink_dist.rename(columns = {'no_internal_links_to_url_
mean': 'avg_int_links',

                                        'log_intlinks_mean': 'logavg_
                                        int_links',
                                        })
intlink_dist
```

This results in the following:

	crawl_depth_	avg_int_links	logavg_int_links
0	0	4139.000000	12.015070
1	1	4808.257576	10.481776
2	2	141.792899	2.081835
3	3	3.928571	0.907680
4	4	2.754941	0.822429
5	5	2.477612	0.789220
6	6	2.319672	0.843477
7	7	1.984375	0.558034
8	8	1.882353	0.508866
9	9	1.500000	-0.433453
10	10	2.000000	1.007196
11	Not Set	0.067304	-6.396314

The averages are in place by site level. Notice how the log column helps make the range of values between crawl depths less extreme and skewed, that is, 4239 to 0.06 for the average vs. 12 to –6.39 for the log average, which makes it easier to normalize the data.

Now we'll set the lower quantile at 35% for all site levels. This will use a customer function quantile_lower:

```
def quantile_lower(x):
    return x.quantile(.35).round(0)

quantiled_intlinks = redir_live_urls.groupby('crawl_depth').agg({'log_
intlinks':
                                                                  [quantile_
                                                                  lower]}).
                                                                  reset_
                                                                  index()
quantiled_intlinks.columns = ['_'.join(col) for col in quantiled_intlinks.
columns.values]
quantiled_intlinks = quantiled_intlinks.rename(columns = {'crawl_depth_':
'crawl_depth',
                                                          'log_intlinks_
                                                          quantile_lower':
                                                          'sd_intlink_
                                                          lowqua'})

quantiled_intlinks
```

This results in the following:

	crawl_depth	sd_intlink_lowqua
0	0	12.0
1	1	12.0
2	2	2.0
3	3	1.0
4	4	1.0
5	5	1.0
6	6	1.0
7	7	1.0
8	8	1.0
9	9	1.0
10	10	1.0
11	Not Set	-7.0

The lower quantile stats are set. Quartiles are limited to the 25th percentile, whereas a quantile means the lower limits can be set to any number, such as 11th, 18th, 24th, etc., which is why we use quantiles instead of quartiles. The next steps are to join the data to the main dataframe, then we'll apply a function to mark URLs that are underlinked for their given site level:

```
redir_live_urls_underidx = redir_live_urls.merge(quantiled_intlinks, on =
'crawl_depth', how = 'left')
```

The following function assesses whether the URL has less links than the lower quantile. If yes, then the value of "sd_int_uidx" is 1, otherwise 0:

```
def sd_intlinkscount_underover(row):
    if row['sd_intlink_lowqua'] > row['log_intlinks']:
        val = 1
    else:
        val = 0
    return val
```

```
redir_live_urls_underidx['sd_int_uidx'] = redir_live_urls_underidx.
apply(sd_intlinkscount_underover, axis=1)
```

There's some code to account for "Not Set" which are effectively orphaned URLs. In this instance, we set these to 1 – meaning they're underlinked:

```
redir_live_urls_underidx['sd_int_uidx'] = np.where(redir_live_urls_
underidx['crawl_depth'] == 'Not Set', 1,

                                          redir_live_urls_
                                          underidx['sd_int_uidx'])
```

```
redir_live_urls_underidx
```

This results in the following:

url	crawl_depth	http_status_code	indexable	indexable_status	no_internal_links_to_url	title	log_intlinks	sd_intlink_lowqua	sd_int_uidx
ttps://www.on24.com/	0	200	Yes	Indexable	4139	Webinar Software & Virtual Event Platform | ON24	12.015070	12.0	0
on24.com/contact-us/	1	200	Yes	Indexable	11189	Contact Us | Global Office Locations | ON24 Teams | ON24	13.449795	12.0	0
/.on24.com/resources/	1	200	Yes	Indexable	11155	Webinar, White Paper, and Video Resources | ON24	13.445404	12.0	0
now-juniper-networks--global-virtual-summit/	1	200	Yes	Indexable	17	How Juniper Networks Set Up a Global Virtual Summit | ON24 Blog	4.088311	12.0	1
lutions/manufacturing/	1	200	Yes	Indexable	7414	Platform for Manufacturing Industry | ON24	12.856038	12.0	0
...	...	...	...	...	...	...	...	...	...

The dataframe shows that the column is in place marking underlinked URLs as 1. With the URLs marked, we're ready to get an overview of how under-linked the URLs are, which will be achieved by aggregating by crawl depth and summing the total number of underlinked URLs:

```
intlinks_agged = redir_live_urls_underidx.groupby('crawl_depth').agg({'sd_
int_uidx': ['sum', 'count']}).reset_index()
```

The following line tidies up the column names by inserting an underscore using a list comprehension:

```
intlinks_agged.columns = ['_'.join(col) for col in intlinks_agged.
columns.values]
intlinks_agged = intlinks_agged.rename(columns = {'crawl_depth_': 'crawl_
depth'})
```

To get a proportion (or percentage), we divide the sum by the count and multiply by 100:

```
intlinks_agged['sd_uidx_prop'] = (intlinks_agged.sd_int_uidx_sum) /
intlinks_agged.sd_int_uidx_count * 100
```

```
print(intlinks_agged)
```

This results in the following:

	crawl_depth	sd_int_uidx_sum	sd_int_uidx_count	sd_uidx_prop
0	0	0	1	0.000000
1	1	38	66	57.575758
2	2	67	169	39.644970
3	3	75	280	26.785714
4	4	57	253	22.529644
5	5	31	201	15.422886
6	6	9	122	7.377049
7	7	9	64	14.062500
8	8	3	17	17.647059
9	9	2	6	33.333333
10	10	0	1	0.000000
11	Not Set	2303	2303	100.000000

So even though the content in levels 1 and 2 have more links than any of the lower levels, they have a higher proportion of underlinked URLs than any other site level (apart from the orphans in Not Set of course).

For example, 57% of pages just below the home page are underlinked.

Let's visualize:

```
# plot the table
depth_uidx_plt = (ggplot(intlinks_agged, aes(x = 'crawl_depth', y = 'sd_
int_uidx_sum')) +
                  geom_bar(stat = 'identity', fill = 'blue', alpha
                  = 0.8) +
                  labs(y = '# Under Linked URLs', x = 'Site Level') +
                  scale_y_log10() +
                  theme_classic() +
                  theme(legend_position = 'none')
                 )

depth_uidx_plt.save(filename = 'images/1_depth_uidx_plt.png', height=5,
width=5, units = 'in', dpi=1000)
depth_uidx_plt
```

It's good to visualize using depth_uidx_plt because we can also see (Figure 3-7) that levels 2, 3, and 4 have the most underlinked URLs by volume.

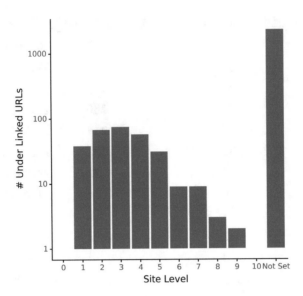

Figure 3-7. *Column chart of the number of internally under-linked URLs by site level*

Let's plot the intlinks_agged table:

```
depth_uidx_prop_plt = (ggplot(intlinks_agged, aes(x = 'crawl_depth', y =
'sd_uidx_prop')) +
                geom_bar(stat = 'identity', fill = 'blue', alpha
                = 0.8) +
                labs(y = '% URLs Under Linked', x = 'Site Level') +
                theme_classic() +
                theme(legend_position = 'none')
                )
depth_uidx_prop_plt.save(filename = 'images/1_depth_uidx_prop_plt.png',
height=5, width=5, units = 'in', dpi=1000)
depth_uidx_prop_plt
```

Plotting depth_uidx_prop_plt (Figure 3-8), we see it just so happens that although level 1 has a lower volume, the proportion is higher. Intuitively, this is indicative of too many pages being linked from the home page but unequally.

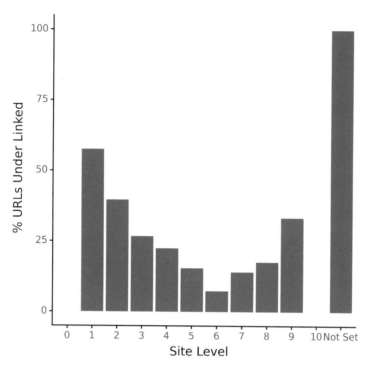

Figure 3-8. *Column chart of the proportion of under internally linked URLs by site level*

It's not a given that URLs in the site level that are underlinked are a problem or perhaps more so by design. However, they are worth reviewing as perhaps they should be at that site level or they do deserve more internal links after all.

The following code exports the underlinked URLs to a CSV which can be viewed in Microsoft Excel:

```
underlinked_urls = redir_live_urls_underidx.loc[redir_live_urls_underidx.
sd_int_uidx == 1]
underlinked_urls = underlinked_urls.sort_values(['crawl_depth', 'no_
internal_links_to_url'])
underlinked_urls.to_csv('exports/underlinked_urls.csv')
```

By Page Authority

Inbound links from external websites are a source of PageRank or, if we're going to be search engine neutral about it, page authority.

Given that not all pages earn inbound links, it is normally desired by SEOs to have pages without backlinks crawled more often. So it would make sense to analyze and explore opportunities to redistribute this PageRank to other pages within the website.

We'll start by tacking on the AHREFs data to the main dataframe so we can see internal links by page authority.

```
intlinks_pageauth = redir_live_urls_underidx.merge(ahrefs_df, on = 'url',
how = 'left')
intlinks_pageauth.head()
```

This results in the following:

url	crawl_depth	http_status_code	indexable_status	no_internal_links_to_url	title	log_intlinks	page_authority	referring_domains
https://www.on24.com/	0	200	Indexable	4139	Webinar Software & Virtual Event Platform | ON24	12.015070	81.0	3215.0
://www.on24.com/contact-us/	1	200	Indexable	11189	Contact Us | Global Office Locations | ON24 Teams | ON24	13.449795	36.0	55.0
s://www.on24.com/resources/	1	200	Indexable	11155	Webinar, White Paper, and Video Resources | ON24	13.445404	28.0	42.0
w.on24.com/blog/how-juniper-.et-up-a-global-virtual-summit/	1	200	Indexable	17	How Juniper Networks Set Up a Global Virtual Summit | ON24 Blog	4.088311	13.0	1.0
.com/solutions/manufacturing/	1	200	Indexable	7414	Platform for Manufacturing Industry | ON24	12.856038	17.0	7.0

We now have page authority and referring domains at the URL level. Predictably, the home page has a lot of referring domains (over 3000) and the most page-level authority at 81.

As usual, we'll perform some aggregations and explore the distribution of the PageRank (interchangeable with page authority).

First, we'll clean up the data to make sure we replace null values with zero:

```
intlinks_pageauth['page_authority'] = np.where(intlinks_pageauth['page_
authority'].isnull(),

                                    0, intlinks_pageauth['page_
                                    authority'])
```

Aggregate by page authority:

```
intlinks_pageauth.groupby('page_authority').agg({'no_internal_links_to_
url': ['describe']})
```

This results in the following:

| page_authority | no_internal_links_to_url | | | | | | | |
| | describe | | | | | | | |
	count	mean	std	min	25%	50%	75%	max
0.0	1320.0	0.034848	0.240667	0.0	0.0	0.0	0.00	3.0
13.0	1077.0	7.839369	120.726053	0.0	0.0	2.0	3.00	3698.0
14.0	148.0	79.763514	524.905542	0.0	0.0	2.0	3.00	3720.0
15.0	725.0	13.477241	200.634714	0.0	0.0	0.0	0.00	3716.0
16.0	67.0	336.134328	1250.082093	0.0	0.0	2.0	5.00	7413.0
17.0	38.0	1082.684211	2095.249390	0.0	2.0	5.0	110.25	7414.0
18.0	22.0	1015.090909	2604.950231	0.0	2.0	3.0	6.75	7436.0
19.0	22.0	510.090909	1304.569147	0.0	2.0	3.5	8.00	3733.0
20.0	15.0	1735.666667	3093.162267	0.0	1.0	4.0	1877.00	7429.0

The preceding table shows the distribution of internal links by different levels of page authority.

At the lower levels, most URLs have around two internal links.

A graph will give us the full picture:

```
# distribution of page_authority
page_authority_dist_plt = (ggplot(intlinks_pageauth, aes(x = 'page_
authority')) +
                geom_histogram(fill = 'blue', alpha = 0.6, bins
                = 30 ) +
                labs(y = '# URLs', x = 'Page Authority') +
                #scale_y_log10() +
                theme_classic() +
                theme(legend_position = 'none')
                )
```

```
page_authority_dist_plt.save(filename = 'images/2_page_authority_dist_
plt.png',
                              height=5, width=5, units = 'in', dpi=1000)
page_authority_dist_plt
```

The distribution, shown in page_authority_dist_plt (Figure 3-9), is heavily negatively skewed when plotting the raw numbers. Most of the site URLs have a PageRank of 15, of which the number of URLs with higher authority shrinks dramatically. A very high number of URLs have no authority, because they are orphaned.

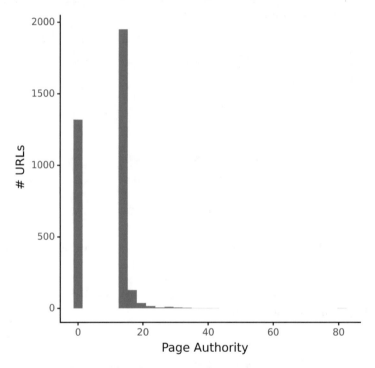

Figure 3-9. *Distribution of URLs by page authority*

Using the log scale, we can see how the higher levels of authority compare:

```
# distribution of page_authority
page_authority_dist_plt = (ggplot(intlinks_pageauth, aes(x = 'page_
authority')) +
                 geom_histogram(fill = 'blue', alpha = 0.6, bins
                 = 30 ) +
                 labs(y = '# URLs (Log)', x = 'Page Authority') +
```

```
            scale_y_log10() +
            theme_classic() +
            theme(legend_position = 'none')
          )
```

```
page_authority_dist_plt.save(filename = 'images/2_page_authority_dist_log_
plt.png',
                            height=5, width=5, units = 'in', dpi=1000)
page_authority_dist_plt
```

Suddenly, the view shown by page_authority_dist_plt (Figure 3-10) is more interesting because as authority increases by an increment of one, there are ten times less URLs than before – a pretty harsh distribution of PageRank.

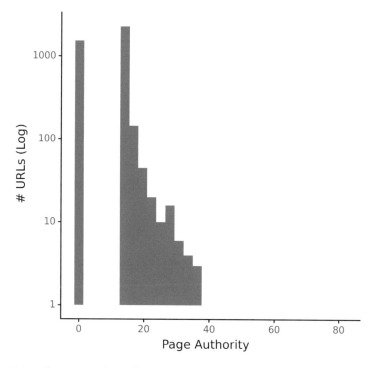

Figure 3-10. *Distribution plot of URLs by logarized scale*

Given this more insightful view, taking a log of "page_authority" to form a new column variable "log_pa" is justified:

```
intlinks_pageauth['page_authority'] = np.where(intlinks_pageauth['page_
authority'] == 0, .1, intlinks_pageauth['page_authority'])
intlinks_pageauth['log_pa'] = np.log2(intlinks_pageauth.page_authority)
intlinks_pageauth.head()
```

url	crawl_depth	http_status_code	indexable_status	no_internal_links_to_url	title	log_intlinks	page_authority	referring_domains	log_pa
://www.on24.com/	0	200	Indexable	4139	Webinar Software & Virtual Event Platform	ON24	12.015070	81.0	3215.0	6.339850
24.com/contact-us/	1	200	Indexable	11189	Contact Us	Global Office Locations	ON24 Teams	ON24	13.449795	36.0	55.0	5.169925
24.com/resources/	1	200	Indexable	11155	Webinar, White Paper, and Video Resources	ON24	13.445404	28.0	42.0	4.807355
/blog/how-juniper-bal-virtual-summit/	1	200	Indexable	17	How Juniper Networks Set Up a Global Virtual Summit	ON24 Blog	4.088311	13.0	1.0	3.700440
ns/manufacturing/	1	200	Indexable	7414	Platform for Manufacturing Industry	ON24	12.856038	17.0	7.0	4.087463

The log_pa column is in place; let's visualize:

```
page_authority_trans_dist_plt = (ggplot(intlinks_pageauth, aes(x =
'log_pa')) +
                    geom_histogram(fill = 'blue', alpha = 0.6, bins
                    = 30 ) +
                    labs(y = '# URLs (Log)', x = 'Log Page Authority') +
                    scale_y_log10() +
                    theme_classic() +
                    theme(legend_position = 'none')
                    )

page_authority_trans_dist_plt.save(filename = 'images/2_page_authority_
trans_dist_plt.png',
                    height=5, width=5, units = 'in', dpi=1000)
page_authority_trans_dist_plt
```

Taking a log has compressed the range of PageRank, as shown by page_authority_ trans_dist_plt (Figure 3-11), by making it less extreme as the home page has a log_pa value of 6, bringing it closer to the rest of the site.

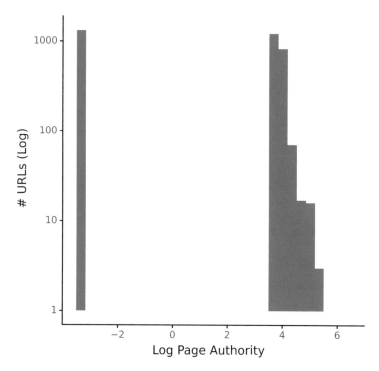

Figure 3-11. *Distribution of URLs by log page authority*

The decimal points will be rounded to make the 3000+ URLs easier to categorize:

```
intlinks_pageauth['pa_band'] = intlinks_pageauth['log_pa'].apply(np.floor)

# display updated DataFrame
intlinks_pageauth
```

wl_depth	http_status_code	indexable	indexable_status	no_internal_links_to_url	title	log_intlinks	page_authority	referring_domains	log_pa	pa_band
0	200	Yes	Indexable	4139	Webinar Software & Virtual Event Platform | ON24	12.015070	81.0	3215.0	6.339850	6.0
1	200	Yes	Indexable	11189	Contact Us | Global Office Locations | ON24 Teams | ON24	13.449795	36.0	55.0	5.169925	5.0
1	200	Yes	Indexable	11155	Webinar, White Paper, and Video Resources | ON24	13.445404	28.0	42.0	4.807355	4.0
1	200	Yes	Indexable	17	How Juniper Networks Set Up a Global Virtual Summit | ON24 Blog	4.088311	13.0	1.0	3.700440	3.0
1	200	Yes	Indexable	7414	Platform for Manufacturing Industry | ON24	12.856038	17.0	7.0	4.087463	4.0

Page Authority URLs That Are Underlinked

With the URLs categorized into PA bands, we want to see if they have less internal links for their authority level than they should. We've set the threshold at 40% so that any URL that has less internal links for their level of PA will be counted as underlinked.

The choice of 40% is not terribly important at this stage as each website (or market even) is different. There are more scientific ways of arriving at the optimal threshold, such as analyzing top-ranking competitors for a search space; however, for now we'll choose 40% as our threshold.

```
def quantile_lower(x):
    return x.quantile(.4).round(0)
```

```
quantiled_pageau = intlinks_pageauth.groupby('pa_band').agg({'no_internal_
links_to_url': [quantile_lower]}).reset_index()
quantiled_pageau.columns = ['_'.join(col) for col in quantiled_pageau.
columns.values]
quantiled_pageau = quantiled_pageau.rename(columns = {'pa_band_':
'pa_band',

                                             'no_internal_links_
                                             to_url_quantile_
                                             lower': 'pa_intlink_
                                             lowqua'})
```

```
quantiled_pageau
```

This results in the following:

	pa_band	pa_intlink_lowqua
0	-4.0	0.0
1	3.0	0.0
2	4.0	3.0
3	5.0	7446.0
4	6.0	4139.0

Going by PageRank, we now have the minimum threshold of inbound internal links we would expect. Time to join the data and mark the URLs that are underlinked for their authority level:

```
intlinks_pageauth_underidx = intlinks_pageauth.merge(quantiled_pageau, on =
'pa_band', how = 'left')

def pa_intlinkscount_underover(row):
    if row['pa_intlink_lowqua'] > row['no_internal_links_to_url']:
        val = 1
    else:
        val = 0
    return val

intlinks_pageauth_underidx['pa_int_uidx'] = intlinks_pageauth_underidx.
apply(pa_intlinkscount_underover, axis=1)
```

This function will allow us to make some aggregations to see how many URLs there are at each PageRank band and how many are under-linked:

```
pageauth_agged = intlinks_pageauth_underidx.groupby('pa_band').agg({'pa_
int_uidx': ['sum', 'count']}).reset_index()
pageauth_agged.columns = ['_'.join(col) for col in pageauth_agged.
columns.values]

pageauth_agged['uidx_prop'] = pageauth_agged.pa_int_uidx_sum / pageauth_
agged.pa_int_uidx_count * 100

print(pageauth_agged)
```

This results in the following:

	pa_band_	pa_int_uidx_sum	pa_int_uidx_count	uidx_prop
0	-4.0	0	1320	0.000000
1	3.0	0	1950	0.000000
2	4.0	77	203	37.931034
3	5.0	4	9	44.444444
4	6.0	0	1	0.000000

Most of the underlinked content appears to be those that have the highest page authority, which is slightly contrary to what the site-level approach suggests (that pages lower down are underlinked). That's assuming most of the high authority pages are closer to the home page.

What is the right answer? It depends on what we're trying to achieve. Let's continue with more analysis for now and visualize the authority stats:

```
# distribution of page_authority
pageauth_agged_plt = (ggplot(intlinks_pageauth_underidx.loc[intlinks_
pageauth_underidx['pa_int_uidx'] == 1],
                        aes(x = 'pa_band')) +
                geom_histogram(fill = 'blue', alpha = 0.6, bins = 10) +
                labs(y = '# URLs Under Linked', x = 'Page Authority
                Level') +
                theme_classic() +
                theme(legend_position = 'none')
                )

pageauth_agged_plt.save(filename = 'images/2_pageauth_agged_hist.png',
                    height=5, width=5, units = 'in', dpi=1000)
pageauth_agged_plt
```

We see in pageauth_agged_plt (Figure 3-12) that there are almost 80 URLs underlinked at PageRank level 4 and a few at PageRank level 5. This is quite an abstract concept admittedly.

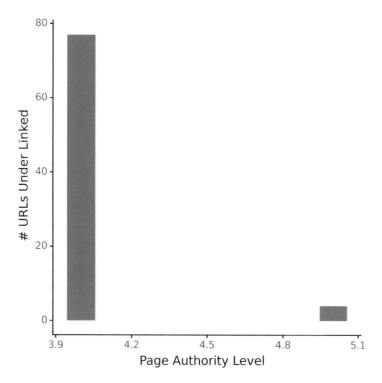

Figure 3-12. *Distribution of under internally linked URLs by page authority level*

Content Type

Perhaps it would be more useful to visualize this by content type just by a "quick and dirty" analysis using the first subdirectory:

```
intlinks_content_underidx = intlinks_depthauth_underidx.copy()
```

To get the first subfolder, we'll define a function that allows the operation to continue in case of a fail (which would happen for the home page URL because there is no subfolder). The k parameter specifies the number of slashes in the URL to find the desired folder and parse the subdirectory name:

```
def get_folder(fp, k=3):
    try:
        return os.path.split(fp)[0].split(os.sep)[k]
    except:
        return 'home'
```

```
intlinks_content_underidx['content'] = intlinks_content_underidx['url'].
apply(lambda x: get_folder(x))
```

Inspect the distribution of links by subfolder:

```
intlinks_content_underidx.groupby('content').agg({'no_internal_links_to_
url': ['describe']})
```

This results in the following:

	no_internal_links_to_url							
	describe							
	count	mean	std	min	25%	50%	75%	max
content								
about-us	6.0	6203.000000	1921.783443	3713.0	4658.50	7442.0	7443.75	7446.0
accelerate-pipeline-on24	1.0	3.000000	NaN	3.0	3.00	3.0	3.00	3.0
accessibility	1.0	0.000000	NaN	0.0	0.00	0.0	0.00	0.0
act-on	1.0	0.000000	NaN	0.0	0.00	0.0	0.00	0.0
add-on-services	1.0	0.000000	NaN	0.0	0.00	0.0	0.00	0.0
...	...	...	...	...	...	...	...	...
webinarworldondemand-london	1.0	0.000000	NaN	0.0	0.00	0.0	0.00	0.0
webinarworldondemand-sydney	1.0	0.000000	NaN	0.0	0.00	0.0	0.00	0.0
webinerd-community	14.0	3.928571	3.452185	0.0	2.00	3.5	5.75	11.0
webinerd-education	2.0	7.500000	10.606602	0.0	3.75	7.5	11.25	15.0
zapier	1.0	0.000000	NaN	0.0	0.00	0.0	0.00	0.0

183 rows × 8 columns

Wow, 183 subfolders! That's way too much for categorical analysis. We could break it down and aggregate it into fewer categories using the ngram techniques described in Chapter 9; feel free to try.

In any case, it looks like the site architecture is too flat and could be better structured to be more hierarchical, that is, more pyramid like.

Also, many of the content folders only have one inbound internal link, so even without the benefit of data science, it's obvious these require SEO attention.

Combining Site Level and Page Authority

Perhaps it would be more useful to visualize by combining site level and page authority?

```
intlinks_depthauth_underidx = intlinks_pageauth_underidx.copy()
intlinks_depthauth_underidx['depthauth_uidx'] = np.where((intlinks_
depthauth_underidx['sd_int_uidx'] +
                                        intlinks_
                                        depthauth_
                                        underidx['pa_
                                        int_uidx'] ==
                                        2), 1, 0)
'''intlinks_depthauth_underidx['depthauth_uidx'] = np.where((intlinks_
depthauth_underidx['sd_int_uidx'] == 1) &
                                        (intlinks_
                                        depthauth_
                                        underidx['pa_int_
                                        uidx'] == 1),
                                        1, 0)'''
depthauth_uidx = intlinks_depthauth_underidx.groupby(['crawl_depth',
'pa_band']).agg({'depthauth_uidx': 'sum'}).reset_index()
depthauth_urls = intlinks_depthauth_underidx.groupby(['crawl_depth',
'pa_band']).agg({'url': 'count'}).reset_index()

depthauth_stats = depthauth_uidx.merge(depthauth_urls,
                                        on = ['crawl_depth',
                                        'pa_band'], how = 'left')
depthauth_stats['depthauth_uidx_prop'] = (depthauth_stats['depthauth_uidx']
/ depthauth_stats['url']).round(2)
depthauth_stats.sort_values('depthauth_uidx', ascending = False)
```

This results in the following:

	crawl_depth	pa_band	depthauth_uidx	url	depthauth_uidx_prop
57	Not Set	4.0	42	44	0.95
12	2	4.0	7	50	0.14
17	3	4.0	5	24	0.21
22	4	4.0	4	18	0.22
27	5	4.0	2	12	0.17
58	Not Set	5.0	1	1	1.00
32	6	4.0	1	4	0.25
47	9	4.0	1	1	1.00
0	0	-4.0	0	0	NaN
40	8	-4.0	0	0	NaN
41	8	3.0	0	15	0.00

Most of the underlinked URLs are orphaned and have page authority (probably from backlinks).

Visualize to get a fuller picture:

```
depthauth_stats_plt = (
    ggplot(depthauth_stats,
           aes(x = 'pa_band', y = 'crawl_depth', fill = 'depthauth_
           uidx')) +
    geom_tile(stat = 'identity', alpha = 0.6) +
    labs(y = '', x = '') +
    theme_classic() +
    theme(legend_position = 'right')
)

depthauth_stats_plt.save(filename = 'images/3_depthauth_stats_plt.png',
                         height=5, width=10, units = 'in', dpi=1000)
depthauth_stats_plt
```

There we have it, depthauth_stats_plt (Figure 3-13) shows most of the focus should go into the orphaned URLs (which they should anyway), but more importantly we know which orphaned URLs to prioritize over others.

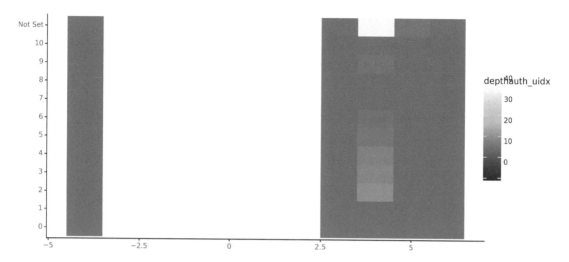

Figure 3-13. *Heatmap of page authority level, site level, and underlinked URLs*

We can also see the extent of the issue. The second highest priority group of underindexed URLs are at site levels 2, 3, and 4.

Anchor Texts

If the count and their distribution represent the quantitative aspect of internal links, then the anchor texts could be said to represent their quality.

Anchor texts signal to search engines and users what content to expect after accessing the hyperlink. This makes anchor texts an important signal and one worth optimizing.

We'll start by aggregating the crawl data from Sitebulb to get an overview of the issues:

```
anchor_issues_agg = crawl_data.agg({'no_anchors_with_empty_href': ['sum'],
              'no_anchors_with_leading_or_trailing_whitespace_in_href':
              ['sum'],
              'no_anchors_with_local_file': ['sum'],
              'no_anchors_with_localhost': ['sum'],
              'no_anchors_with_malformed_href': ['sum'],
              'no_anchors_with_no_text': ['sum'],
              'no_anchors_with_non_descriptive_text': ['sum'],
              'no_anchors_with_non-http_protocol_in_href': ['sum'],
```

```
                    'no_anchors_with_url_in_onclick': ['sum'],
                    'no_anchors_with_username_and_password_in_href': ['sum'],
                    'no_image_anchors_with_no_alt_text': ['sum']
                }).reset_index()

anchor_issues_agg = pd.melt(anchor_issues_agg, var_name=['issues'],
                            value_vars=['no_anchors_with_empty_href',
                                        'no_anchors_with_leading_or_
                                        trailing_whitespace_in_href',
                                        'no_anchors_with_local_file','no_
                                        anchors_with_localhost',
                                        'no_anchors_with_malformed_href',
                                        'no_anchors_with_no_text',
                                        'no_anchors_with_non_
                                        descriptive_text',
                                        'no_anchors_with_non-http_protocol_
                                        in_href',
                                        'no_anchors_with_url_in_onclick',
                                        'no_anchors_with_username_and_
                                        password_in_href',
                                        'no_image_anchors_with_no_
                                        alt_text'],
                            value_name='instances'
                        )

anchor_issues_agg
```

This results in the following:

	issues	instances
0	no_anchors_with_empty_href	19
1	no_anchors_with_leading_or_trailing_whitespace_in_href	3724
2	no_anchors_with_local_file	0
3	no_anchors_with_localhost	0
4	no_anchors_with_malformed_href	11
5	no_anchors_with_no_text	297
6	no_anchors_with_non_descriptive_text	4047
7	no_anchors_with_non-http_protocol_in_href	0
8	no_anchors_with_url_in_onclick	0
9	no_anchors_with_username_and_password_in_href	0
10	no_image_anchors_with_no_alt_text	112

Over 4000 links with no descriptive anchor text jump out as the most common issue, not to mention the 19 anchors with empty HREF (albeit very low in number).

To visualize

```
anchor_issues_count_plt = (ggplot(anchor_issues_agg, aes(x =
'reorder(issues, instances)', y = 'instances')) +
                om_bar(stat = 'identity', fill = 'blue', alpha = 0.6) +
                labs(y = '# instances of Anchor Text Issues', x = '') +
                theme_classic() +
                coord_flip() +
                theme(legend_position = 'none')
                )
```

```
anchor_issues_count_plt.save(filename = 'images/4_anchor_issues_count_
plt.png',
                height=5, width=5, units = 'in', dpi=1000)
anchor_issues_count_plt
```

anchor_issues_count_plt (Figure 3-14) visually confirms the number of internal links with nondescriptive anchor text.

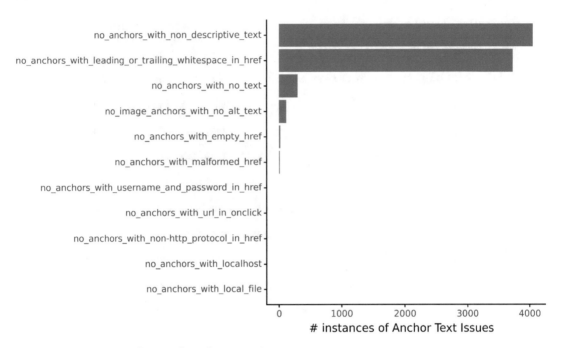

Figure 3-14. *Bar chart of anchor text issues*

Anchor Issues by Site Level

We'll drill down on the preceding example by site level to get a bit more insight to see where the problems are happening:

```
anchor_issues_levels = crawl_data.groupby('crawl_depth').agg({'no_anchors_
with_empty_href': ['sum'],
                'no_anchors_with_leading_or_trailing_whitespace_in_href':
                ['sum'],
                'no_anchors_with_local_file': ['sum'],
                'no_anchors_with_localhost': ['sum'],
                'no_anchors_with_malformed_href': ['sum'],
                'no_anchors_with_no_text': ['sum'],
                'no_anchors_with_non_descriptive_text': ['sum'],
                'no_anchors_with_non-http_protocol_in_href': ['sum'],
                'no_anchors_with_url_in_onclick': ['sum'],
                'no_anchors_with_username_and_password_in_href': ['sum'],
                'no_image_anchors_with_no_alt_text': ['sum']
            }).reset_index()
```

```
anchor_issues_levels.columns = ['_'.join(col) for col in anchor_issues_
levels.columns.values]
anchor_issues_levels.columns = [str.replace(col, '_sum', '') for col in
anchor_issues_levels.columns.values]
anchor_issues_levels.columns = [str.replace(col, 'no_anchors_with_', '')
for col in anchor_issues_levels.columns.values]
anchor_issues_levels = anchor_issues_levels.rename(columns = {'crawl_
depth_': 'crawl_depth'})

anchor_issues_levels = pd.melt(anchor_issues_levels, id_vars=['crawl_
depth'], var_name=['issues'],
                        value_vars=['empty_href',
                                    'leading_or_trailing_whitespace_
                                    in_href',
                                    'local_file','localhost',
                                    'malformed_href', 'no_text',
                                    'non_descriptive_text',
                                    'non-http_protocol_in_href',
                                    'url_in_onclick',
                                    'username_and_password_in_href',
                                    'no_image_anchors_with_no_
                                    alt_text'],
                        value_name='instances'
                    )

print(anchor_issues_levels)
```

This results in the following:

	crawl_depth	issues	instances
111	Not Set	non_descriptive_text	2458
31	Not Set	leading_or_trailing_whitespace_in_href	2295
104	3	non_descriptive_text	350
24	3	leading_or_trailing_whitespace_in_href	328
105	4	non_descriptive_text	307
..	...	...	...
85	13	no_text	0

84	12	no_text	0
83	11	no_text	0
82	10	no_text	0
0	0	empty_href	0

```
[176 rows x 3 columns]
```

Most of the issues are on orphaned pages followed by URLs three to four levels deep. To visualize

```
anchor_levels_issues_count_plt = (ggplot(anchor_issues_levels, aes
(x = 'crawl_depth',
                                              y = 'issues',
                                              fill =
                                              'instances'
                                              )) +
              geom_tile() +
              labs(y = '# instances of Anchor Text Issues', x = '') +
              scale_fill_cmap(cmap_name='viridis') +
              theme_classic()
              )
anchor_levels_issues_count_plt.save(filename = 'images/4_anchor_levels_
issues_count_plt.png',
                    height=5, width=5, units = 'in', dpi=1000)
anchor_levels_issues_count_plt
```

The anchor_levels_issues_count_plt graphic (Figure 3-15) makes it clearer; the technical issues with anchor text lay with the orphaned pages.

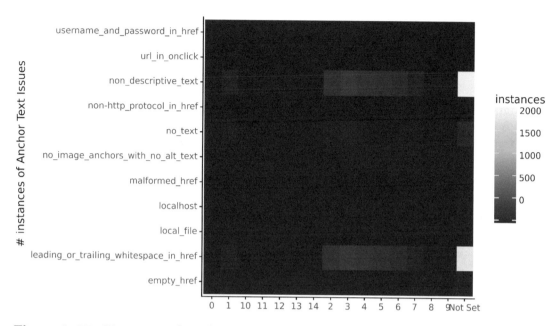

Figure 3-15. *Heatmap of site level, anchor text issues, and instances*

Anchor Text Relevance

Of course, that's not the only aspect of anchor text that SEOs are interested in. SEOs want to know the extent of the relevance between the anchor text and the destination URL.

For that task, we'll use string matching techniques on the Sitebulb link report to measure that relevance and then aggregate to see the overall picture:

```
link_df = link_data[['target_url', 'referring_url', 'anchor_text',
'location']]
link_df = link_df.rename(columns = {'target_url':'url'})
```

Merge with the crawl data using the URL as the primary key and then filter for indexable URLs only:

```
anchor_merge = crawl_data.merge(link_df, on = 'url', how = 'left')
anchor_merge = anchor_merge.loc[anchor_merge['host'] == website]

anchor_merge = anchor_merge.loc[anchor_merge['indexable'] == 'Yes']

anchor_merge['crawl_depth'] = anchor_merge['crawl_depth'].
astype('category')
```

```
anchor_merge['crawl_depth'] = anchor_merge['crawl_depth'].cat
.reorder_categories(['0', '1', '2', '3', '4', '5', '6', '7', '8', '9',
'10', 'Not Set'])
```

Then we compare the string similarity of the anchor text and title tag of the destination URLs:

```
anchor_merge['anchor_relevance'] = anchor_merge.loc[:, ['title',
                                                'anchor_text']].
apply(lambda x: sorensen_dice(*x), axis=1)
```

And any URLs with less than 70% relevance score will be marked as irrelevant under the new column "irrel_anchors" as a 1.

Why 70%? This is from experience, and you're more than welcome to try different thresholds.

With Sorensen-Dice, which is not only fast but meets SEO needs for measuring relevance, 70% seems to be the right limit between relevance and irrelevance, especially when accounting for the site markers in the title tag string:

```
anchor_merge['irrel_anchors'] = np.where(anchor_merge['anchor_relevance'] <
.7, 1, 0)
```

Having a single factor makes it easier to aggregate the entire dataframe by column although there are alternative methods to this:

```
anchor_merge['project'] = target_name
```

```
anchor_merge
```

This results in the following:

rect_url	redirect_url_status	redirect_url_status_code	referring_url	anchor_text	anchor_relevance	irrel_anchors	project
No Data	Not Set	Not Set	https://www.on24.com/about-us/	ON24	0.153846	1	ON24
No Data	Not Set	Not Set	https://www.on24.com/about-us/board-of-directors/	ON24	0.153846	1	ON24
No Data	Not Set	Not Set	https://www.on24.com/about-us/careers/	ON24	0.153846	1	ON24
No Data	Not Set	Not Set	https://www.on24.com/about-us/careers/	https://www.on24.com	0.289855	1	ON24
No Data	Not Set	Not Set	https://www.on24.com/about-us/careers/	https://www.on24.com	0.289855	1	ON24
...	...	...	...	...	...	...	...
No Data	Not Set	Not Set	https://www.on24.com/blog/business_types/enter...	Shell expands global brand awareness and drive...	0.965517	0	ON24
No Data	Not Set	Not Set	NaN	nan	0.060606	1	ON24
No Data	Not Set	Not Set	NaN	nan	0.125000	1	ON24
No Data	Not Set	Not Set	https://www.on24.com/blog/solutions/manufactur...	Agilent Optimizes its Global Digital Marketing...	0.956522	0	ON24
No Data	Not Set	Not Set	NaN	nan	0.148148	1	ON24

Because there is a many-to-many relationship between referring pages and destination URLs (i.e., a destination URL can receive links from multiple URLs, and the former can link to multiple URLs), the dataframe has expanded to over 350,000 rows from 8611.

Let's aggregate by counting the number of URLs per referring URL:

```
anchor_rel_stats_site_agg = anchor_merge.groupby('project').agg({'irrel_
anchors': 'sum'}).reset_index()
anchor_rel_stats_site_agg['total_urls'] = anchor_merge.shape[0]
anchor_rel_stats_site_agg['irrel_anchors_prop'] = anchor_rel_stats_site_
agg['irrel_anchors'] /anchor_rel_stats_site_agg['total_urls']
print(anchor_rel_stats_site_agg)
```

```
 project   irrel_anchors   total_urls   irrel_anchors_prop
0    ON24          333946       350643             0.952382
```

About 95% of anchor texts on this site are irrelevant. How does this compare to their competitors? That's your homework.

Let's go slightly deeper and analyze this by site depth:

```
anchor_rel_depth_irrels = anchor_merge.groupby(['crawl_depth']).
agg({'irrel_anchors': 'sum'}).reset_index()
anchor_rel_depth_urls = anchor_merge.groupby(['crawl_depth']).
agg({'project': 'count'}).reset_index()
anchor_rel_depth_stats = anchor_rel_depth_irrels.merge(anchor_rel_depth_
urls, on = 'crawl_depth', how = 'left')
```

```
anchor_rel_depth_stats['irrel_anchors_prop'] = anchor_rel_depth_
stats['irrel_anchors'] / anchor_rel_depth_stats['project']

anchor_rel_depth_stats
```

This results in the following:

	crawl_depth	irrel_anchors	project	irrel_anchors_prop
0	0	4139	4139	1.000000
1	1	306156	317345	0.964742
2	2	19162	23974	0.799283
3	3	655	1117	0.586392
4	4	608	713	0.852735
5	5	467	510	0.915686
6	6	278	289	0.961938
7	7	128	131	0.977099
8	8	33	33	1.000000
9	9	10	10	1.000000
10	10	2	2	1.000000
11	Not Set	2308	2380	0.969748

Virtually, all content at all site levels with the exception of those three clicks away from the home page (probably blog posts) have irrelevant anchors.

Let's visualize:

```
# anchor issues text
anchor_rel_stats_site_agg_plt = (ggplot(anchor_rel_depth_stats,
                                  aes(x = 'crawl_depth', y = 'irrel_
                                  anchors_prop')) +
            geom_bar(stat = 'identity', fill = 'blue', alpha
            = 0.6) +
            labs(y = '# irrel_anchors', x = '') +
            #scale_y_log10() +
            theme_classic() +
            coord_flip() +
```

```
                theme(legend_position = 'none')
                )
anchor_rel_stats_site_agg_plt.save(filename = 'images/3_anchor_rel_stats_
site_agg_plt.png',
                        height=5, width=5, units = 'in', dpi=1000)
anchor_rel_stats_site_agg_plt
```

Irrelevant anchors by site level are shown in the anchor_rel_stats_site_agg_plt plot (Figure 3-16), where we can see it is pretty much sitewide with less instances on URLs in site level 3.

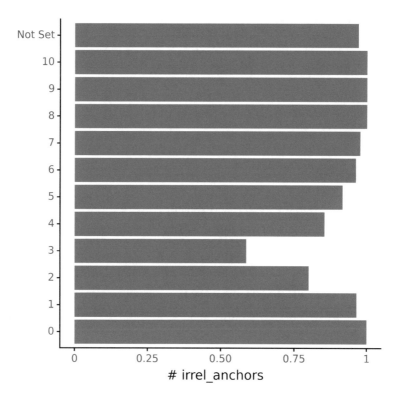

Figure 3-16. *Bar chart of irrelevant anchor texts by site level*

Location

More insight could be gained by looking at the location of the anchors:

```
anchor_rel_locat_irrels = anchor_merge.groupby(['location']).agg({'irrel_
anchors': 'sum'}).reset_index()
```

```
anchor_rel_locat_urls = anchor_merge.groupby(['location']).agg({'project':
'count'}).reset_index()
anchor_rel_locat_stats = anchor_rel_locat_irrels.merge(anchor_rel_locat_
urls, on = 'location', how = 'left')
anchor_rel_locat_stats['irrel_anchors_prop'] = anchor_rel_locat_
stats['irrel_anchors'] / anchor_rel_locat_stats['project']

anchor_rel_locat_stats
```

This results in the following:

	location	irrel_anchors	project	irrel_anchors_prop
0	Footer	137470	148609	0.925045
1	Header	194183	199741	0.972174

The irrelevant anchors are within the header or footer which make these relatively easy to solve.

Anchor Text Words

Let's look at the anchor texts themselves. Anchor texts are the words that make up the HTML hyperlinks. Search engines use these words to assign some meaning to the page that is being linked to.

Naturally, search engines will score anchor texts that accurately describe the content of the page they're linking to, because if a user does click the link, then they will receive a good experience of the content such that it matches their expectations created by the anchor text.

We'll start by looking at the most common words anchor texts used in the website:

```
anchor_count = anchor_merge[['anchor_text']].copy()
anchor_count['count'] = 1

anchor_count_agg = anchor_count.groupby('anchor_text').agg({'count':
'sum'}).reset_index()
anchor_count_agg = anchor_count_agg.sort_values('count', ascending = False)

anchor_count_agg
```

This results in the following:

	anchor_text	count
203	Contact Us	7427
551	Live Demo Discover how to create engaging webi...	7426
876	Resources	7426
550	Live Demo	7426
724	ON24 Webcast Elite	3851
...	...	...
916	Selling to Tech Means Investing in Customer Co...	1
915	Seismic shifted its APAC marketing strategy	1
176	Call-to-Action	1
181	Cassandra Clark Senior Manager of Webinar Prog...	1
1807	"The Ultimate Guide to Planning the Perfect We...	1

1808 rows × 2 columns

There are over 1,808 variations of anchor texts of which "Contact Us" is the most popular along with "Live Demo" and "Resources."

Let's visualize using a word cloud. We'll have to import the WordCloud package and convert the dataframe into a dictionary:

```
from wordcloud import WordCloud

data = anchor_count_agg.set_index('anchor_text').to_dict()['count']
data
```

```
{'Contact Us ': 7427,
 'Live Demo Discover how to create engaging webinar experiences designed to
 cativate and convert your audience. ': 7426,
 'Resources ': 7426,
 'Live Demo ': 7426,
 'ON24 Webcast Elite ': 3851,
 'ON24 platform ': 3806,
```

'Press Releases ': 3799, …}

Once converted, we feed this into the wordcloud function, limiting the data to the 200 most popular anchors:

```
wc = WordCloud(background_color='white',
               width=800, height=400,
               max_words=30).generate_from_frequencies(anchor_count_agg)

import matplotlib.pyplot as plt

plt.figure(figsize=(10, 10))
plt.imshow(wc, interpolation='bilinear')
plt.axis('off')

# Save image
wc.to_file("images/wordcloud.png")

plt.show()
```

The word cloud (Figure 3-17) could be used in a management presentation. There are some pretty long anchors there!

Figure 3-17. *Word cloud of the most commonly used anchor texts*

The activation from this point would be to see about finding semiautomated rules to improve the relevance of anchor texts, which is made easier by virtue of the fact that these are within the header or footer.

Core Web Vitals (CWV)

Core Web Vitals (CWV) is a Google initiative to help websites deliver a better UX. This includes speed, page stability during load, and the time it takes for the web page to become user interactive. So if CWV is about users, why is this in the technical section?

The technical SEO benefits which are less advertised help Google (and other search engines) mainly conserve computing resources to crawl and render websites. So it's a massive win-win-win for search engines, users, and webmasters.

So by pursuing CWV, you're effectively increasing your crawl and render budget which benefits your technical SEO.

However, technical SEO doesn't hold great appeal to marketing teams, whereas it's a much easier sell to marketing teams if you can imply the ranking benefits to justify web developments of improving CWV. And that is what we'll aim to do in this section.

We'll start with the landscape to show the overall competitive picture before drilling down on the website itself for the purpose of using data to prioritize development.

Landscape

```
import re
import time
import random
import pandas as pd
import numpy as np
import requests
import json
import plotnine
import tldextract
from plotnine import *
from mizani.transforms import trans
from client import RestClient

target_bu = 'boundless'
```

```
target_site = 'https://boundlesshq.com/'
target_name = target_bu
```

We start by obtaining the SERPs for your target keywords using the pandas read_csv function. We're interested in the URL which will form the input for querying the Google PageSpeed API which gives us the CWV metric values:

```
desktop_serps_df = pd.read_csv('data/1_desktop' + client_name +
'_serps.csv')
desktop_serps_df
```

This results in the following:

	keyword	rank_absolute	url	device
0	permanent establishment risk	1	https://papayaglobal.com/blog/how-to-avoid-per...	desktop
1	permanent establishment risk	2	None	desktop
2	permanent establishment risk	3	https://www.pwc.co.uk/assets/pdf/permanent-est...	desktop
3	permanent establishment risk	4	https://www.safeguardglobal.com/resources/blog...	desktop
4	permanent establishment risk	5	https://www.omnipresent.com/resources/permanen...	desktop
5	permanent establishment risk	6	https://www.airswift.com/blog/permanent-establ...	desktop
6	permanent establishment risk	7	https://www.gtn.com/resources/newsletters/perm...	desktop
7	permanent establishment risk	8	https://www.letsdeel.com/blog/permanent-establ...	desktop
8	permanent establishment risk	9	https://freemanlaw.com/the-tax-risk-of-a-perma...	desktop
9	permanent establishment risk	10	https://www.oysterhr.com/library/avoid-permane...	desktop
10	permanent establishment risk	11	https://nhglobalpartners.com/what-is-permanent...	desktop

The SERPs data can get a bit noisy, and ultimately the business is only interested in their direct competitors, so we'll create a list of them to filter the SERPs accordingly:

```
selected_sites = [target_site, 'https://papayaglobal.com/', 'https://www.
airswift.com/', 'https://shieldgeo.com/',
                  'https://remote.com/', 'https://www.letsdeel.com/',
'https://www.omnipresent.com/']
```

```
desktop_serps_select = desktop_serps_df[~desktop_serps_df['url'].
isnull()].copy()
```

```
desktop_serps_select = desktop_serps_select[desktop_serps_select['url'].
str.contains('|'.join(selected_sites))]
desktop_serps_select
```

	keyword	rank_absolute	url	device
0	permanent establishment risk	1	https://papayaglobal.com/blog/how-to-avoid-permanent-establishment-risk/	desktop
4	permanent establishment risk	5	https://www.omnipresent.com/resources/permanent-establishment-risk-a-remote-workforce	desktop
5	permanent establishment risk	6	https://www.airswift.com/blog/permanent-establishment-risks	desktop
7	permanent establishment risk	8	https://www.letsdeel.com/blog/permanent-establishment-risk	desktop
14	permanent establishment risk	15	https://shieldgeo.com/ultimate-guide-permanent-establishment/	desktop
...	...	...	...	...
2355	eor country	32	https://papayaglobal.com/blog/differences-between-icps-why-it-matters/	desktop
2370	eor country	47	https://www.omnipresent.com/resources/peo-vs-eor-tapping-into-global-talent	desktop
2374	eor country	51	https://boundlesshq.com/guides/croatia/	desktop
2376	eor country	53	https://www.letsdeel.com/blog/international-employee-experience	desktop
2403	eor country	80	https://www.airswift.com/blog/what-is-an-employer-of-record	desktop

114 rows × 4 columns

There are much less rows as a result, which means less API queries and less time required to get the data.

Note the data is just for desktop, so this process would need to be repeated for mobile SERPs also.

To query the PageSpeed API efficiently and avoid duplicate requests, we want a unique set of URLs. We achieve this by

Exporting the URL column to a list

```
desktop_serps_urls = desktop_serps_select['url'].to_list()
```

Deduplicating the list

```
desktop_serps_urls = list(dict.fromkeys(desktop_serps_urls))
desktop_serps_urls
```

```
['https://papayaglobal.com/blog/how-to-avoid-permanent-
establishment-risk/',
 'https://www.omnipresent.com/resources/permanent-establishment-risk-a-
remote-workforce',
 'https://www.airswift.com/blog/permanent-establishment-risks',
 'https://www.letsdeel.com/blog/permanent-establishment-risk',
```

```
 'https://shieldgeo.com/ultimate-guide-permanent-establishment/',
 'https://remote.com/blog/what-is-permanent-establishment',
 'https://remote.com/lp/global-payroll',
 'https://remote.com/services/global-payroll?nextInternalLocale=
en-us', . . . ]
```

With the list, we query the API, starting by setting the parameters for the API itself, the device, and the API key (obtained by getting a Google Cloud Platform account which is free):

```
base_url = 'https://www.googleapis.com/pagespeedonline/v5/
runPagespeed?url='
strategy = '&strategy=desktop'
api_key = '&key=[Your PageSpeed API key]'
```

Initialize an empty dictionary and set i to zero which will be used as a counter to help us keep track of how many API calls have been made and how many to go:

```
desktop_cwv = {}
i = 1

for url in desktop_serps_urls:
    request_url = base_url + url + strategy + api_key
    response = json.loads(requests.get(request_url).text)
    i += 1
    print(i, " ", request_url)
    desktop_cwv[url] = response
```

The result is a dictionary containing the API response. To get this output into a usable format, we iterate through the dictionary to pull out the actual CWV scores as the API has a lot of other micro measurement data which doesn't serve our immediate objectives.

Initialize an empty list which will store the API response data:

```
desktop_psi_lst = []
```

Loop through the API output which is a JSON dictionary, so we need to pull out the relevant "keys" and add them to the list initialized earlier:

```
for key, data in desktop_cwv.items():
```

```
if 'lighthouseResult' in data:
    FCP = data['lighthouseResult']['audits']['first-contentful-paint']
    ['numericValue']
    LCP = data['lighthouseResult']['audits']['largest-contentful-
    paint']['numericValue']
    CLS = data['lighthouseResult']['audits']['cumulative-layout-shift']
    ['numericValue']
    FID = data['lighthouseResult']['audits']['max-potential-fid']
    ['numericValue']
    SIS = data['lighthouseResult']['audits']['speed-index']
    ['score'] * 100

    desktop_psi_lst.append([key, FCP, LCP, CLS, FID, SIS])
```

Convert the list into a dataframe:

```
desktop_psi_df = pd.DataFrame(desktop_psi_lst, columns = ['url', 'FCP',
'LCP', 'CLS', 'FID', 'SIS'])
desktop_psi_df
```

This results in the following:

	url	FCP	LCP	CLS	FID	SIS
0	https://papayaglobal.com/blog/how-to-avoid-permanent-establishment-risk/	890.00000	5241.5	0.140066	103	17.0
1	https://www.omnipresent.com/resources/permanent-establishment-risk-a-remote-workforce	1010.00000	1720.0	0.144360	138	92.0
2	https://www.airswift.com/blog/permanent-establishment-risks	662.13298	2734.0	0.118718	99	28.0
3	https://www.letsdeel.com/blog/permanent-establishment-risk	1093.00000	1736.5	0.000000	81	67.0
4	https://shieldgeo.com/ultimate-guide-permanent-establishment/	1030.00000	1254.5	0.121591	81	96.0
...	...	...	...	...	...	...
72	https://www.letsdeel.com/blog/remote-interview-guide	942.00000	1320.0	0.000000	122	68.0
73	https://www.omnipresent.com/resources/how-to-recruit-and-hire-remote-employees	1070.00000	1640.0	0.138055	94	72.0
74	https://papayaglobal.com/blog/differences-between-icps-why-it-matters/	930.00000	3432.0	0.089238	129	33.0
75	https://boundlesshq.com/guides/croatia/	730.00000	2050.0	0.051751	78	99.0
76	https://www.letsdeel.com/blog/international-employee-experience	972.00000	1411.0	0.000000	129	67.0

77 rows × 6 columns

The PageSpeed data on all of the ranking URLs is in a dataframe with all of the CWV metrics:

- *FCP*: First Contentful Paint

- *LCP*: Largest Contentful Paint

- *CLS*: Cumulative Layout Shift

- *SIS*: Speed Index Score

To show the relevance of the ranking (and hopefully the benefit to ranking by improving CWV), we want to merge this with the rank data:

```
dtp_psi_serps = desktop_serps_select.merge(desktop_psi_df, on = 'url', how
= 'left')
dtp_psi_serps_bu = dtp_psi_serps.merge(target_keywords_df, on = 'keyword',
how = 'left')
dtp_psi_serps_bu.to_csv('data/'+ target_bu +'_dtp_psi_serps_bu.csv')
dtp_psi_serps_bu
```

This results in the following:

	keyword	rank_absolute	url	device	FCP	LCP	CLS	FID	SIS	bu
0	permanent establishment risk	1	https://papayaglobal.com/blog/how-to-avoid-permanent-establishment-risk/	desktop	890.00000	5241.5	0.140066	103.0	17.0	boundless
1	permanent establishment risk	5	https://www.omnipresent.com/resources/permanent-establishment-risk-a-remote-workforce	desktop	1010.00000	1720.0	0.144360	138.0	92.0	boundless
2	permanent establishment risk	6	https://www.airswift.com/blog/permanent-establishment-risks	desktop	662.13298	2734.0	0.118718	99.0	28.0	boundless
3	permanent establishment risk	8	https://www.letsdeel.com/blog/permanent-establishment-risk	desktop	1093.00000	1736.5	0.000000	81.0	67.0	boundless
4	permanent establishment risk	15	https://shieldgeo.com/ultimate-guide-permanent-establishment/	desktop	1030.00000	1254.5	0.121591	81.0	96.0	boundless
...	...	...	...	...	...	...	...	...	...	...
112	eor country	32	https://papayaglobal.com/blog/differences-between-icps-why-it-matters/	desktop	930.00000	3432.0	0.089238	129.0	33.0	boundless
113	eor country	47	https://www.omnipresent.com/resources/peo-vs-eor-tapping-into-global-talent	desktop	1070.00000	1740.0	0.138055	128.0	94.0	boundless
114	eor country	51	https://boundlesshq.com/guides/croatia/	desktop	730.00000	2050.0	0.051751	78.0	99.0	boundless
115	eor country	53	https://www.letsdeel.com/blog/international-employee-experience	desktop	972.00000	1411.0	0.000000	129.0	67.0	boundless
116	eor country	80	https://www.airswift.com/blog/what-is-an-employer-of-record	desktop	662.13298	2986.0	0.120042	94.0	64.0	boundless

117 rows × 10 columns

The dataframe is complete with the keyword, its rank, URL, device, and CWV metrics.

At this point, rather than repeat near identical code for mobile, you can assume we have the data for mobile which we have combined into a single dataframe using the pandas concat function (same headings).

To add some additional features, we have added another column is_target indicating whether the ranking URL is the client or not:

```
overall_psi_serps_bu['is_target'] = np.where(overall_psi_serps_bu['url'].
str.contains(target_site), '1', '0')
```

Parse the site name:

```
overall_psi_serps_bu['site'] = overall_psi_serps_bu['url'].apply(lambda
url: tldextract.extract(url).domain)
```

Count the column for easy aggregation:

```
overall_psi_serps_bu['count'] = 1
```

The resultant dataframe is overall_psi_serps_bu shown as follows:

keyword	rank_absolute	url	device	FCP	LCP	CLS	FID	SIS	bu	is_target	site
permanent stablishment risk	1	https://papayaglobal.com/blog/how-to-avoid-permanent-establishment-risk/	mobile	3990.00000	4140.0	0.051357	349.0	4.0	boundless	0	papayaglobal
permanent stablishment risk	4	https://www.omnipresent.com/resources/permanent-establishment-risk-a-remote-workforce	mobile	4065.00000	4515.0	0.040469	471.0	30.0	boundless	0	omnipresent
permanent stablishment risk	5	https://www.airswift.com/blog/permanent-establishment-risks?hs_amp=true	mobile	927.00000	2495.0	0.000000	316.0	97.0	boundless	0	airswift
permanent stablishment risk	8	https://www.letsdeel.com/blog/permanent-establishment-risk	mobile	5763.00000	7131.0	0.000000	286.0	25.0	boundless	0	letsdeel
permanent stablishment risk	16	https://shieldgeo.com/ultimate-guide-permanent-establishment/	mobile	4623.00000	6695.0	0.000343	299.0	70.0	boundless	0	shieldgeo
...	...	...	...	...	...	...	...	...	...	...	...
eor country	32	https://papayaglobal.com/blog/differences-between-icps-why-it-matters/	desktop	930.00000	3432.0	0.089238	129.0	33.0	boundless	0	papayaglobal
eor country	47	https://www.omnipresent.com/resources/peo-vs-eor-tapping-into-global-talent	desktop	1070.00000	1740.0	0.138055	128.0	94.0	boundless	0	omnipresent
eor country	51	https://boundlesshq.com/guides/croatia/	desktop	730.00000	2050.0	0.051751	78.0	99.0	boundless	1	boundlesshq
eor country	53	https://www.letsdeel.com/blog/international-employee-experience	desktop	972.00000	1411.0	0.000000	129.0	67.0	boundless	0	letsdeel
eor country	80	https://www.airswift.com/blog/what-is-an-employer-of-record	desktop	662.13298	2986.0	0.120042	94.0	64.0	boundless	0	airswift

The aggregation will be executed at the site level so we can compare how each site scores on average for their CWV metrics and correlate that with performance:

```
overall_psi_serps_agg = overall_psi_serps_bu.groupby('site').
agg({'LCP': 'mean',
                                              'FCP': 'mean',
                                              'CLS': 'mean',
                                              'FID': 'mean',
                                              'SIS': 'mean',
```

```
                                                            'rank_
                                                            absolute':
                                                            'mean',
                                                            'count':
                                                            'sum'}).
                                                            reset_
                                                            index()
overall_psi_serps_agg = overall_psi_serps_agg.rename(columns = {'count':
'reach'})
```

Here are some operations to make the site names shorter for the graphs later:

```
overall_psi_serps_agg['site'] = np.where(overall_psi_serps_agg['site'] ==
'papayaglobal', 'papaya',
                                    overall_psi_serps_agg['site'])
overall_psi_serps_agg['site'] = np.where(overall_psi_serps_agg['site'] ==
'boundlesshq', 'boundless',
                                    overall_psi_serps_agg['site'])
overall_psi_serps_agg
```

This results in the following:

	site	LCP	FCP	CLS	FID	SIS	rank_absolute	reach
0	airswift	2850.282609	952.014339	0.060763	150.347826	72.521739	30.043478	23
1	boundless	4563.175000	2043.200000	0.186265	158.900000	92.450000	35.900000	20
2	letsdeel	5153.855556	3340.166667	0.000000	229.955556	38.755556	36.044444	45
3	omnipresent	3809.153846	2380.897436	0.064785	242.000000	72.820513	34.948718	39
4	papaya	10835.166753	2640.666667	0.057782	312.800000	8.833333	42.562500	32
5	remote	5358.801887	2169.452830	0.074436	845.716981	89.660377	17.641509	53
6	shieldgeo	5756.100000	3017.533333	0.056910	234.833333	71.500000	33.333333	30

That's the summary which is not so easy to discern trends, and now we're ready to plot the data, starting with the overall speed index. The Speed Index Score (SIS) is scaled between 0 and 100, 100 being the fastest and therefore best.

Note that in all of the charts that will compare Google rank with the individual CWV metrics, the vertical axis will be inverted such that the higher the position, the higher the ranking. This is to make the charts more intuitive and easier to understand.

```
SIS_cwv_landscape_plt = (
    ggplot(overall_psi_serps_agg,
            aes(x = 'SIS', y = 'rank_absolute', fill = 'site', colour = 'site',
                        size = 'reach')) +
    geom_point(alpha = 0.8) +
    geom_text(overall_psi_serps_agg, aes(label = 'site'),
    position=position_stack(vjust=-0.08)) +
    labs(y = 'Google Rank', x = 'Speed Score') +
    scale_y_reverse() +
  scale_size_continuous(range = [7, 17]) +
    theme(legend_position = 'none', axis_text_x=element_text(rotation=0,
    hjust=1, size = 12))
)

SIS_cwv_landscape_plt.save(filename = 'images/0_SIS_cwv_landscape.png',
                            height=5, width=8, units = 'in', dpi=1000)
SIS_cwv_landscape_plt
```

Already we can see in SIS_cwv_landscape_plt (Figure 3-18) that the higher your speed score, the higher you rank in general which is a nice easy sell to the stakeholders, acting as motivation to invest resources into improving CWV.

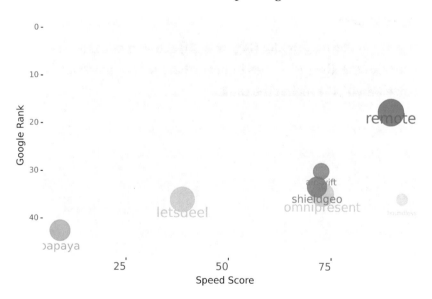

Figure 3-18. *Scatterplot comparing speed scores and Google rank of different websites*

Boundless in this instance are doing relatively well. Although they don't rank the highest, this could indicate that either some aspects of CWV are not being attended to or something non-CWV related or more likely a combination of both.

```
LCP_cwv_landscape_plt = (
    ggplot(overall_psi_serps_agg,
            aes(x = 'LCP', y = 'rank_absolute', fill = 'site', colour
            = 'site',
                                    size = 'reach')) +
    geom_point(alpha = 0.8) +
    geom_text(overall_psi_serps_agg, aes(label = 'site'),
    position=position_stack(vjust=-0.08)) +
    labs(y = 'Google Rank', x = 'Largest Contentful Paint') +
    scale_y_reverse() +
  scale_size_continuous(range = [7, 17]) +
    theme(legend_position = 'none', axis_text_x=element_text(rotation=0,
    hjust=1, size = 12))
)

LCP_cwv_landscape_plt.save(filename = 'images/0_LCP_cwv_landscape.png',
                            height=5, width=8, units = 'in', dpi=1000)
LCP_cwv_landscape_plt
```

The LCP_cwv_landscape_plt plot (Figure 3-19) shows that Papaya and Remote look like outliers; in any case, the trend does indicate that the less time it takes to load the largest content element, the higher the rank.

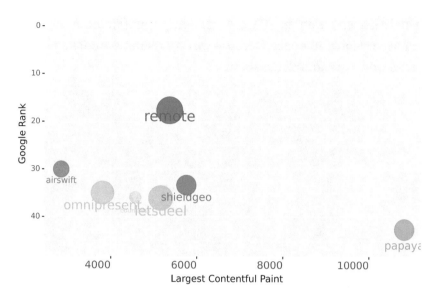

Figure 3-19. *Scatterplot comparing Largest Contentful Paint (LCP) and Google rank by website*

```
FID_cwv_landscape_plt = (
    ggplot(overall_psi_serps_agg,
           aes(x = 'FID', y = 'rank_absolute', fill = 'site', colour
           = 'site',
                                 size = 'reach')) +
    geom_point(alpha = 0.8) +
    geom_text(overall_psi_serps_agg, aes(label = 'site'),
    position=position_stack(vjust=-0.08)) +
    labs(y = 'Google Rank', x = 'First Input Delay') +
    scale_y_reverse() +
    scale_x_log10() +
  scale_size_continuous(range = [7, 17]) +
    theme(legend_position = 'none', axis_text_x=element_text(rotation=0,
    hjust=1, size = 12))
)

FID_cwv_landscape_plt.save(filename = 'images/0_FID_cwv_landscape.png',
                          height=5, width=8, units = 'in', dpi=1000)
FID_cwv_landscape_plt
```

Remote looks like an outlier in FID_cwv_landscape_plt (Figure 3-20). Should the outlier be removed? Not in this case, because we don't remove outliers just because it doesn't show us what we wanted it to show.

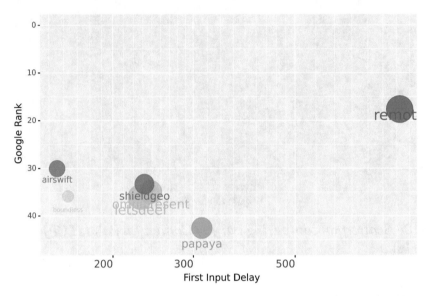

Figure 3-20. *Scatterplot comparing First Input Delay (FID) and Google rank by website*

The trend indicates that the less time it takes to make the page interactive for users, the higher the rank.

Boundless are doing well in this respect.

```
CLS_cwv_landscape_plt = (
    ggplot(overall_psi_serps_agg,
           aes(x = 'CLS', y = 'rank_absolute', fill = 'site', colour
           = 'site',
                            size = 'reach')) +
    geom_point(alpha = 0.8) +
    geom_text(overall_psi_serps_agg, aes(label = 'site'),
    position=position_stack(vjust=-0.08)) +
    labs(y = 'Google Rank', x = 'Cumulative Layout Shift') +
    scale_y_reverse() +
  scale_size_continuous(range = [7, 17]) +
```

```
    theme(legend_position = 'none', axis_text_x=element_text(rotation=0,
    hjust=1, size = 12))
)
```

```
CLS_cwv_landscape_plt.save(filename = 'images/0_CLS_cwv_landscape.png',
                            height=5, width=8, units = 'in', dpi=1000)
CLS_cwv_landscape_plt
```

Okay, CLS where Boundless don't perform as well is shown in CLS_cwv_landscape_ plt (Figure 3-21). The impact on improving rank is quite unclear too.

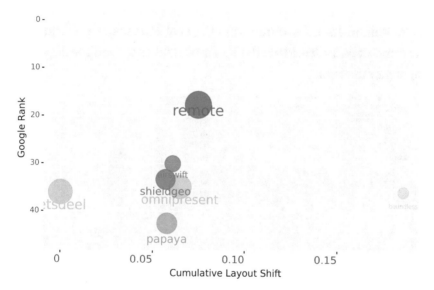

Figure 3-21. *Scatterplot comparing Cumulative Layout Shift (CLS) and Google rank by website*

```
FCP_cwv_landscape_plt = (
    ggplot(overall_psi_serps_agg,
          aes(x = 'FCP', y = 'rank_absolute', fill = 'site', colour
          = 'site',
                            size = 'reach')) +
    geom_point(alpha = 0.8) +
    geom_text(overall_psi_serps_agg, aes(label = 'site'),
    position=position_stack(vjust=-0.08)) +
```

```
    labs(y = 'Google Rank', x = 'First Contentful Paint') +
    scale_y_reverse() +
  scale_size_continuous(range = [7, 17]) +
    theme(legend_position = 'none', axis_text_x=element_text(rotation=0,
    hjust=1, size = 12))
)

FCP_cwv_landscape_plt.save(filename = 'images/0_FCP_cwv_landscape.png',
                           height=5, width=8, units = 'in', dpi=1000)
FCP_cwv_landscape_plt
```

Papaya and Remote look like outliers in FCP_cwv_landscape_plt (Figure 3-22); in any case, the trend does indicate that the less time it takes to load the largest content element, the higher the rank.

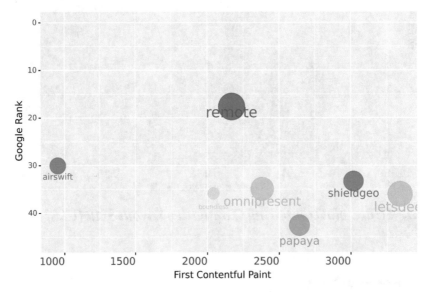

Figure 3-22. *Scatterplot comparing First Contentful Paint (FCP) and Google rank by website*

That's the deep dive into the overall scores. The preceding example can be repeated for both desktop and mobile scores to drill down into, showing which specific CWV metrics should be prioritized. Overall, for boundless, CLS appears to be its weakest point.

In the following, we'll summarize the analysis on a single chart by pivoting the data in a format that can be used to power the single chart:

```
overall_psi_serps_long = overall_psi_serps_agg.copy()
```

We select the columns we want:

```
overall_psi_serps_long = overall_psi_serps_long[['site', 'LCP', 'FCP',
'CLS', 'FID', 'SIS']]
```

and use the melt function to pivot the table:

```
overall_psi_serps_long = overall_psi_serps_long.melt(id_vars=['site'],
                                            value_vars=['LCP',
'FCP', 'CLS', 'FID', 'SIS'],

                                            var_name='Metric',
value_name='Index')
overall_psi_serps_long['x_axis'] = overall_psi_serps_long['Metric']
overall_psi_serps_long['site'] = np.where(overall_psi_serps_long['site'] ==
'papayaglobal', 'papaya',

                                    overall_psi_serps_long['site'])
overall_psi_serps_long['site'] = np.where(overall_psi_serps_long['site'] ==
'boundlesshq', 'boundless',

                                    overall_psi_serps_long['site'])

overall_psi_serps_long
```

This results in the following:

	site	Metric	Index	x_axis
0	airswift	LCP	2807.090909	LCP
1	airswift	LCP	2889.875000	LCP
2	boundless	LCP	1935.050000	LCP
3	boundless	LCP	7191.300000	LCP
4	letsdeel	LCP	1926.391304	LCP
...	...	...	...	...
65	papaya	SIS	2.333333	SIS
66	remote	SIS	96.227273	SIS
67	remote	SIS	85.000000	SIS
68	shieldgeo	SIS	88.785714	SIS
69	shieldgeo	SIS	56.375000	SIS

70 rows × 4 columns

That's the long format in place, ready to plot.

```
speed_ex_plt = (
    ggplot(overall_psi_serps_long,
            aes(x = 'site', y = 'Index', fill = 'site')) +
    geom_bar(stat = 'identity', alpha = 0.8) +
    labs(y = '', x = '') +
    theme(legend_position = 'right',
          axis_text_x =element_text(rotation=90, hjust=1, size = 12),
          legend_title = element_blank()
          ) +
    facet_grid('Metric ~ .', scales = 'free')
)

speed_ex_plt.save(filename = 'images/0_CWV_Metrics_plt.png',
                        height=5, width=8, units = 'in', dpi=1000)
speed_ex_plt
```

The speed_ex_plt chart (Figure 3-23) shows the competitors being compared for each metric. Remote seem to perform the worst on average, so their prominent rankings are probably due to non-CWV factors.

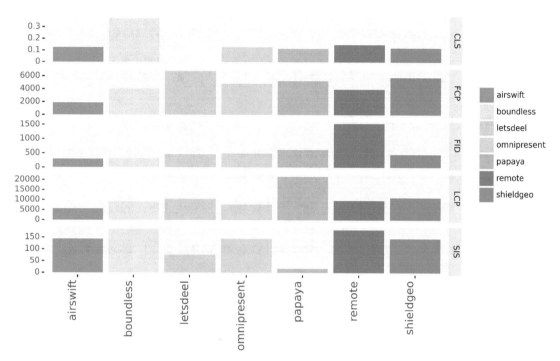

Figure 3-23. *Faceted column chart of different sites by CWV metric*

Onsite CWV

The purpose of the landscape was to use data to motivate the client, colleagues, and stakeholders of the SEO benefits that would follow CWV improvement. In this section, we're going to drill into the site itself to see where the improvements could be made.

We'll start by importing the data and cleaning up the columns as usual:

```
target_crawl_raw = pd.read_csv('data/boundlesshq_com_all_urls__excluding_
uncrawled__filtered_20220427203402.csv')

target_crawl_raw.columns = [col.lower() for col in target_crawl_raw.
columns]
target_crawl_raw.columns = [col.replace('(', '') for col in target_crawl_
raw.columns]
```

```
target_crawl_raw.columns = [col.replace(')', '') for col in target_crawl_
raw.columns]
target_crawl_raw.columns = [col.replace('@', '') for col in target_crawl_
raw.columns]
target_crawl_raw.columns = [col.replace('/', '') for col in target_crawl_
raw.columns]
target_crawl_raw.columns = [col.replace(' ', '_') for col in target_crawl_
raw.columns]
print(target_crawl_raw.columns)
```

We're using Sitebulb crawl data, and we want to only include onsite indexable URLs since those are the ones that rank, which we will filter as follows:

```
target_crawl_raw = target_crawl_raw.loc[target_crawl_raw['host'] ==
target_host]
target_crawl_raw = target_crawl_raw.loc[target_crawl_raw['indexable_
status'] == 'Indexable']
target_crawl_raw = target_crawl_raw.loc[target_crawl_raw['content_type']
== 'HTML']
```

```
target_crawl_raw
```

This results in the following:

	url	crawl_depth	crawl_status	host	is_subdomain	scheme	crawl_source	first_parent_url	h
0	https://boundlesshq.com/	0	Success	boundlesshq.com	No	https	Crawler	None	
13	https://boundlesshq.com/pricing/	1	Success	boundlesshq.com	No	https	Crawler	https://boundlesshq.com/	
17	https://boundlesshq.com/how-it-works/countries/	1	Success	boundlesshq.com	No	https	Crawler	https://boundlesshq.com/	
23	https://boundlesshq.com/borderless-benefits/	1	Success	boundlesshq.com	No	https	Crawler	https://boundlesshq.com/	
36	https://boundlesshq.com/blog/employment/what-is-an-employer-of-record/	1	Success	boundlesshq.com	No	https	Crawler	https://boundlesshq.com/	
...	...	...	...	...	...	...	...	...	
4404	https://boundlesshq.com/guides/united-arab-emirates/	Not Set	Success	boundlesshq.com	No	https	Google Search Analytics	None	
4407	https://boundlesshq.com/guides/vietnam/	Not Set	Success	boundlesshq.com	No	https	Google Search Analytics	None	
4477	https://boundlesshq.com/hr-tech-ireland/	Not Set	Success	boundlesshq.com	No	https	Google Search Analytics	None	
4486	https://boundlesshq.com/guides/uruguay/	Not Set	Success	boundlesshq.com	No	https	Google Search Analytics	None	
4498	https://boundlesshq.com/guides/slovenia/	Not Set	Success	boundlesshq.com	No	https	Google Search Analytics	None	

279 rows × 71 columns

With 279 rows, it's a small website. The next step is to select the desired columns which will comprise the CWV measures and anything that could possibly explain it:

```
target_speedDist_df = target_crawl_raw[['url', 'cumulative_layout_shift',
'first_contentful_paint',
                                        'largest_contentful_paint',
                                        'performance_score', 'time_to_
                                        interactive',
                                        'total_blocking_time', 'images_
                                        without_dimensions', 'perf_
                                        budget_fonts',
                                        'font_transfer_size_kib', 'fonts_
                                        files', 'images_files',
                                        'images_not_efficiently_encoded',
                                        'images_size_kib',
                                        'images_transfer_size_kib',
                                        'images_without_dimensions',
                                        'media_files',
                                        'media_size_kib', 'media_transfer_
                                        size_kib',
                                        'next-gen_format_savings_kib',
                                        'offscreen_images_not_deferred',
                                        'other_files', 'other_size_kib',
                                        'other_transfer_size_kib',
                                        'passed_font-face_display_urls',
                                        'render_blocking_savings',
                                        'resources_not_http2', 'scaled_
                                        images', 'perf_budget_total']]

target_speedDist_df
```

This results in the following:

	url	cumulative_layout_shift	first_contentful_paint	largest_contentful_paint	performance_score	time_to_interactive
0	https://boundlesshq.com/	0.339	1757	3518	83	2111
13	https://boundlesshq.com/pricing/	0.096	2459	2635	84	3273
17	https://boundlesshq.com/how-it-works/countries/	0.104	2395	2616	75	3825
23	https://boundlesshq.com/borderless-benefits/	0.139	1952	2578	90	2629
36	https://boundlesshq.com/blog/employment/what-is-an-employer-of-record/	0.236	2052	5380	59	3648
...	...	...	...	...	...	...
4404	https://boundlesshq.com/guides/united-arab-emirates/	0.103	2103	9684	64	2711
4407	https://boundlesshq.com/guides/vietnam/	0.104	1868	9427	71	2529
4477	https://boundlesshq.com/hr-tech-ireland/	0.214	2190	2616	83	3297
4486	https://boundlesshq.com/guides/uruguay/	0.104	2057	9460	66	2551
4498	https://boundlesshq.com/guides/slovenia/	0.104	2113	8651	64	2760

279 rows × 29 columns

The dataframe columns have reduced from 71 to 29, and the CWV scores are more apparent.

Attempting to analyze the sites at the URL will not be terribly useful, so to make pattern identification easier, we will classify the content by folder location:

```
section_conds = [
    target_speedDist_df['url'] == 'https://boundlesshq.com/',
    target_speedDist_df['url'].str.contains('/guides/'),
    target_speedDist_df['url'].str.contains('/how-it-works/')
]
```

```
section_vals = ['home', 'guides', 'commercial']
```

```
target_speedDist_df['content'] = np.select(section_conds, section_vals,
default = 'blog')
```

We'll also convert the main metrics to a number:

```
cols = ['cumulative_layout_shift', 'first_contentful_paint', 'largest_
contentful_paint', 'performance_score',
        'time_to_interactive', 'total_blocking_time']
```

```
target_speedDist_df[cols] = pd.to_numeric(target_speedDist_df[cols].
stack(), errors='coerce').unstack()
```

```
target_speedDist_df
```

This results in the following:

I	other_files	other_size_kib	other_transfer_size_kib	passed_font-face_display_urls	render_blocking_savings	resources_not_http2	scaled_images	perf_budget_total	content
	0	0	0	0	10550	0	20	Yes	home
	0	0	0	0	8348	0	0	Yes	blog
	0	0	0	0	8805	0	5	Yes	commercial
	0	0	0	0	8572	0	0	Yes	blog
!	0	0	0	0	8895	0	0	Yes	blog
.	...	...	...	...	...	...	...	...	...
!	0	0	0	0	8415	0	0	Yes	guides
!	0	0	0	0	8327	0	0	Yes	guides
	0	0	0	0	8501	0	1	Yes	blog
!	0	0	0	0	8501	0	0	Yes	guides
!	0	0	0	0	8660	0	0	Yes	guides

A new column has been created in which each indexable URL is labeled by their content category.

Time for some aggregation using groupby on "content":

```
speed_dist_agg = target_speedDist_df.groupby('content').agg({'url':
'count', 'performance_score'}).reset_index()
speed_dist_agg
```

This results in the following:

	content	url	performance_score
0	blog	66	68.636364
1	commercial	3	66.333333
2	guides	209	76.631579
3	home	1	83.000000

Most of the content are guides followed by blog posts with three offer pages. To visualize, we're going to use a histogram showing the distribution of the overall performance score and color code the URLs in the score columns by their segment.

The home page and the guides are by far the fastest.

```
target_speedDist_plt = (
    ggplot(target_speedDist_df,
          aes(x = 'performance_score', fill = 'content')) +
    geom_histogram(alpha = 0.8, bins = 20) +
    labs(y = 'Page Count', x = '\nSpeed Score') +
```

```
    #scale_x_continuous(breaks=range(0, 100, 20)) +
    theme(legend_position = 'right',
        axis_text_x = element_text(rotation=90, hjust=1, size = 7))
)

target_speedDist_plt.save(filename = 'images/3_target_speedDist_plt.png',
                            height=5, width=8, units = 'in', dpi=1000)
target_speedDist_plt
```

The target_speedDist_plt plot (Figure 3-24) shows the home page (in purple) performs reasonably well with a speed score of 84. The guides vary, but most of these have a speed above 80, and the majority of blog posts are in the 70s.

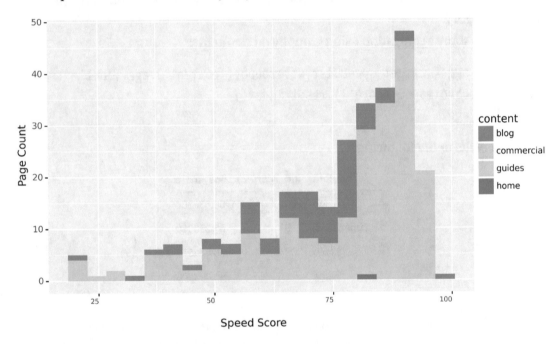

Figure 3-24. *Distribution of speed score by content type*

Let's drill down by CWV score category, starting with CLS:

```
target_CLS_plt = (
    ggplot(target_speedDist_df,
        aes(x = 'cumulative_layout_shift', fill = 'content')) +
    geom_histogram(alpha = 0.8, bins = 20) +
    labs(y = 'Page Count', x = '\ncumulative_layout_shift') +
```

```
    #scale_x_continuous(breaks=range(0, 100, 20)) +
    theme(legend_position = 'right',
         axis_text_x = element_text(rotation=90, hjust=1, size = 7))
)

target_CLS_plt.save(filename = 'images/3_target_CLS_plt.png',
                                height=5, width=8, units = 'in', dpi=1000)
target_CLS_plt
```

As shown in target_CLS_plt (Figure 3-25), guides have the least amount of shifting during browser rendering, whereas the blogs and the home page shift the most.

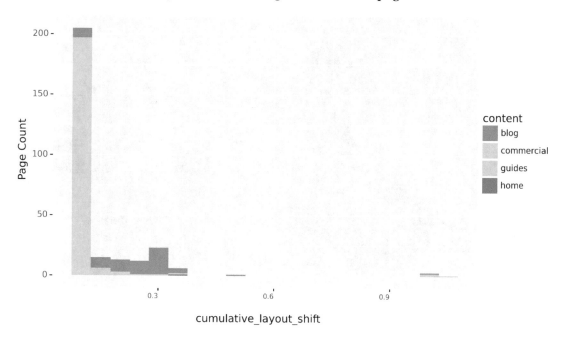

Figure 3-25. *Distribution of CLS by content type*

So we now know which content templates to focus our CLS development efforts.

```
target_FCP_plt = (
    ggplot(target_speedDist_df,
         aes(x = 'first_contentful_paint', fill = 'content')) +
    geom_histogram(alpha = 0.8, bins = 30) +
    labs(y = 'Page Count', x = '\nContentful paint') +
    theme(legend_position = 'right',
```

```
               axis_text_x = element_text(rotation=90, hjust=1, size = 7))
)

target_FCP_plt.save(filename = 'images/3_target_FCP_plt.png',
                             height=5, width=8, units = 'in', dpi=1000)
target_FCP_plt
```

In this area, target_FCP_plt (Figure 3-26) shows no discernible trends here which indicates it's an overall site problem. So digging into the Chrome Developer Tools and looking into the network logs would be the obvious next step.

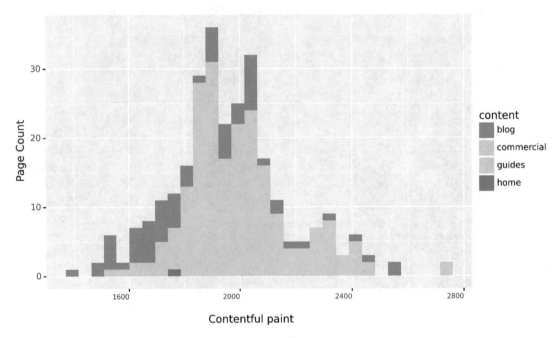

Figure 3-26. *Distribution of FCP by content type*

```
target_LCP_plt = (
    ggplot(target_speedDist_df,
           aes(x = 'largest_contentful_paint', fill = 'content')) +
    geom_histogram(alpha = 0.8, bins = 20) +
    labs(y = 'Page Count', x = '\nlargest_contentful_paint') +
    theme(legend_position = 'right',
          axis_text_x = element_text(rotation=90, hjust=1, size = 7))
)
```

```
target_LCP_plt.save(filename = 'images/3_target_LCP_plt.png',
                              height=5, width=8, units = 'in', dpi=1000)
target_LCP_plt
```

target_LCP_plt (Figure 3-27) shows most guides and some blogs have the fastest LCP scores; in any case, the blog template and the rogue guides would be the areas of focus.

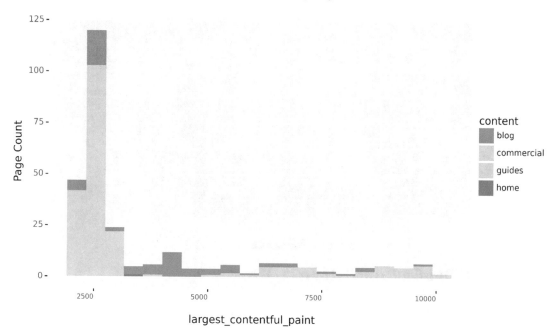

Figure 3-27. *Distribution of LCP by content type*

```
target_FID_plt = (
    ggplot(target_speedDist_df,
           aes(x = 'time_to_interactive', fill = 'content')) +
    geom_histogram(alpha = 0.8, bins = 20) +
    labs(y = 'Page Count', x = '\ntime_to_interactive') +
    theme(legend_position = 'right',
          axis_text_x = element_text(rotation=90, hjust=1, size = 7))
)

target_FID_plt.save(filename = 'images/3_target_FID_plt.png',
                              height=5, width=8, units = 'in', dpi=1000)
target_FID_plt
```

The majority of the site appears in target_FID_plt (Figure 3-28) to enjoy fast FID times, so this would be the least priority for CWV improvement.

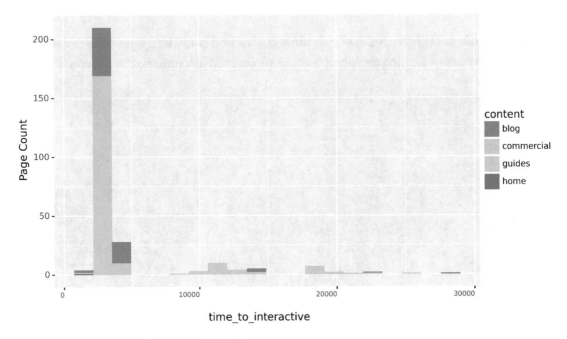

Figure 3-28. *Distribution of FID by content type*

Summary

In this chapter, we covered how data-driven approach could be taken toward technical SEO by way of

- Modeling page authority to estimate the benefit of technical SEO recommendations to colleagues and clients

- Internal link optimization analyzed in different ways to improve content discoverability and labeling via anchor text

- Core Web Vitals to see which metrics require improvement and by content type

The next chapter will focus on using data to improve content and UX.

CHAPTER 4

Content and UX

Content and UX for SEO is about the quality of the experience you're delivering to your website users, especially when they are referred from search engines. This means a number of things including but not limited to

- Having the content your target audiences are searching for

- Content that best satisfies the user query

  - *Content creation*: Planning landing page content

  - *Content consolidation*: (I) Splitting content (in instances where "too much" content might be impacting user satisfaction or hindering search engines from understanding the search intent the content is targeting) and (II) merging content (in instances where multiple pages are competing for the same intent)

- Fast to load – ensuring you're delivering a good user experience (UX)

- Renders well on different device types

By no means do we claim that this is the final word on data-driven SEO from a content and UX perspective. What we will do is expose data-driven ways of solving the most important SEO challenge using data science techniques, as not all require data science.

For example, getting scientific evidence that fast page speeds are indicative of higher ranked pages uses similar code from Chapter 6. Our focus will be on the various flavors of content that best satisfies the user query: keyword mapping, content gap analysis, and content creation.

A. Voniatis, *Data-Driven SEO with Python*, https://doi.org/10.1007/978-1-4842-9175-7_4

Content That Best Satisfies the User Query

An obvious challenge of SEO is deciding which content should go on which pages. Arguably, getting this right means you're optimizing for Google's RankBrain (a component of Google's core algorithm which uses machine learning to help understand and process user search queries).

While many crawling tools provide visuals of the distributions of pages by site depth or by segment, for example, data science enables you to benefit from a richer level of detail. To help you work out the content that best satisfies the user query, you need to

- Map keywords to content

- Plan content sections for those landing pages

- Decide what content to create for target keywords that will satisfy users searching for them

Data Sources

Your most likely data sources will be a combination of

- Site auditor URL exports

- SERPs tracking tools

Keyword Mapping

While there is so much to be gained from creating value-adding content, there is also much to be gained from retiring or consolidating content. This is achieved by merging it with another on the basis that they share the same search intent. Assuming the keywords have been grouped together by search intent, the next stage is to map them.

Keyword mapping is the process of mapping target keywords to pages and then optimizing the page toward these – as a result, maximizing a site's rank position potential in the search result. There are a number of approaches to achieve this:

- TF-IDF

- String matching

- Third-party neural network models (BERT, GPT-3)

- Build your own AI

We recommend string matching as it's fast, reasonably accurate, and the easiest to deploy.

String Matching

String matching works to see how many strings overlap and is used in DNA sequencing. String matching can work in two ways, which are to either treat strings as one object or strings made up of tokens (i.e., words within a string). We're opting for the latter because words mean something to humans and are not serial numbers. For that reason, we'll be using Sorensen-Dice which is fast and accurate compared to others we've tested.

The following code extract shows how we use string distance to map keywords to content by seeking the most similar URL titles to the target keyword. Let's go, importing libraries:

```
import requests
from requests.exceptions import ReadTimeout
from json.decoder import JSONDecodeError
import re
import time
import random
import pandas as pd
import numpy as np
import datetime
from client import RestClient
import json
import py_stringmatching as sm
from textdistance import sorensen_dice
from plotnine import *
import matplotlib.pyplot as plt

target = 'wella'
```

We'll start by importing the crawl data, which is a CSV export of website auditing software, in this case from "Sitebulb":

```
crawl_raw = pd.read_csv('data/www_wella_com_internal_html_urls_by_
indexable_status_filtered_20220629220833.csv')
```

Clean up the column heading title texts using a list comprehension:

```
crawl_raw.columns = [col.lower().replace('(','').replace(')','').
replace('%','').replace(' ', '_')
  for col in crawl_raw.columns]
```

```
crawl_df = crawl_raw.copy()
```

We're only interested in indexable pages as those are the URLs available for mapping:

```
crawl_df = crawl_df.loc[crawl_df['indexable'] == 'Yes']
crawl_df
```

This results in the following:

	url	ur	crawl_depth	scheme	crawl_source	first_parent_url	url_source	http_status	http_status_code	indexable
0	https://www.wella.com/international/wella-tuto...	7	Not Set	https	XML Sitemap	None	XML Sitemap	OK	200	Yes
1	https://www.wella.com/international/hair-style...	1	Not Set	https	XML Sitemap	None	XML Sitemap	OK	200	Yes
2	https://www.wella.com/international/hair-style...	4	Not Set	https	XML Sitemap	None	XML Sitemap	OK	200	Yes

The crawl import is complete. However, we're only interested in the URL and title as that's all we need for mapping keywords to URLs. Still it's good to import the whole file to visually inspect it, to be more familiar with the data.

```
urls_titles = crawl_df[['url', 'title']].copy()
urls_titles
```

This results in the following:

	url	title
0	https://www.wella.com/international/wella-tuto...	How to Style a Faux Hawk Like a Headline Act |...
1	https://www.wella.com/international/hair-style...	Silvikrin Classic Voluminous Hold Hairspray 75...
2	https://www.wella.com/international/hair-style...	Wellaflex 2nd Day Volume Strong Hold Mousse, H...
3	https://www.wella.com/international/hair-color...	Wella Koleston Permanent Hair Color Cream Fore...
4	https://www.wella.com/international/hair-color...	Wella Koleston Permanent Hair Color Cream With...
5	https://www.wella.com/international/hair-style...	Wellaflex Mega Strong Hold Hairspray, Hold: 5+...
6	https://www.wella.com/international/hair-style...	Wella Deluxe 24 Hour Wonder Volume Mousse 75 m...
7	https://www.wella.com/international/hair-color...	Wella Koleston Permanent Hair Color Cream With...
8	https://www.wella.com/international/hair-style...	Wellaflex Hydro Style Extra Strong Hold Hairsp...
9	https://www.wella.com/international/hair-style...	Wella Shockwaves Ultra Strong Power Hold Gel S...
10	https://www.wella.com/international/hair-color...	Wella Koleston Permanent Hair Color Cream With...

The dataframe is showing the URLs and titles. Let's load the keywords we want to map that have been clustered using techniques in Chapter 2:

```
keyword_discovery = pd.read_csv('data/keyword_discovery.csv')
```

This results in the following:

	topic	keyword	se_results_count	topic_results	topic_group
7	black hair	black hair	7340000000	7340000000	1
14	brown hair	brown hair	5170000000	5170000000	2
8	blonde hair	blonde hair	2730000000	2730000000	3
20	color perfect	color perfect	2270000000	2270000000	4
104	virtual try on	virtual try on	1670000000	2270000000	4
103	virtual try on	virtual try on tool	600000000	2270000000	4
61	how hair coloring works	how hair coloring works	2140000000	2140000000	5
26	cover roots	quick hair root touch ups	816000000	1941000000	6
27	cover roots	the best root touch	169000000	1941000000	6
28	cover roots	cover roots at home fast	197000000	1941000000	6
29	cover roots	cover roots	390000000	1941000000	6

155

The dataframe shows the topics, keywords, number of search engine results for the keywords, topic web search results, and the topic group. Note these were clustered using the methods disclosed in Chapter 2.

We'll map the topic as this is the central keyword that would also rank for their topic group keywords. This means we only require the topic column.

```
total_mapping_simi = keyword_discovery[['topic']].copy().drop_duplicates()
```

We want all the combinations of topics and URL titles before we can test each combination for string similarity. We achieve this using the cross-product merge:

```
total_mapping_simi = total_mapping_simi.merge(urls_titles, how = 'cross')
```

A new column "test" is created which will be formatted to remove boilerplate brand strings and force lowercase. This will make the string matching values more accurate.

```
total_mapping_simi['test'] = total_mapping_simi['title']
total_mapping_simi['test'] = total_mapping_simi['test'].str.lower()
total_mapping_simi['test'] = total_mapping_simi['test'].str.replace(' \|
wella', '')
```

```
total_mapping_simi
```

This results in the following:

	topic	url	title	test
0	black hair	https://www.wella.com/international/wella-tuto...	How to Style a Faux Hawk Like a Headline Act |...	how to style a faux hawk like a headline act
1	black hair	https://www.wella.com/international/hair-style...	Silvikrin Classic Voluminous Hold Hairspray 75...	silvikrin classic voluminous hold hairspray 75ml
2	black hair	https://www.wella.com/international/hair-style...	Wellaflex 2nd Day Volume Strong Hold Mousse, H...	wellaflex 2nd day volume strong hold mousse, h...
3	black hair	https://www.wella.com/international/hair-color...	Wella Koleston Permanent Hair Color Cream Fore...	wella koleston permanent hair color cream fore...
4	black hair	https://www.wella.com/international/hair-color...	Wella Koleston Permanent Hair Color Cream With...	wella koleston permanent hair color cream with...
...	...	...	...	...
30815	half up top knot	https://www.wella.com/international/curly-and-...	Style Wavy & Curly Hair | Curly Hair Products ...	style wavy & curly hair | curly hair products
30816	half up top knot	https://www.wella.com/international/wella-maga...	Coloring at home for the first time? | Wella	coloring at home for the first time?
30817	half up top knot	https://www.wella.com/international/styling	Wella - Hair passion and expertise, shared wit...	wella - hair passion and expertise, shared wit...
30818	half up top knot	https://www.wella.com/international/about-well...	Wella Color by You | Mix, match and wear the h...	wella color by you | mix, match and wear the h...
30819	half up top knot	https://www.wella.com/international/wella-x-you	Wella X You LP | Wella	wella x you lp

30820 rows × 4 columns

Now we're ready to compare strings by creating a new column "simi," meaning string similarity. The scores will take the topic and test columns as inputs and feed the sorensen_dice function imported earlier:

```
total_mapping_simi['simi'] = total_mapping_simi.loc[:, ['topic',
    'test']].apply(lambda x: sorensen_dice(*x), axis=1)
total_mapping_simi
```

	topic	url	title	test	simi
0	black hair	https://www.wella.com/international/wella-tuto...	How to Style a Faux Hawk Like a Headline Act |...	how to style a faux hawk like a headline act	0.296296
1	black hair	https://www.wella.com/international/hair-style...	Silvikrin Classic Voluminous Hold Hairspray 75...	silvikrin classic voluminous hold hairspray 75ml	0.310345
2	black hair	https://www.wella.com/international/hair-style...	Wellaflex 2nd Day Volume Strong Hold Mousse, H...	wellaflex 2nd day volume strong hold mousse, h...	0.166667
3	black hair	https://www.wella.com/international/hair-color...	Wella Koleston Permanent Hair Color Cream Fore...	wella koleston permanent hair color cream fore...	0.227273
4	black hair	https://www.wella.com/international/hair-color...	Wella Koleston Permanent Hair Color Cream With...	wella koleston permanent hair color cream with...	0.202020
...	...	...	...	...	...
30815	half up top knot	https://www.wella.com/international/curly-and-...	Style Wavy & Curly Hair | Curly Hair Products	style wavy & curly hair | curly hair products	0.360656
30816	half up top knot	https://www.wella.com/international/wella-maga...	Coloring at home for the first time? | Wella	coloring at home for the first time?	0.461538
30817	half up top knot	https://www.wella.com/international/styling	Wella - Hair passion and expertise, shared wit...	wella - hair passion and expertise, shared wit...	0.382353
30818	half up top knot	https://www.wella.com/international/about-well...	Wella Color by You | Mix, match and wear the h...	wella color by you | mix, match and wear the h...	0.296296
30819	half up top knot	https://www.wella.com/international/wella-x-you	Wella X You LP | Wella	wella x you lp	0.533333

30820 rows × 5 columns

The simi column has been added complete with scores. A score of 1 is identical, and 0 is completely dissimilar. The next stage is to select the closest matching URLs to topic keywords:

```
keyword_mapping_grp = total_mapping_simi.copy()
```

The dataframe is first sorted by similarity score and topic in descending order so that the first row by topic is the closest matching:

```
keyword_mapping_grp = keyword_mapping_grp.sort_values(['simi', 'topic'],
ascending = False)
```

After sorting, we use the first() function to select the top matching URL for each topic using the groupby() function:

```
keyword_mapping_grp = keyword_mapping_grp.groupby('topic').first().
reset_index()
```

```
keyword_mapping_grp
```

This results in the following:

	topic	url	title	test	simi
0	80s rock'n'roll hairstyle	https://www.wella.com/international/wella-tuto...	Get the 80's rock'n'roll hairstyle.	get the 80's rock'n'roll hairstyle.	0.833333
1	accessorize braided hairstyle	https://www.wella.com/international/wella-tuto...	Accessorize your braided hairstyle for a chic ...	accessorize your braided hairstyle for a chic ...	0.725000
2	back comb safely	https://www.wella.com/international/wella-tuto...	Back comb safely for stylish results!	back comb safely for stylish results!	0.603774
3	bardot look	https://www.wella.com/international/wella-tuto...	BARDOT LOOK | Wella	bardot look	1.000000
4	beachy waves hairstyle	https://www.wella.com/international/wella-maga...	Hair color safety tests | Wella	hair color safety tests	0.666667
5	best hair color results	https://www.wella.com/international/wella-maga...	Hair color safety tests | Wella	hair color safety tests	0.826087
6	big hair volume	https://www.wella.com/international/blonde-hair	Blonde Hair | Wella	blonde hair	0.692308
7	black hair	https://www.wella.com/international/black-hair	Black Hair | Wella	black hair	1.000000
8	blonde hair	https://www.wella.com/international/blonde-hair	Blonde Hair | Wella	blonde hair	1.000000

Each topic now has its closest matching URL. The next stage is to decide whether these matches are good enough or not:

```
keyword_mapping = keyword_mapping_grp[['topic', 'url', 'title',
'simi']].copy()
```

At this point, we eyeball the data to see what threshold number is good enough. I've gone with 0.7 or 70% as it seems to do the job mostly correctly, which is to act as the natural threshold for matching test content to URLs.

Using np.where(), which is equivalent to Excel's IF formula, we'll make any rows exceeding 0.7 as "mapped" and the rest as "unmatched":

```
keyword_mapping['url'] = np.where(keyword_mapping['simi'] < 0.7,
'unmatched', keyword_mapping['url'])
keyword_mapping['mapped'] = np.where(keyword_mapping['simi'] =< 0.7,
'No', 'Yes')
```

```
keyword_mapping
```

This results in the following:

	topic	url	title	simi	mapped
0	80s rock'n'roll hairstyle	https://www.wella.com/international/wella-tuto...	Get the 80's rock'n'roll hairstyle.	0.833333	Yes
1	accessorize braided hairstyle	https://www.wella.com/international/wella-tuto...	Accessorize your braided hairstyle for a chic ...	0.725000	Yes
2	back comb safely	unmatched	Back comb safely for stylish results!	0.603774	No
3	bardot look	https://www.wella.com/international/wella-tuto...	BARDOT LOOK	Wella	1.000000	Yes
4	beachy waves hairstyle	unmatched	Hair color safety tests	Wella	0.666667	No
5	best hair color results	https://www.wella.com/international/wella-maga...	Hair color safety tests	Wella	0.826087	Yes
6	big hair volume	unmatched	Blonde Hair	Wella	0.692308	No
7	black hair	https://www.wella.com/international/black-hair	Black Hair	Wella	1.000000	Yes
8	blonde hair	https://www.wella.com/international/blonde-hair	Blonde Hair	Wella	1.000000	Yes
9	blunt cut bob	unmatched	Fun and functional braids.	0.512821	No
10	boost fine & thinning hair	https://www.wella.com/international/fine-and-t...	Boost Fine & Thinning Hair	Styling Essential...	0.712329	Yes

Finally, we have keywords mapped to URLs and some stats on the overall exercise.

```
keyword_mapping_aggs = keyword_mapping.copy()
keyword_mapping_aggs = keyword_mapping_aggs.groupby('mapped').count().
reset_index()

Keyword_mapping_aggs
```

This results in the following:

	mapped	topic	url	title	simi
0	No	32	32	32	32
1	Yes	60	60	60	60

String Distance to Map Keyword Evaluation

So 65% of the 92 URLs got mapped – not bad and for the minimum code too. Those unmapped will have to be done manually, probably because

- Existing unmapped URL titles are not optimized.

- New content needs to be created.

Content Gap Analysis

Search engines require content to rank as a response to a keyword search by their users. Content gap analysis helps your site extend its reach to your target audiences by identifying keywords (and topics) where your direct competitors are visible, and your site is not.

The analysis is achieved by using search analytics data sources such as SEMRush overlaying your site data with your competitors to find

- *Core content set*: Of which keywords are common to multiple competitors

- *Content gaps*: The extent to which the brand is not visible for keywords that form the content set

Without this analysis, your site risks being left behind in terms of audience reach and also appearing less authoritative because your site appears less knowledgeable about the topics covered by your existing content. This is particularly important when considering the buying cycle. Let's imagine you're booking a holiday, and now imagine the variety of search queries that you might use as you carry out that search, perhaps searching by destination ("beach holidays to Spain"), perhaps refining by a specific requirement ("family beach holidays in Spain"), and then more specific including a destination (Majorca), and perhaps ("family holidays with pool in Majorca"). Savvy SEOs think deeply about mapping customer demand (right across the search journey) to compelling landing page (and website) experiences that can satisfy this demand. Data science enables you to manage this opportunity at a significant scale.

Warnings and motivations over, let's roll starting with the usual package loading:

```
import re
import time
import random
import pandas as pd
import numpy as np
```

OS and Glob allow the environment to read the SEMRush files from a folder:

```
import os
import glob

from pandas.api.types import is_string_dtype
from pandas.api.types import is_numeric_dtype
import uritools
```

Combinations is particularly useful for generating combinations of list elements which will be used to work out which datasets to intersect and in a given order:

```
from itertools import combinations
```

To see all columns of a dataframe and without truncation:

```
pd.set_option('display.max_colwidth', None)
```

These variables are set in advance so that when copying this script over for another site, the script can be run with minimal changes to the code:

```
root_domain = 'wella.com'
hostdomain = 'www.wella.com'
hostname = 'wella'
full_domain = 'https://www.wella.com'
target_name = 'Wella'
```

With the variables set, we're now ready to start importing data.

Getting the Data

We set the directory path where all of the SEMRush files are stored:

```
data_dir = os.path.join('data/semrush/')
```

Glob reads all of the files in the folder, and we store the output in a variable "semrush_csvs":

```
semrush_csvs = glob.glob(data_dir + "/*.csv")
Semrush_csvs
```

Print out the files in the folder:

```
['data/hair.com-organic.Positions-uk-20220704-2022-07-05T14_04_59Z.csv',
 'data/johnfrieda.com-organic.Positions-
 uk-20220704-2022-07-05T13_29_57Z.csv',
 'data/madison-reed.com-organic.Positions-
 uk-20220704-2022-07-05T13_38_32Z.csv',
 'data/sebastianprofessional.com-organic.Positions-
 uk-20220704-2022-07-05T13_39_13Z.csv',
 'data/matrix.com-organic.Positions-uk-20220704-2022-07-05T14_04_12Z.csv',
 'data/wella.com-organic.Positions-uk-20220704-2022-07-05T13_30_29Z.csv',
 'data/redken.com-organic.Positions-uk-20220704-2022-07-05T13_37_31Z.csv',
 'data/schwarzkopf.com-organic.Positions-
 uk-20220704-2022-07-05T13_29_03Z.csv',
 'data/garnier.co.uk-organic.Positions-
 uk-20220704-2022-07-05T14_07_16Z.csv']
```

Initialize the final dataframe where we'll be storing the imported SEMRush data:

```
semrush_raw_df = pd.DataFrame()
```

Initialize a list where we'll be storing the imported SEMRush data:

```
semrush_li = []
```

The for loop uses the pandas read_csv() function to read the SEMRush CSV file and extract the filename which is put into a new column "filename." A bit superfluous to requirements but it will help us know where the data came from.

Once the data is read, it is added to the semrush_li list we initialized earlier:

```
for cf in semrush_csvs:
    df = pd.read_csv(cf, index_col=None, header=0)
    df['filename'] = os.path.basename(cf)
    df['filename'] = df['filename'].str.replace('.csv', '')
    df['filename'] = df['filename'].str.replace('_', '.')
    semrush_li.append(df)

semrush_raw_df = pd.concat(semrush_li, axis=0, ignore_index=True)
```

Clean up the columns to make these lowercase and data-friendly. A list comprehension can also be used, but we used a different approach to show an alternative.

```
semrush_raw_df.columns = semrush_raw_df.columns.str.strip().str.lower().
str.replace(' ', '_').str.replace('(', '').str.replace(')', '')
```

A site column is created so we know which content the site belongs to. Here, we used regex on the filename column, but we could have easily derived this from the URL also:

```
semrush_raw_df['site'] = semrush_raw_df['filename'].str.extract('(.*?)\-')
semrush_raw_df.head()
```

This results in the following:

	keyword	position	previous_position	search_volume	keyword_difficulty	cpc	url	traffic	traffic_%	traffic_cost	competition	number_
0	blonde balayage	6	6	14800	47	3.55	https://www.hair.com/blonde-balayage-ideas.html	518	2.24	1838.0	0.51	
1	ginger hair color	2	2	2400	47	2.11	https://www.hair.com/ginger-hair-color.html	316	1.36	668.0	0.99	2
2	hair colors for pale skin	1	1	1000	51	0.94	https://www.hair.com/hair-color-for-pale-skin.html	248	1.07	233.0	0.83	
3	dark brown hair	9	12	9900	62	3.72	https://www.hair.com/dark-brown-hair-ideas.html	237	1.02	883.0	0.66	2
4	best purple shampoo	5	5	4400	66	0.40	https://www.hair.com/best-purple-shampoo.html	193	0.83	77.0	1.00	

That's the dataframe, although we're more interested in the keywords and the site it belongs to.

```
semrush_raw_presect = semrush_raw_sited.copy()
semrush_raw_presect = semrush_raw_presect[['keyword', 'site']]
semrush_raw_presect
```

This results in the following:

	keyword	site
0	blonde balayage	hair.com
1	ginger hair color	hair.com
2	hair colors for pale skin	hair.com
3	dark brown hair	hair.com
4	best purple shampoo	hair.com
...	...	...
118676	honey blonde hair dye for dark hair	garnier.co.uk
118677	dye hair blonde	garnier.co.uk
118678	colour touch shade chart	garnier.co.uk
118679	red and blue hair dye	garnier.co.uk
118680	garnier 3-1	garnier.co.uk

118681 rows × 2 columns

The aim of the exercise is to find keywords to two or more competitors which will define the core content set.

To achieve this, we will use a list comprehension to split the semrush_raw_presect dataframe by site into unnamed dataframes:

```
df1, df2, df3, df4, df5, df6, df7, df8, df9 = [x for _, x in semrush_raw_
presect.groupby(semrush_raw_presect['site'])]
```

Now that each dataframe has the site and keywords, we can dispense with the site column as we're only interested in the keywords and not where they come from.

We start by defining a list of dataframes, df_list:

```
df_list = [df1, df2, df3, df4, df5, df6, df7, df8, df9]
```

Here's an example; df1 is Garnier:

`df1`

This results in the following:

	keyword	site
100596	garnier	garnier.co.uk
100597	hair colour	garnier.co.uk
100598	garnier.co.uk	garnier.co.uk
100599	garnier hair color	garnier.co.uk
100600	garnier hair colour	garnier.co.uk
...	...	...
118676	honey blonde hair dye for dark hair	garnier.co.uk
118677	dye hair blonde	garnier.co.uk
118678	colour touch shade chart	garnier.co.uk
118679	red and blue hair dye	garnier.co.uk
118680	garnier 3-1	garnier.co.uk

18085 rows × 2 columns

Define the function drop_col, which as the name suggests

1. Drops the column (col) of the dataframe (df)

2. Takes the desired column (list_col)

3. Converts the desired column to a list

4. Adds the column to a big list (master_list)

```
def drop_col(df, col, listcol, master_list):
    df.drop(col, axis = 1, inplace = True)
    df_tolist = df[listcol].tolist()
    master_list.append(df_tolist)
```

Our master list is initiated as follows:

```
keywords_lists = []
```

List comprehension which will go through all of the keyword sets in df_list, and these as lists to get a list of keyword lists.

```
_ = [drop_col(x, 'site', 'keyword', keywords_lists) for x in df_list]
```

The lists within the list of lists are too long to print here; however, the double bracket at the beginning should show this is indeed a list of lists.

```
keywords_lists
```

This results in the following:

```
[['garnier',
  'hair colour',
  'garnier.co.uk',
  'garnier hair color',
  'garnier hair colour',
  'garnier micellar water',
  'garnier hair food',
  'garnier bb cream',
  'garnier face mask',
  'bb cream from garnier',
  'garnier hair mask',
  'garnier shampoo',
  'hair dye',
```

The list of keyword lists is exported into separated lists:

```
lst_1, lst_2, lst_3, lst_4, lst_5, lst_6, lst_7, lst_8, lst_9 =
keywords_lists
```

List 1 is shown as follows:

```
lst_1
```

This results in the following:

```
['garnier',
 'hair colour',
 'garnier.co.uk',
 'garnier hair color',
 'garnier hair colour',
 'garnier micellar water',
 'garnier hair food',
 'garnier bb cream',
 'garnier face mask',
 'bb cream from garnier',
 'garnier hair mask',
 'garnier shampoo',
 'hair dye',
 'garnier hair dye',
 'garnier shampoo bar',
 'garnier vitamin c serum',
```

Now we want to generate combinations of lists so we can control how each of the site's keywords get intersected:

```
values_list = [lst_1, lst_2, lst_3, lst_4, lst_5, lst_6, lst_7,
lst_8, lst_9]
```

The dictionary comprehension will append each list into a dictionary we create called keywords_dict, where the key (index) is the number of the list:

```
keywords_dict = {listo: values_list[listo]  for listo in
range(len(values_list))}
```

When we print the keywords_dict keys

```
keywords_dict.keys()
```

we get the list numbers. The reason it goes from 0 to 8 and not 1 to 9 is because Python uses zero indexing which means it starts from zero:

```
dict_keys([0, 1, 2, 3, 4, 5, 6, 7, 8])
```

Now we'll convert the keys to a list for ease of manipulation shortly:

```
keys_list = list(keywords_dict.keys())
keys_list
```

This results in the following:

```
[0, 1, 2, 3, 4, 5, 6, 7, 8]
```

With the list, we can construct combinations of the site's keywords to intersect. The intersection of the website keyword lists will be the words that are common to the websites.

Creating the Combinations

Initialize list_combos which will be a list of the combinations generated:

```
list_combos = []
```

List comprehension using the combinations function picking four site keywords at random and storing it in list combos using the append() function:

```
_ = [list_combos.append(comb) for comb in combinations(keys_list, 4)]
```

This line converts the combination into a list so that list_combos will be a list of lists:

```
list_combos = [list(combo) for combo in list_combos]

list_combos
```

This results in the following:

```
[[0, 1, 2, 3],
 [0, 1, 2, 4],
 [0, 1, 2, 5],
 [0, 1, 2, 6],
 [0, 1, 2, 7],
 [0, 1, 2, 8],
 [0, 1, 3, 4],
 [0, 1, 3, 5],
 [0, 1, 3, 6], ...
```

With the list of lists, we're ready to start intersecting the keyword lists to build the core content (keyword) set.

Finding the Content Intersection

Initialize an empty list keywords_intersected:

```
keywords_intersected = []
```

Define the multi_intersect function which takes a list of dictionaries and their keys, then finds the common keywords (i.e., intersection), and adds it to the keywords_ intersected list.

The function can be adapted to just compare two sites, three sites, and so on. Just ensure you rerun the combinations function with the number of lists desired and edit the function as follows:

```
def multi_intersect(list_dict, combo):
    a = list_dict[combo[0]]
    b = list_dict[combo[1]]
    c = list_dict[combo[2]]
    d = list_dict[combo[3]]
    intersection = list(set(a) & set(b) & set(c) & set(d))
    keywords_intersected.append(intersection)
```

Using the list comprehension, we loop through the list of combinations list_combos to run the multi_intersect function which takes the dictionary containing all the site keywords (keywords_dict), pulls the appropriate keywords, and finds the common ones, before adding to keywords_intersected:

```
_ = [multi_intersect(keywords_dict, combo) for combo in list_combos]
```

And we get a list of lists, because each list is an iteration of the function for each combination:

```
keywords_intersected
```

This results in the following:

```
[['best way to cover grey hair',
  'rich red hair colour',
  'hair dye colors chart',
  'different shades of blonde hair',
  'adding colour to grey hair',
  'cool hair colors',
  'dark red hair',
  'light brown toner',
  'medium light brown hair',
  'hair color on brown skin',
  'highlights to cover grey in dark brown hair',
  'auburn color swatch', ..
```

Let's turn the list of lists into a single list:

```
flat_keywords_intersected = [elem for sublist in keywords_intersected for
elem in sublist]
```

Then deduplicate it. list(set(the_list_you_want_to_de-duplicate)) is a really helpful technique to deduplicate lists.

```
unique_keywords_intersected = list(set(flat_keywords_intersected))
print(len(flat_keywords_intersected), len(unique_keywords_intersected))
```

This results in the following:

```
87031 8380
```

There were 87K keywords originally and 8380 keywords post deduplication.

```
unique_keywords_intersected
```

This results in the following:

```
['hairspray for holding curls',
 'burgundy colour hair',
 'cool hair colors',
 'dark red hair',
 'color stripes hair',
```

```
'for frizzy hair products',
'blue purple hair',
'autumn balayage 2021',
'ash brown hair color',
'blonde highlights in black hair',
'what hair colour will suit me',
'hair gloss treatment at home',
'dark roots with red hair',
'silver shoulder length hair',
'mens curly hair',
'ash brunette hair',
'toners for grey hair',
```

That's the list, but it's not over yet as we need to establish the gap, which we all want to know.

Establishing Gap

The question is which keywords are "Wella" not targeting and how many are there?

We'll start by filtering the SEMRush site for the target site Wella.com:

```
target_semrush = semrush_raw_sited.loc[semrush_raw_sited['site'] ==
root_domain]
```

And then we include only the keywords in the core content set:

```
target_on = target_semrush.loc[target_semrush['keyword'].isin(unique_
keywords_intersected)]
target_on
```

This results in the following:

	keyword	position	previous_position	search_volume	keyword_difficulty	cpc	url	traffic	traffic_%	traffic_cost	comp
60526	balayage	9	11	60500	72	3.32	https://blog.wella.com/gb/what-is-balayage	1452	1.04	4820.0	
60537	brown to blonde hair	1	1	2400	48	1.49	https://blog.wella.com/gb/foolproof-way-go-brown-blonde-hair	595	0.42	886.0	
60542	copper hair	4	6	8100	40	2.78	https://blog.wella.com/gb/copper-red-hair-color	526	0.37	1463.0	
60545	auburn hair	10	9	22200	54	2.24	https://blog.wella.com/gb/auburn-hair-color-ideas-and-formulas	488	0.34	1094.0	
60546	yellow hair	1	1	1900	30	0.00	https://blog.wella.com/gb/how-to-tone-yellow-hair	471	0.33	0.0	

Let's get some stats starting with the number of keywords in the preceding dataframe and the number of keywords in the core content set:

```
print(target_on[['keyword'].drop_duplicates().shape[0], len(unique_
keywords_intersected))
```

This results in the following:

```
6936 8380
```

So just under 70% of Wella's keyword content is in the core content set, which is about 1.4K keywords short.

To find the 6.9K intersect keywords, we can use the list and set functions:

```
target_on_list = list(set(target_semrush['keyword'].tolist()) & set(unique_
keywords_intersected))
target_on_list[:10]
```

This results in the following:

```
['hairspray for holding curls',
 'burgundy colour hair',
 'cool hair colors',
 'dark red hair',
 'blue purple hair',
```

```
'autumn balayage 2021',
'ash brown hair color',
'blonde highlights in black hair',
'what hair colour will suit me',
'hair gloss treatment at home']
```

To find the keywords that are not in the core content set, that is, the content gap, we'll remove the target SEMRush keywords from the core content set:

```
target_gap = list(set(unique_keywords_intersected) - set(target_
semrush['keyword'].tolist()))
print(len(target_gap), len(unique_keywords_intersected))
target_gap[:10]
```

This results in the following:

```
['bleaching hair with toner',
 'color stripes hair',
 'for frizzy hair products',
 'air dry beach waves short hair',
 'does semi permanent black dye wash out',
 'balayage for dark skin',
 'matte hairspray',
 'mens curly hair',
 'how to change hair color',
 'ginger and pink hair']
```

Now that we know what these gap keywords are, we can filter the dataframe by listing keywords:

```
cga_semrush = semrush_raw_sited.loc[semrush_raw_sited['keyword'].
isin(target_gap)]

cga_semrush
```

This results in the following:

	keyword	position	previous_position	search_volume	keyword_difficulty	cpc	url	traffic	traffic_%	traffic_cost	competition
42	hair dye ideas	8	7	2900	64	0.94	https://www.hair.com/dark-brown-hair-ideas.html	69	0.29	65.0	0.56
78	best hair color for pale skin	3	3	590	53	0.52	https://www.hair.com/hair-color-for-pale-skin.html	48	0.20	25.0	0.64
92	best hair color for pale skin and blue eyes	1	1	170	51	0.52	https://www.hair.com/best-hair-colors-blue-eyes.html	42	0.18	21.0	0.51
109	color stripes hair	11	11	1900	38	0.26	https://www.hair.com/skunk-stripe-hair.html	36	0.15	9.0	1.00
112	best hair color for blue eyes and fair skin	1	1	140	49	0.87	https://www.hair.com/best-hair-colors-blue-eyes.html	34	0.14	30.0	0.95
...	...	...	...	...	...	...	...	...	...	...	...

We only want the highest ranked target URLs per keyword, which we'll achieve with a combination of sort_values(), groupby(), and first():

```
cga_unique = cga_semrush.sort_values('position').groupby('keyword').
first().reset_index()
cga_unique['project'] = target_name
```

To make the dataframe more user-friendly, we'll prioritize keywords by

```
cga_unique = cga_unique.sort_values('search_volume', ascending = False)
```

Ready to export:

```
cga_unique.to_csv('exports/cga_unique.csv')
cga_unique
```

Now it's time to decide what content should be on these pages.

Content Creation: Planning Landing Page Content

Of course, now that you know which keywords belong together and which ones don't, and which keywords to pursue thanks to the content gap analysis, the question becomes what content should be on these pages?

One strategy we're pursuing is to

1. Look at the top 10 ranking URLs for each keyword

2. Extract the headings (<h1>, <h2>) from each ranking URL

3. Check the search results for each heading as writers can phrase the intent differently

4. Cluster the headings and label them

5. Count the frequency of the clustered headings for a given keyword, to see which ones are most popular and are being rewarded by Google (in terms of rankings)

6. Export the results for each search phrase

This strategy won't work for all verticals as there's a lot of noise in some market sectors compared to others. For example, with hair styling articles, a lot of the headings (and their sections) are celebrity names which will not have the same detectable search intent as another celebrity.

In contrast, in other verticals this method works really well because there aren't endless lists with the same HTML heading tags shared with related article titles (e.g., "Drew Barrymore" and "54 ways to wear the modern Marilyn").

Instead, the headings are fewer in number and have a meaning in common, for example, "What is account-based marketing?" and "Defining ABM," which is something Google is likely to understand.

With those caveats in mind, let's go.

```
import requests
from requests.exceptions import ReadTimeout
from json.decoder import JSONDecodeError
import re
import time
import random
import pandas as pd
import numpy as np
import datetime
import requests
import json
```

```
from datetime import timedelta
from glob import glob
import os
from client import RestClient
from textdistance import sorensen_dice
from plotnine import *
import matplotlib.pyplot as plt
from mizani.transforms import trans
from pandas.api.types import is_string_dtype
from pandas.api.types import is_numeric_dtype
import uritools
```

This is the website we're creating content for:

```
target = 'on24'
```

These are the keywords the target website wants to rank for. There's only eight keywords, but as you'll see, this process generates a lot of noisy data, which will need cleaning up:

```
queries = ['webinar best practices',
           'webinar marketing guide',
           'webinar guide',
           'funnel marketing guide',
           'scrappy marketing guide',
           'b2b marketing guide',
           'how to run virtual events',
           'webinar benchmarks']
```

Getting SERP Data

Import the SERP data which will form the basis of finding out what content is Google rewarding for the sites to rank in the top 10:

```
serps_input = pd.read_csv('data/serps_input_' + target + '.csv')

serps_input
```

This results in the following:

	keyword	rank	url	se_results_count	domain	title	is_video
0	funnel marketing guide	1	https://www.mageplaza.com/blog/marketing-funnel.html	14900000	www.mageplaza.com	What Is A Marketing Funnel? A Step-By-Step Guide - Mageplaza	None
1	funnel marketing guide	2	https://www.singlegrain.com/blog-posts/content-marketing/how-to-create-marketing-funnel/	14900000	www.singlegrain.com	How to Create a Powerful Marketing Funnel Step-by-Step	False
2	funnel marketing guide	3	None	14900000	None	None	None
3	funnel marketing guide	4	https://ahrefs.com/blog/marketing-funnels/	14900000	ahrefs.com	Marketing Funnels for Beginners: A Comprehensive Guide	False
4	funnel marketing guide	5	https://www.hotjar.com/blog/marketing-funnel/	14900000	www.hotjar.com	The Marketing Funnel: Stages, Strategies, & How to Optimize	False
...	...	...	...	...	...	...	...

The extract function from the TLD extract package is useful for extracting the hostname and domain name from URLs:

```
from tldextract import extract

serps_input_clean = serps_input.copy()
```

Set the URL column as a string:

```
serps_input_clean['url'] = serps_input_clean['url'].astype(str)
```

Use lambda to apply the extract function to the URL column:

```
serps_input_clean['host'] = serps_input_clean['url'].apply(lambda x: extract(x))
```

Convert the function output (which is a tuple) to a list:

```
serps_input_clean['host'] = [list(lst) for lst in serps_input_clean['host']]
```

Extract the hostname by taking the penultimate list element from the list using the string get method:

```
serps_input_clean['host'] = serps_input_clean['host'].str.get(-2)
```

The site uses a similar logic as before:

```
serps_input_clean['site'] = serps_input_clean['url'].apply(lambda x:
extract(x))
serps_input_clean['site'] = [list(lst) for lst in serps_input_
clean['site']]
```

Only this time, we want both the hostname and the top-level domain (TLD) which we will join to form the site or domain name:

```
serps_input_clean['site'] = serps_input_clean['site'].str.get(-2) + '.'
+serps_input_clean['site'].str.get(-1)


serps_input_clean
```

This results in the following:

eyword	rank	url	se_results_count	domain	title	is_video	host	site
funnel arketing guide	1	https://www.mageplaza.com/blog/marketing-funnel.html	14900000	www.mageplaza.com	What Is A Marketing Funnel? A Step-By-Step Guide - Mageplaza	NaN	mageplaza	mageplaza.com
funnel arketing guide	2	https://www.singlegrain.com/blog-posts/content-marketing/how-to-create-marketing-funnel/	14900000	www.singlegrain.com	How to Create a Powerful Marketing Funnel Step-by-Step	False	singlegrain	singlegrain.com
funnel arketing guide	3	nan	14900000	NaN	NaN	NaN	nan	nan.
funnel arketing guide	4	https://ahrefs.com/blog/marketing-funnels/	14900000	ahrefs.com	Marketing Funnels for Beginners: A Comprehensive Guide	False	ahrefs	ahrefs.com
funnel arketing guide	5	https://www.hotjar.com/blog/marketing-funnel/	14900000	www.hotjar.com	The Marketing Funnel: Stages, Strategies, & How to Optimize	False	hotjar	hotjar.com
...	...	...	...	...	...	...	...	...

The augmented dataframe shows the host and site columns added.

This line allows the column values to be read by setting the column widths to their maximum value:

```
pd.set_option('display.max_colwidth', None)
```

Crawling the Content

The next step is to get a list of top ranking URLs that we'll crawl for their content sections:

```
serps_to_crawl_df = serps_input_clean.copy()
```

There are some sites not worth crawling because they won't let you, which are defined in the following list:

```
dont_crawl = ['wikipedia', 'google', 'youtube', 'linkedin', 'foursquare',
'amazon', 'twitter', 'facebook', 'pinterest', 'tiktok', 'quora',
'reddit', 'None']
```

The dataframe is filtered to exclude sites in the don't crawl list:

```
serps_to_crawl_df = serps_to_crawl_df.loc[~serps_to_crawl_df['host'].
isin(dont_crawl)]
```

We'll also remove nulls and sites outside the top 10:

```
serps_to_crawl_df = serps_to_crawl_df.loc[~serps_to_crawl_df['domain'].
isnull()]
serps_to_crawl_df = serps_to_crawl_df.loc[serps_to_crawl_df['rank'] < 10]

serps_to_crawl_df.head(10)
```

This results in the following:

url	se_results_count	domain	title	is_video	host	site
https://www.mageplaza.com/blog/marketing-funnel.html	14900000	www.mageplaza.com	What Is A Marketing Funnel? A Step-By-Step Guide - Mageplaza	NaN	mageplaza	mageplaza.com
https://www.singlegrain.com/blog-posts/content-marketing/how-to-create-marketing-funnel/	14900000	www.singlegrain.com	How to Create a Powerful Marketing Funnel Step-by-Step	False	singlegrain	singlegrain.com
https://ahrefs.com/blog/marketing-funnels/	14900000	ahrefs.com	Marketing Funnels for Beginners: A Comprehensive Guide	False	ahrefs	ahrefs.com
https://www.hotjar.com/blog/marketing-funnel/	14900000	www.hotjar.com	The Marketing Funnel: Stages, Strategies, & How to Optimize	False	hotjar	hotjar.com
https://sproutsocial.com/insights/social-media-marketing-funnel/	14900000	sproutsocial.com	How to Build a Social Media Marketing Funnel That Converts	False	sproutsocial	sproutsocial.com

With the dataframe filtered, we just want the URLs to export to our desktop crawler.

Some URLs may rank for multiple search phrases. To avoid crawling the same URL multiple times, we'll use drop_duplicates() to make the URL list unique:

```
serps_to_crawl_upload = serps_to_crawl_df[['url']].drop_duplicates()
serps_to_crawl_upload.to_csv('data/serps_to_crawl_upload.csv', index=False)

serps_to_crawl_upload
```

This results in the following:

	url
0	https://www.mageplaza.com/blog/marketing-funnel.html
1	https://www.singlegrain.com/blog-posts/content-marketing/how-to-create-marketing-funnel/
3	https://ahrefs.com/blog/marketing-funnels/
4	https://www.hotjar.com/blog/marketing-funnel/
5	https://sproutsocial.com/insights/social-media-marketing-funnel/
...	...
719	https://www.netline.com/netline003h/?d=on24scrappymarketer&k=190815nlw24sm
721	https://blog.marketo.com/2016/08/get-scrappy-7-tips-for-smarter-digital-marketing.html
722	https://www.scootermediaco.com/2021/05/scrappy-marketing-strategies/
723	https://www.slideshare.net/kflanagan/the-scrappy-guide-to-marketing
724	https://www.bookdepository.com/Scrappy-Marketing-Handbook-Ann-Handley/9781118929636

62 rows × 1 columns

Now we have a list of 62 URLs to crawl, which cover the eight target keywords. Let's import the results of the crawl:

```
crawl_raw = pd.read_csv('data/all_inlinks.csv')
pd.set_option('display.max_columns', None)
```

Using a list comprehension, we'll clean up the column names to make it easier to work with:

```
crawl_raw.columns = [col.lower().replace(' ', '_') for col in crawl_raw.columns]
```

Print out the column names to see how many extractor fields were extracted:

```
print(crawl_raw.columns)
```

This results in the following:

```
Index(['type', 'source', 'destination', 'form_action_link', 'indexability',
       'indexability_status', 'hreflang', 'size_(bytes)', 'alt_text',
       'length',
       'anchor', 'status_code', 'status', 'follow', 'target', 'rel',
       'path_type', 'unlinked', 'link_path', 'link_position', 'link_
       origin',
       'extractor_1_1', 'extractor_1_2', 'extractor_1_3', 'extractor_1_4',
       'extractor_1_5', 'extractor_1_6', 'extractor_1_7', 'extractor_2_1',
       'extractor_2_2', 'extractor_2_3', 'extractor_2_4', 'extractor_2_5',
       'extractor_2_6', 'extractor_2_7', 'extractor_2_8', 'extractor_2_9',
       'extractor_2_10', 'extractor_2_11', 'extractor_2_12',
       'extractor_2_13',
       'extractor_2_14', 'extractor_2_15', 'extractor_2_16',
       'extractor_2_17',
       'extractor_2_18', 'extractor_2_19', 'extractor_2_20',
       'extractor_2_21',
       'extractor_2_22', 'extractor_2_23', 'extractor_2_24',
       'extractor_2_25',
       'extractor_2_26', 'extractor_2_27', 'extractor_2_28',
       'extractor_2_29',
       'extractor_2_30', 'extractor_2_31', 'extractor_2_32',
       'extractor_2_33',
       'extractor_2_34', 'extractor_2_35', 'extractor_2_36',
       'extractor_2_37',
       'extractor_2_38', 'extractor_2_39', 'extractor_2_40',
       'extractor_2_41',
       'extractor_2_42', 'extractor_2_43', 'extractor_2_44',
       'extractor_2_45',
       'extractor_2_46', 'extractor_2_47', 'extractor_2_48',
       'extractor_2_49',
```

```
    'extractor_2_50', 'extractor_2_51', 'extractor_2_52',
    'extractor_2_53',
    'extractor_2_54', 'extractor_2_55', 'extractor_2_56',
    'extractor_2_57',
    'extractor_2_58', 'extractor_2_59', 'extractor_2_60',
    'extractor_2_61',
    'extractor_2_62', 'extractor_2_63', 'extractor_2_64',
    'extractor_2_65'],
  dtype='object')
```

There are 6 primary headings (H1 in HTML) and 65 H2 headings altogether. These will form the basis of our content sections which tell us what content should be on those pages.

```
crawl_raw
```

This results in the following:

link_position	link_origin	extractor_1_1	extractor_1_2	extractor_1_3	extractor_1_4	extractor_1_5	extractor_1_6	extractor_1_7	extractor_2_1	extractor_2_2	ext
Header	HTML	The ultimate guide to creating engaging webinars.	NaN	NaN	NaN	NaN	NaN	NaN	Webinar guide contents	Presentations vs Webinars	
Header	HTML	The ultimate guide to creating engaging webinars.	NaN	NaN	NaN	NaN	NaN	NaN	Webinar guide contents	Presentations vs Webinars	
Header	HTML	The ultimate guide to creating engaging webinars.	NaN	NaN	NaN	NaN	NaN	NaN	Webinar guide contents	Presentations vs Webinars	
Content	HTML	The ultimate guide to creating engaging webinars.	NaN	NaN	NaN	NaN	NaN	NaN	Webinar guide contents	Presentations vs Webinars	
Content	HTML	The ultimate guide to creating engaging webinars.	NaN	NaN	NaN	NaN	NaN	NaN	Webinar guide contents	Presentations vs Webinars	
...	...	...	...	...	...	...	...	...	...	...	

Extracting the Headings

Since we're only interested in the content, we'll filter for it:

```
crawl_headings = crawl_raw.loc[crawl_raw['link_position'] ==
'Content'].copy()
```

The dataframe also contains columns that are superfluous to our requirements such as link_position and link_origin. We can remove these by listing the columns by position (saves space and typing out the names of which there are many!).

```
drop_cols = [0, 2, 3, 4, 5, 6, 7, 8, 9, 10, 11, 12, 13, 14, 15, 16, 17, 18, 19, 20]
```

Using the .drop() method, we can drop multiple columns in place (i.e., without having to copy the result onto itself):

```
crawl_headings.drop(crawl_headings.columns[drop_cols], axis = 1, inplace = True)
```

Rename the columns from source to URL, which will be useful for joining later:

```
crawl_headings = crawl_headings.rename(columns = {'source': 'url'})

crawl_headings
```

This results in the following:

		url	extractor_1_1	extractor_1_2	extractor_1_3	extractor_1_4	extractor_1_5	extractor_1_6	extractor_1_7	extra
3		https://buffalo7.co.uk/blog/webinar-guide/	The ultimate guide to creating engaging webinars.	NaN	NaN	NaN	NaN	NaN	NaN	Web
4		https://buffalo7.co.uk/blog/webinar-guide/	The ultimate guide to creating engaging webinars.	NaN	NaN	NaN	NaN	NaN	NaN	Web
5		https://buffalo7.co.uk/blog/webinar-guide/	The ultimate guide to creating engaging webinars.	NaN	NaN	NaN	NaN	NaN	NaN	Web
6		https://buffalo7.co.uk/blog/webinar-guide/	The ultimate guide to creating engaging webinars.	NaN	NaN	NaN	NaN	NaN	NaN	Web
7		https://buffalo7.co.uk/blog/webinar-guide/	The ultimate guide to creating engaging webinars.	NaN	NaN	NaN	NaN	NaN	NaN	Web
...		...	...	...	...	...	...	...	...	

With the desired columns of URL and their content section columns, these need to be converted to long format, where all of the sections will be in a single column called "heading":

```
crawl_headings_long = crawl_headings.copy()
```

We'll want a list of the extractor column names (again to save typing) by subsetting the dataframe from the second column onward using .iloc and extracting the column names (.columns.values):

```
heading_cols = crawl_headings_long.iloc[:, 1:].columns.values.tolist()
```

Using the .melt() function, we'll pivot the dataframe to reshape the content sections into a single column "heading" using the preceding list:

```
crawl_headings_long = pd.melt(crawl_headings_long, id_vars='url', value_
name = 'heading', var_name = 'position',
        value_vars= heading_cols)
```

Remove the null values:

```
crawl_headings_long = crawl_headings_long.loc[~crawl_headings_
long['heading'].isnull()]
```

Remove the duplicates:

```
crawl_headings_long = crawl_headings_long.drop_duplicates()

crawl_headings_long
```

This results in the following:

	url	position	heading
0	https://buffalo7.co.uk/blog/webinar-guide/	extractor_1_1	The ultimate guide to creating engaging webinars.
6	https://blog.hubspot.com/marketing/what-is-a-webinar	extractor_1_1	The Ultimate Guide to Creating Compelling Webinars
16	https://blog.hubspot.com/marketing/are-webinars-dead-how-to-make-a-webinar	extractor_1_1	The Ultimate Guide to Creating Compelling Webinars
18	https://blog.hubspot.com/blog/tabid/6307/bid/2391/10-best-practices-for-webinars-or-webcasts.aspx	extractor_1_1	The Ultimate Guide to Creating Compelling Webinars
20	https://www.creativebloq.com/advice/virtual-event-tips	extractor_1_1	How to host a virtual event: 10 expert tips
...	...	...	...
16179	https://surveysparrow.com/blog/how-to-conduct-a-webinar-guide/	extractor_2_61	Company
16420	https://surveysparrow.com/blog/how-to-conduct-a-webinar-guide/	extractor_2_62	Resources
16661	https://surveysparrow.com/blog/how-to-conduct-a-webinar-guide/	extractor_2_63	Free Tools
16902	https://surveysparrow.com/blog/how-to-conduct-a-webinar-guide/	extractor_2_64	Sales
17143	https://surveysparrow.com/blog/how-to-conduct-a-webinar-guide/	extractor_2_65	Connect

647 rows × 3 columns

The resulting dataframe shows the URL, the heading, and the position where the first number denotes whether it was an h1 or h2 and the second number indicates the order of the heading on the page. The heading is the text value.

You may observe that the heading contains some values that are not strictly content but boilerplate content that is sitewide, such as Company, Resources, etc. These will require removal at some point.

```
serps_headings = serps_to_crawl_df.copy()
```

Let's join the headings to the SERPs data:

```
serps_headings = serps_headings.merge(crawl_headings_long, on = 'url',
how = 'left')
```

Replace null headings with '' so that these can be aggregated:

```
serps_headings['heading'] = np.where(serps_headings['heading'].isnull(),
'', serps_headings['heading'])
```

```
serps_headings['project'] = 'target'
```

```
serps_headings
```

This results in the following:

url	se_results_count	domain	title	is_video	host	site	position	heading	project
eplaza.com/blog/marketing-funnel.html	14900000	www.mageplaza.com	What Is A Marketing Funnel? A Step-By-Step Guide - Mageplaza	NaN	mageplaza	mageplaza.com	extractor_1_1	What Is A Marketing Funnel? A Step-By-Step Guide!	target
eplaza.com/blog/marketing-funnel.html	14900000	www.mageplaza.com	What Is A Marketing Funnel? A Step-By-Step Guide - Mageplaza	NaN	mageplaza	mageplaza.com	extractor_2_1	What is Marketing funnel?	target
eplaza.com/blog/marketing-funnel.html	14900000	www.mageplaza.com	What Is A Marketing Funnel? A Step-By-Step Guide - Mageplaza	NaN	mageplaza	mageplaza.com	extractor_2_2	Understanding the stages of a marketing funnel	target
eplaza.com/blog/marketing-funnel.html	14900000	www.mageplaza.com	What Is A Marketing Funnel? A Step-By-Step Guide - Mageplaza	NaN	mageplaza	mageplaza.com	extractor_2_3	Marketing and sales funnel: What's the difference?	target
eplaza.com/blog/marketing-funnel.html	14900000	www.mageplaza.com	What Is A Marketing Funnel? A Step-By-Step Guide - Mageplaza	NaN	mageplaza	mageplaza.com	extractor_2_4	Do you need a Marketing funnel?	target
...	...	...	...	...	...	...	...	...	...

With the data joined, we'll take the domain, heading, and the position:

```
headings_tosum = serps_headings[['domain', 'heading', 'position']].copy()
```

Split position by underscore and extract the last number in the list (using -1) to get the order the heading appears on the page:

```
headings_tosum['pos_n'] = headings_tosum['position'].str.split('_').str[-1]
```

Convert the data type into a number:

```
headings_tosum['pos_n'] = headings_tosum['pos_n'].astype(float)
```

Add a count column for easy aggregation:

```
headings_tosum['count'] = 1
headings_tosum
```

This results in the following:

	domain	heading	position	pos_n	count
0	www.mageplaza.com	What Is A Marketing Funnel? A Step-By-Step Guide!	extractor_1_1	1.0	1
1	www.mageplaza.com	What is Marketing funnel?	extractor_2_1	1.0	1
2	www.mageplaza.com	Understanding the stages of a marketing funnel	extractor_2_2	2.0	1
3	www.mageplaza.com	Marketing and sales funnel: What's the difference?	extractor_2_3	3.0	1
4	www.mageplaza.com	Do you need a Marketing funnel?	extractor_2_4	4.0	1
...	...	...	...	...	...
674	www.bookdepository.com	Books By Language	extractor_2_9	9.0	1
675	www.bookdepository.com	Description	extractor_2_10	10.0	1
676	www.bookdepository.com	\n Product details	extractor_2_11	11.0	1
677	www.bookdepository.com	People who viewed this also viewed	extractor_2_12	12.0	1
678	www.bookdepository.com	Bestsellers in Sales & Marketing Management	extractor_2_13	13.0	1

679 rows × 5 columns

Cleaning and Selecting Headings

We're ready to aggregate and start removing nonsense headings.

We'll start by removing boilerplate headings that are particular to each site. This is achieved by summing the number of times a heading appears by domain and removing any that appear more than once as that will theoretically mean the heading is not unique.

```
domsheadings_tosum_agg = headings_tosum.groupby(['domain', 'heading']).
agg({'count': sum,
'pos_n': 'mean'
        }).reset_index().sort_values(['domain', 'count'],
        ascending = False)
domsheadings_tosum_agg['heading'] = domsheadings_tosum_agg['heading'].
str.lower()
domsheadings_tosum_agg.head(50)
```

Stop headings is a list containing headings that we want to remove.

Include those that appear more than once:

```
stop_headings = domsheadings_tosum_agg.loc[domsheadings_tosum_
agg['count'] > 1]
```

and contain line break characters like "\n":

```
stop_headings = stop_headings.loc[stop_headings['heading'].str.
contains('\n')]
stop_headings = stop_headings['heading'].tolist()

stop_headings
```

This results in the following:

```
['\n  \n     the scrappy guide to marketing\n  \n',
 '\n                \n                          danny goodwin
\n              ',
 '\n          \n                  how to forecast seo with better precision &
transparency            \n      ',
 '\n          \n                    should you switch to ga4 now? what you need to
know           \n      ',
 '\n             the ultimate guide to webinars: 41 tips for successful
webinars         ',
 '\n          \n          \n            \n              \n                \n
\n            \n          \n            \n                    \n get timely
updates and fresh ideas delivered to your inbox. \n              \n
\n                \n             \n          \n          \n
\n          \n      ',
 '4 best webinar practices for marketing and promotion in 2020\n',
 '\n    company\n  ',
 '\n    customers\n  ',
 '\n    free tools\n  ',
 '\n    partners\n  ',
 '\n    popular features\n  ']
```

The list of boilerplate has been reasonably successful on a domain level, but there is more work to do.

We'll now analyze the headings per se, starting by counting the number of headings:

```
headings_tosum_agg = headings_tosum.groupby(['heading']).agg({'count': sum,
'pos_n': 'mean'
            }).reset_index().sort_values('count',
            ascending = False)
headings_tosum_agg['heading'] = headings_tosum_agg['heading'].str.lower()
```

Remove the headings containing the boilerplate items:

```
headings_tosum_agg = headings_tosum_agg.loc[~headings_tosum_agg['heading'].
isin(stop_headings)]
```

Subset away from headings containing nothing (''):

```
headings_tosum_agg = headings_tosum_agg.loc[headings_tosum_
agg['heading'] != '']
```

```
headings_tosum_agg.head(10)
```

This results in the following:

	heading	count	pos_n
195	company	4	25.000000
467	webinar marketing strategy	3	2.000000
281	how to record a webinar	3	5.000000
161	b2b marketing examples	3	5.666667
507	what is a webinar?	3	1.000000
163	b2b marketing strategies	3	5.333333
494	what is b2b marketing?	3	1.000000
476	webinar statistics	3	5.000000
263	how does a webinar work?	3	4.000000
273	how to create a webinar	3	1.000000

The dataframe looks to contain more sensible content headings with the exception of "company," which also is much further down the order of the page at 25.

Let's filter further:

```
headings_tosum_filtered = headings_tosum_agg.copy()
```

Remove headings with a position of 10 or above as these are unlikely to contain actual content sections. Note 10 is an arbitrary number and could be more or less depending on the nature of content.

```
headings_tosum_filtered = headings_tosum_filtered.loc[headings_tosum_
filtered['count'] < 10 ]
```

Measure the number of words in the heading:

```
headings_tosum_filtered['tokens'] = headings_tosum_filtered['heading'].str.
count(' ') + 1
```

Clean up the headings by removing spaces on either side of the text:

```
headings_tosum_filtered['heading'] = headings_tosum_filtered['heading'].
str.strip()
```

Split heading using colons as a punctuation mark and extract the right-hand side of the colon:

```
headings_tosum_filtered['heading'] =  headings_tosum_filtered['heading'].
str.split(':').str[-1]
```

Apply the same principle to the full stop:

```
headings_tosum_filtered['heading'] =  headings_tosum_filtered['heading'].
str.split('.').str[-1]
```

Remove headings containing pagination, for example, 1 of 9:

```
headings_tosum_filtered = headings_tosum_filtered.loc[~headings_tosum_
filtered['heading'].str.contains('[0-9] of [0-9]', regex = True)]
```

Remove headings that are less than 5 words long or more than 12:

```
headings_tosum_filtered = headings_tosum_filtered.loc[headings_tosum_
filtered['tokens'].between(5, 12)]
headings_tosum_filtered = headings_tosum_filtered.sort_values('count',
ascending = False)
```

```
headings_tosum_filtered = headings_tosum_filtered.loc[headings_tosum_
filtered['heading'] != '' ]
```

```
headings_tosum_filtered.head(10)
```

This results in the following:

	heading	count	pos_n	tokens
281	how to record a webinar	3	5.0	5
273	how to create a webinar	3	1.0	5
274	how to create an amazing webinar in 2022	3	1.0	8
276	how to host a webinar for free	3	7.0	7
409	the ultimate guide to creating compelling webinars	3	1.0	7
356	reach your target audience with webinars	3	8.0	6
263	how does a webinar work?	3	4.0	5
166	b2b marketing trends to watch in 2022 [new data]	2	4.0	9
509	what is scrappy marketing – and is it an answer?	2	1.0	10
499	what is scrappy marketing and why is it beneficial?	2	1.0	9

Now we have headings that look more like actual content sections. These are now ready for clustering.

Cluster Headings

The reason for clustering is that writers will describe the same section heading using different words and deliberately so as to avoid copyright infringement and plagiarism. However, Google is smart enough to know that "webinar best practices" and "best practices for webinars" are the same.

To make use of Google's knowledge, we'll make use of the SERPs to see if the search results of each heading are similar enough to know if they mean the same thing or not (i.e., whether the underlying meaning or intent is the same).

We'll create a list and use the search intent clustering code (see Chapter 2) to categorize the headings into topics:

```
headings_to_cluster = headings_tosum_filtered[['heading']].drop_
duplicates()
headings_to_cluster = headings_to_cluster.loc[~headings_to_
cluster['heading'].isnull()]
headings_to_cluster = headings_to_cluster.rename(columns = {'heading':
'keyword'})
```

```
headings_to_cluster
```

This results in the following:

	keyword
281	how to record a webinar
273	how to create a webinar
274	how to create an amazing webinar in 2022
276	how to host a webinar for free
409	the ultimate guide to creating compelling webinars
...	...
402	the difference between sales and marketing
401	the complete guide to virtual events in 2022
400	how to host a virtual event
429	understanding the difference between b2b and b2c marketing
280	how to protect your virtual events from cyberattacks

206 rows × 1 columns

With the headings clustered by search intent, we'll import the results:

```
topic_keyw_map = pd.read_csv('data/topic_keyw_map.csv')
```

Let's rename the keyword column to heading, which we can use to join to the SERP dataframe later:

```
topic_keyw_map = topic_keyw_map.rename(columns = {'keyword': 'heading'})

topic_keyw_map
```

This results in the following:

	topic	heading	topic_results
0	what is a virtual event	what is a virtual event?	6130000000
1	what is a virtual event	what is a virtual event	6130000000
2	how to create your own effective webinar	how to create your own effective webinar	2384900000
3	how to create your own effective webinar	how to design a webinar	2384900000
4	how to create your own effective webinar	how to create an amazing webinar in 2022	2384900000
5	how to create your own effective webinar	is creating a webinar right for you?	2384900000
6	how to create your own effective webinar	how to create a webinar	2384900000
7	how to create your own effective webinar	what's the best time to host a webinar?	2384900000
8	how to create your own effective webinar	how to create your webinar content	2384900000
9	the complete guide to virtual events in 2022	in-person or virtual - the fundamentals matter	1863000000

The dataframe shows the heading and the meaning of the heading as "topic." The next stage is to get some statistics and see how many headings constitute a topic. As the topics are the central meaning of the headings, this will form the core content sections per target keyword.

```
topic_keyw_map_agg = topic_keyw_map.copy()
topic_keyw_map_agg['count'] = 1
topic_keyw_map_agg = topic_keyw_map_agg.groupby('topic').agg({'count':
'sum'}).reset_index()
topic_keyw_map_agg = topic_keyw_map_agg.sort_values('count',
ascending = False)

topic_keyw_map_agg
```

This results in the following:

	topic	count
7	how to create your own effective webinar	7
17	what is scrappy marketing and why is it beneficial?	4
14	webinar attendance facts and statistic	4
3	building your webinar using virtual event software	4
8	how to optimize your marketing funnel for the customer journey	3
0	5 best webinar presentation design practices for 2020	3
5	how to conduct a webinar – the ultimate guide	3
2	b2b marketing the ultimate guide to b2b marketing	3
6	how to create a powerful marketing funnel?	2
4	how does this apply to webinars?	2
1	8 tactics for your b2b marketing strategy	2
10	the essential guide to webinar marketing for 2022 [with best practices]	2
11	understanding the stages of a marketing funnel	2
12	use webinar best practices to host a great webinar	2
13	webinar / webcast best practices	2
15	what is a social media marketing funnel?	2
16	what is a virtual event	2
9	the complete guide to virtual events in 2022	2

"Creating effective webinars" was the most popular content section.

These will now be merged with the SERPs so we can map suggested content to target keywords:

```
serps_topics_merge = serps_headings.copy()
```

For a successful merge, we'll require the heading to be in lowercase:

```
serps_topics_merge['heading'] = serps_topics_merge['heading'].str.lower()
```

```
serps_topics_merge = serps_topics_merge.merge(topic_keyw_map, on =
'heading', how = 'left')
```

```
serps_topics_merge
```

This results in the following:

_count	domain	title	is_video	host	site	position	heading	project	topic	topic_results	count
900000	www.mageplaza.com	What Is A Marketing Funnel? A Step-By-Step Guide - Mageplaza	NaN	mageplaza	mageplaza.com	extractor_1_1	what is a marketing funnel? a step-by-step guide!	target	NaN	NaN	1
900000	www.mageplaza.com	What Is A Marketing Funnel? A Step-By-Step Guide - Mageplaza	NaN	mageplaza	mageplaza.com	extractor_2_1	what is marketing funnel?	target	NaN	NaN	1
900000	www.mageplaza.com	What Is A Marketing Funnel? A Step-By-Step Guide - Mageplaza	NaN	mageplaza	mageplaza.com	extractor_2_2	understanding the stages of a marketing funnel	target	understanding the stages of a marketing funnel	33890000.0	1
900000	www.mageplaza.com	What Is A Marketing Funnel? A Step-By-Step Guide - Mageplaza	NaN	mageplaza	mageplaza.com	extractor_2_3	marketing and sales funnel: what's the difference?	target	NaN	NaN	1
900000	www.mageplaza.com	What Is A Marketing Funnel? A Step-By-Step Guide - Mageplaza	NaN	mageplaza	mageplaza.com	extractor_2_4	do you need a marketing funnel?	target	NaN	NaN	1
...	...	...	...	...	...	...	...	...	...	...	...

```
keyword_topics_summary = serps_topics_merge.groupby(['keyword', 'topic']).
agg({'count': 'sum'}).reset_index().sort_values(['keyword', 'count'],
ascending = False)
```

The count will be reset to 1, so we can count the number of suggested content sections per target keyword:

```
keyword_topics_summary['count'] = 1
```

```
keyword_topics_summary
```

This results in the following:

	keyword	topic	count
24	webinar marketing guide	how to create your own effective webinar	1
25	webinar marketing guide	the essential guide to webinar marketing for 2022 [with best practices]	1
22	webinar guide	how to conduct a webinar – the ultimate guide	1
23	webinar guide	how to create your own effective webinar	1
20	webinar guide	building your webinar using virtual event software	1
21	webinar guide	how does this apply to webinars?	1
17	webinar best practices	how to create your own effective webinar	1
13	webinar best practices	5 best webinar presentation design practices for 2020	1
15	webinar best practices	how does this apply to webinars?	1
19	webinar best practices	webinar / webcast best practices	1
14	webinar best practices	b2b marketing the ultimate guide to b2b marketing	1
16	webinar best practices	how to conduct a webinar – the ultimate guide	1
18	webinar best practices	use webinar best practices to host a great webinar	1
12	webinar benchmarks	webinar attendance facts and statistic	1
11	webinar benchmarks	use webinar best practices to host a great webinar	1
10	scrappy marketing guide	what is scrappy marketing and why is it beneficial?	1
9	scrappy marketing guide	how does this apply to webinars?	1
6	how to run virtual events	building your webinar using virtual event software	1
8	how to run virtual events	what is a virtual event	1
7	how to run virtual events	the complete guide to virtual events in 2022	1
3	funnel marketing guide	how to optimize your marketing funnel for the customer journey	1
2	funnel marketing guide	how to create a powerful marketing funnel?	1
4	funnel marketing guide	understanding the stages of a marketing funnel	1
5	funnel marketing guide	what is a social media marketing funnel?	1
0	b2b marketing guide	8 tactics for your b2b marketing strategy	1
1	b2b marketing guide	b2b marketing the ultimate guide to b2b marketing	1

The preceding dataframe shows the content sections (topic) that should be written for each target keyword.

```
keyword_topics_summary.groupby(['keyword']).agg({'count': 'sum'}).
reset_index()
```

This results in the following:

	keyword	count
0	b2b marketing guide	2
1	funnel marketing guide	4
2	how to run virtual events	3
3	scrappy marketing guide	2
4	webinar benchmarks	2
5	webinar best practices	7
6	webinar guide	4
7	webinar marketing guide	2

Webinar best practices will have the most content, while other target keywords will have around two core content sections on average.

Reflections

For B2B marketing, it works really well as it's a good way of automating a manual process most SEOs go through (i.e., seeing what content the top 10 ranking pages cover) especially when you have a lot of keywords to create content for.

We used the H1 and H2 because using even more copy from the body (such as H3 or <p> paragraphs even after filtering out stop words) would introduce more noise into the string distance calculations.

Sometimes, you get some reliable suggestions that are actually quite good; however, the output should be reviewed first before raising content requests from your creative team or agency.

Summary

There are many aspects of SEO that go into delivering content and UX better than your competitors. This chapter focused on

- *Keyword mapping*: Assigning keywords to existing content and identifying opportunities for new content creation

- *Content gap analysis*: Identifying critical content and the gaps in your website

- *Content creation*: Finding the core content common to top ranking articles for your target search phrases

The next chapter deals with the third major pillar of SEO: authority.

CHAPTER 5

Authority

Authority is arguably 50% of the Google algorithm. You could optimize your site to your heart's content by creating the perfect content and deliver it with the perfect UX that's hosted on a site with the most perfect information architecture, only to find it's nowhere in Google's search results when searching by the title of the page – assuming it's not a unique search phrase, so what gives?

You'll find out about this and the following in this chapter:

- What site authority is and how it impacts SEO

- How brand searches could impact search visibility

- Review single and multiple site analysis

Some SEO History

To answer the question, one must appreciate the evolution of search engines and just how wild things were before Google came along in 1998. And even when Google did come along, things were still wild and evolving quickly.

Before Google, most of the search engines like AltaVista, Yahoo!, and Ask (Jeeves) were primarily focused on the keywords embedded within the content on the page. This made search engines relatively easy to game using all kinds of tricks including hiding keywords in white text on white backgrounds or substantial repetition of keywords.

When Google arrived, they did a couple of things differently, which essentially turned competing search engines on their heads.

© Andreas Voniatis 2023
A. Voniatis, *Data-Driven SEO with Python*, https://doi.org/10.1007/978-1-4842-9175-7_5

The first thing is that their algorithm ranked pages based on their authority, in other words, how trustworthy the document (or website) was, as opposed to only matching a document on keyword relevance. Authority in those days was measured by Google as the amount of links from other sites linking to your site. This was much in the same way as citations in a doctoral dissertation. The more links (or citations), the higher the probability a random surfer on the Web would find your content. This made SEO harder to game and the results (temporarily yet significantly) more reliable relative to the competition.

The second thing they did was partner with Yahoo! which openly credited Google for powering their search results. So what happened next? Instead of using Yahoo!, people went straight to Google, bypassing the intermediary Yahoo! Search engine, and the rest is history – or not quite.

A Little More History

Although Google got the lion's share of searches, the SEO industry worked out the gist of Google's algorithm and started engineering link popularity schemes such as swapping links (known as reciprocal linking) and creating/renting links from private networks (still alive and well today, unfortunately). Google responded with antispam algorithms, such as Panda and Penguin, which more or less decimated these schemes to the point that most businesses in the brand space resorted to advertising and digital PR. And it works.

Authority, Links, and Other

While there is a widespread confusion in that back links are authority. We've seen plenty of evidence to show that authority is the effect of links and advertising, that is, authority is not only measured in links. Refer to Figure 5-1.

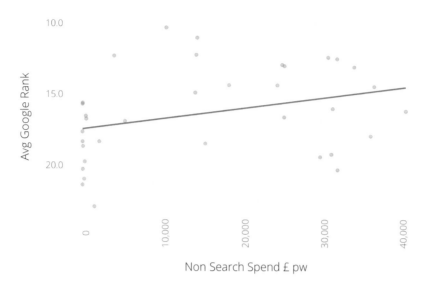

Figure 5-1. *Positive relationship between rankings and authority*

Figure 5-1 is just one example of many showing a positive relationship between rankings and authority. In this case, the authority is the product of nonsearch advertising. And why is that? It's because good links and effective advertising drive brand impressions, which are also positively linked.

What we will set out to do is show how data science can help you:

- Examine your own links

- Analyze your competitor's links

- Find power networks

- Determine the key ingredients for a good link

Examining Your Own Links

If you've ever wanted to analyze your site's backlinks, the chances are you'd use one of the more popular tools like AHREFs and SEMRush. These services trawl the Web to get a list of sites linking to your website with a domain rating and other info describing the quality of your backlinks, which they store in vast indexes which can be queried.

It's no secret that backlinks play a big part in Google's algorithm so it makes sense as a minimum to understand your own site before comparing it with the competition, of which the former is what we will do today.

While most of the analysis can be done on a spreadsheet, Python has certain advantages. Other than the sheer number of rows it can handle, it can also look at the statistical side more readily such as distributions.

Importing and Cleaning the Target Link Data

We're going to pick a small website from the UK furniture sector (for no particular reason) and walk through some basic analysis using Python.

So what is the value of a site's backlinks for SEO? At its simplest, I'd say quality and quantity. Quality is subjective to the expert yet definitive to Google by way of metrics such as authority and content relevance.

We'll start by evaluating the link quality with the available data before evaluating the quantity. Time to code.

```python
import re
import time
import random
import pandas as pd
import numpy as np
import datetime
from datetime import timedelta
from plotnine import *
import matplotlib.pyplot as plt
from pandas.api.types import is_string_dtype
from pandas.api.types import is_numeric_dtype
import uritools

pd.set_option('display.max_colwidth', None)
%matplotlib inline

root_domain = 'johnsankey.co.uk'
hostdomain = 'www.johnsankey.co.uk'
hostname = 'johnsankey'
full_domain = 'https://www.johnsankey.co.uk'
target_name = 'John Sankey'
```

We start by importing the data and cleaning up the column names to make it easier to handle and quicker to type, for the later stages.

```
target_ahrefs_raw = pd.read_csv(
    'data/johnsankey.co.uk-refdomains-subdomains__2022-03-18_15-15-47.csv')
```

List comprehensions are a powerful and less intensive way to clean up the column names.

```
target_ahrefs_raw.columns = [col.lower() for col in target_ahrefs_raw.
columns]
```

The list comprehension instructs Python to convert the column name to lowercase for each column ("col") in the dataframe columns.

```
target_ahrefs_raw.columns = [col.replace(' ','_') for col in target_ahrefs_
raw.columns]
target_ahrefs_raw.columns = [col.replace('.','_') for col in target_ahrefs_
raw.columns]
target_ahrefs_raw.columns = [col.replace('__','_') for col in target_
ahrefs_raw.columns]
target_ahrefs_raw.columns = [col.replace('(','') for col in target_ahrefs_
raw.columns]
target_ahrefs_raw.columns = [col.replace(')','') for col in target_ahrefs_
raw.columns]
target_ahrefs_raw.columns = [col.replace('%','') for col in target_ahrefs_
raw.columns]
```

An alternative to repeating the preceding lines of code would be to chain the function calls to process the columns in a single line:

```
target_ahrefs_raw.columns = [col.lower().replace(' ','_').replace('.','_').
replace('__','_').replace('(','').replace(')','').replace('%','') for col
in target_ahrefs_raw.columns]
```

Though not strictly necessary, I like having a count column as standard for aggregations and a single value column "project" should I need to group the entire table:

```
target_ahrefs_raw['rd_count'] = 1
target_ahrefs_raw['project'] = target_name
Target_ahrefs_raw
```

This results in the following:

	domain	dr	dofollow_ref_domains	dofollow_linked_domains	traffic_	links_to_target	new_links	lost_links	dofollow_links
0	dribbble.com	93.0	598872	14388	3460486	1	0	0	0
1	msn.com	92.0	748243	265009	111799543	9	0	0	0
2	thetimes.co.uk	91.0	448410	28603	4977287	1	0	0	1
3	ow.ly	90.0	118913	2	42910	2	2	0	2
4	10times.com	78.0	19109	19346	522290	2	0	0	0
...	...	...	...	...	...	...	...	...	...
102	ikhatotherescue.blogspot.com	0.0	73	2314	0	1	0	0	1
103	peakedgehotel.blogspot.com	0.0	0	0	0	8	0	0	8
104	thelibertyscale.blogspot.com	0.0	0	0	1	1	0	0	1
105	plums-rhombus-lrg3.squarespace.com	0.0	0	0	0	1	1	1	1
106	upholsterycleaningdozaowa.blogspot.com	0.0	0	0	0	1	0	0	1

107 rows × 13 columns

Now we have a dataframe with clean column names. The next step is to clean the actual table values and make them more useful for analysis.

Make a copy of the previous dataframe and give it a new name:

```
target_ahrefs_clean_dtypes = target_ahrefs_raw.copy()
```

Clean the dofollow_ref_domains column which tells us how many ref domains the sitelinking has. In this case, we'll convert the dashes to zeros and then cast the whole column as a whole number.

Start with referring domains:

```
target_ahrefs_clean_dtypes['dofollow_ref_domains'] = np.where(target_
ahrefs_clean_dtypes['dofollow_ref_domains'] == '-',
                    0, target_ahrefs_clean_dtypes['dofollow_ref_
                    domains'])
target_ahrefs_clean_dtypes['dofollow_ref_domains'] = target_ahrefs_clean_
dtypes['dofollow_ref_domains'].astype(int)
```

then linked domains:

```
target_ahrefs_clean_dtypes['dofollow_linked_domains'] = np.where(target_
ahrefs_clean_dtypes['dofollow_linked_domains'] == '-',
                        0, target_ahrefs_clean_dtypes['dofollow_linked_
                        domains'])
target_ahrefs_clean_dtypes['dofollow_linked_domains'] = target_ahrefs_
clean_dtypes['dofollow_linked_domains'].astype(int)
```

"First seen" tells us the date when the link was first found (i.e., discovered and then added to the index of ahrefs). We'll convert the string to a date format that Python can process and then use this to derive the age of the links later on:

```
target_ahrefs_clean_dtypes['first_seen'] = pd.to_datetime(target_ahrefs_
clean_dtypes['first_seen'], format='%d/%m/%Y %H:%M')
```

Converting first_seen to a date also means we can perform time aggregations by month year, as it's not always the case that links for a site will get acquired on a daily basis:

```
target_ahrefs_clean_dtypes['month_year'] = target_ahrefs_clean_
dtypes['first_seen'].dt.to_period('M')
```

The link age is calculated by taking today's date and subtracting the first seen date. Then it's converted to a number format and divided by a huge number to get the number of days:

```
target_ahrefs_clean_dtypes['link_age'] = dt.datetime.now() - target_ahrefs_
clean_dtypes['first_seen']
target_ahrefs_clean_dtypes['link_age'] = target_ahrefs_clean_
dtypes['link_age']
target_ahrefs_clean_dtypes['link_age'] = target_ahrefs_clean_dtypes
['link_age'].astype(int)
target_ahrefs_clean_dtypes['link_age'] = (target_ahrefs_clean_dtypes
['link_age']/(3600 * 24 * 1000000000)).round(0)
```

```
target_ahrefs_clean_dtypes
```

This results in the following:

lomains	dofollow_linked_domains	traffic_	links_to_target	new_links	lost_links	dofollow_links	first_seen	lost	rd_count	project	month_year	link_age
598872	14388	3460486	1	0	0	0	2021-04-16 03:06:00	NaN	1	John Sankey	2021-04	403.0
748243	265009	111799543	9	0	0	0	2021-09-28 11:39:00	NaN	1	John Sankey	2021-09	238.0
448410	28603	4977287	1	0	0	1	2017-09-17 07:13:00	NaN	1	John Sankey	2017-09	1710.0
118913	2	42910	2	2	0	2	2022-02-16 21:31:00	NaN	1	John Sankey	2022-02	97.0
19109	19346	522290	2	0	0	0	2018-10-06 01:03:00	NaN	1	John Sankey	2018-10	1326.0
...	...	...	...	...	...	...	...	...	...	...	...	...
73	2314	0	1	0	0	1	2021-11-18 05:37:00	NaN	1	John Sankey	2021-11	187.0
0	0	0	8	0	0	8	2020-06-17 23:30:00	NaN	1	John Sankey	2020-06	705.0
0	0	1	1	0	0	1	2021-04-11 06:21:00	NaN	1	John Sankey	2021-04	408.0
0	0	0	1	1	1	1	2022-03-14 11:18:00	16/03/2022 16:51	1	John Sankey	2022-03	71.0
0	0	0	1	0	0	1	2019-04-04 13:29:00	NaN	1	John Sankey	2019-04	1146.0

With the data types cleaned, and some new data features created (note columns added earlier), the fun can begin.

Targeting Domain Authority

The first part of our analysis evaluates the link quality, which starts by summarizing the whole dataframe using the describe function to get descriptive statistics of all the columns:

```
target_ahrefs_analysis = target_ahrefs_clean_dtypes
target_ahrefs_analysis.describe()
```

This results in the following:

	dr	dofollow_ref_domains	dofollow_linked_domains	traffic_	links_to_target	new_links	lost_links	dofollow_links	rd_count	link_age
count	107.000000	107.000000	1.070000e+02	1.070000e+02	107.000000	107.000000	107.000000	107.000000	107.0	107.000000
mean	26.593458	22557.794393	3.661365e+05	1.334358e+06	3.383178	0.644860	0.093458	2.411215	1.0	492.121495
std	28.092862	103659.472991	1.477562e+06	1.088568e+07	6.180683	3.629679	0.445792	5.606521	0.0	555.336286
min	0.000000	0.000000	0.000000e+00	0.000000e+00	1.000000	0.000000	0.000000	0.000000	1.0	6.000000
25%	0.100000	29.000000	2.700000e+01	0.000000e+00	1.000000	0.000000	0.000000	1.000000	1.0	123.500000
50%	16.000000	184.000000	1.373000e+03	1.380000e+02	2.000000	0.000000	0.000000	1.000000	1.0	234.000000
75%	47.500000	1973.000000	7.001000e+03	7.329500e+03	2.500000	0.000000	0.000000	2.000000	1.0	645.500000
max	93.000000	748243.000000	7.923115e+06	1.117995e+08	42.000000	34.000000	4.000000	42.000000	1.0	2504.000000

So from the preceding table, we can see the average (mean), the number of referring domains (107), and the variation (the 25th percentiles and so on).

The average domain rating (equivalent to Moz's Domain Authority) of referring domains is 27. Is that a good thing? In the absence of competitor data to compare in this market sector, it's hard to know, which is where your experience as an SEO practitioner comes in. However, I'm certain we could all agree that it could be much higher – given that it falls on a scale between 0 and 100. How much higher to make a shift is another question.

The preceding table can be a bit dry and hard to visualize, so we'll plot a histogram to get more of an intuitive understanding of the referring domain authority:

```
dr_dist_plt = (
    ggplot(target_ahrefs_analysis,
          aes(x = 'dr')) +
    geom_histogram(alpha = 0.6, fill = 'blue', bins = 100) +
    scale_y_continuous() +
    theme(legend_position = 'right'))

dr_dist_plt
```

The distribution is heavily skewed, showing that most of the referring domains have an authority rating of zero (Figure 5-2). Beyond zero, the distribution looks fairly uniform with an equal amount of domains across different levels of authority.

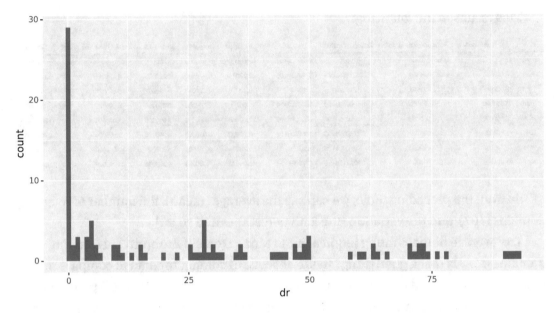

Figure 5-2. *Distribution of domain rating in the backlink profile*

Domain Authority Over Time

We'll now look at the domain authority as a proxy for the link quality as a time series. If we were to plot the number of links by date, the time series would look rather messy and less useful as follows:

```
dr_firstseen_plt = (
    ggplot(target_ahrefs_analysis, aes(x = 'first_seen', y = 'dr',
    group = 1)) +
    geom_line(alpha = 0.6, colour = 'blue', size = 2) +
    labs(y = 'Domain Rating', x = 'Month Year') +
    scale_y_continuous() +
    scale_x_date() +
    theme(legend_position = 'right',
        axis_text_x=element_text(rotation=90, hjust=1)
        )
)
```

```
dr_firstseen_plt.save(filename = 'images/1_dr_firstseen_plt.png',
                          height=5, width=10, units = 'in', dpi=1000)

dr_firstseen_plt
```

The plot looks very noisy as you'd expect and only really shows you what the DR (domain rating) of a referring domain was at a point in time (Figure 5-3). The utility of this chart is that if you have a team tasked with acquiring links, you can monitor the link quality over time in general.

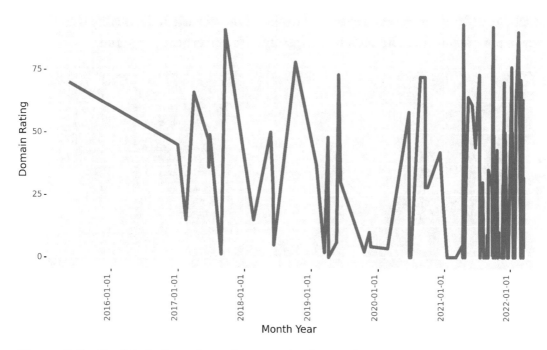

Figure 5-3. *Backlink domain rating acquired over time*

For a more smoother view:

```
dr_firstseen_smooth_plt = (
    ggplot(target_ahrefs_analysis, aes(x = 'first_seen', y = 'dr',
    group = 1)) +
    geom_smooth(alpha = 0.6, colour = 'blue', size = 3, se = False) +
    labs(y = 'Domain Rating', x = 'Month Year') +
    scale_y_continuous() +
    scale_x_date() +
```

```
    theme(legend_position = 'right',
        axis_text_x=element_text(rotation=90, hjust=1)
    ))
```

dr_firstseen_smooth_plt.save(filename = 'images/1_dr_firstseen_smooth_plt.
png', height=5, width=10, units = 'in', dpi=1000)

dr_firstseen_smooth_plt

The use of geom_smooth() gives a somewhat less noisy view and shows the variability of the domain rating over time to show how consistent the quality is (Figure 5-4). Again, this correlates to the quality of the links being acquired.

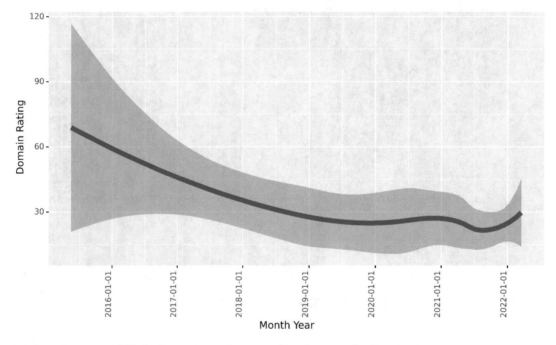

Figure 5-4. *Backlink domain rating acquired smoothed over time*

What this doesn't quite describe is the overall site authority over time, because the value of links acquired is retained over time; therefore, a different math approach is required.

To see the site's authority over time, we will calculate a running average of the domain rating by month of the year. Note the use of the expanding() function which instructs Pandas to include all previous rows with each new row:

```
target_rd_cummean_df = target_ahrefs_analysis
target_rd_mean_df = target_rd_cummean_df.groupby(['month_year'])['dr'].
sum().reset_index()

target_rd_mean_df['dr_runavg'] = target_rd_mean_df['dr'].expanding().mean()

target_rd_mean_df.head(10)
```

This results in the following:

	month_year	dr	dr_runavg
0	2015-05	70.0	70.000000
1	2016-12	45.0	57.500000
2	2017-02	15.0	43.333333
3	2017-03	66.0	49.000000
4	2017-06	132.0	65.600000
5	2017-08	1.3	54.883333
6	2017-09	91.0	60.042857
7	2018-02	15.0	54.412500
8	2018-05	50.0	53.922222
9	2018-06	32.9	51.820000

We now have a table which we can use to feed the graph and visualize.

```
dr_cummean_smooth_plt = (
    ggplot(target_rd_mean_df, aes(x = 'month_year', y = 'dr_runavg',
    group = 1)) +
    geom_line(alpha = 0.6, colour = 'blue', size = 2) +
    #labs(y = 'GA Sessions', x = 'Date') +
    scale_y_continuous() +
    scale_x_date() +
    theme(legend_position = 'right',
        axis_text_x=element_text(rotation=90, hjust=1)
        ))

dr_cummean_smooth_plt
```

So the target site started with high authority links (which may have been a PR campaign announcing the business brand), which faded soon after for four years and then rebooted with new acquisition of high authority links again (Figure 5-5).

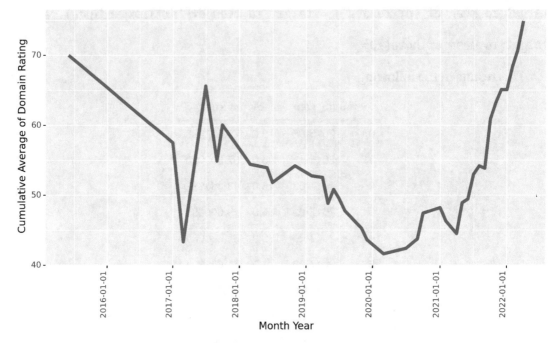

Figure 5-5. *Cumulative average domain rating of backlinks over time*

Most importantly, we can see the site's general authority over time, which is how a search engine like Google may see it too.

A really good extension to this analysis would be to regenerate the dataframe so that we would plot the distribution over time on a cumulative basis. Then we could not only see the median quality but also the variation over time too.

That's the link quality, what about quantity?

Targeting Link Volumes

Quality is one thing; the volume of quality links is quite another, which is what we'll analyze next.

We'll use the expanding function like the previous operation to calculate a cumulative sum of the links acquired to date:

```
target_count_cumsum_df = target_ahrefs_analysis
print(target_count_cumsum_df.columns)
target_count_cumsum_df = target_count_cumsum_df.groupby(['month_year'])
['rd_count'].sum().reset_index()
```

```
target_count_cumsum_df['count_runsum'] = target_count_cumsum_df['rd_
count'].expanding().sum()
target_count_cumsum_df['link_velocity'] = target_count_cumsum_df['rd_
count'].diff()
```

```
target_count_cumsum_df
```

This results in the following:

	month_year	rd_count	count_runsum	link_velocity
0	2015-05	1	1.0	NaN
1	2016-12	1	2.0	0.0
2	2017-02	1	3.0	0.0
3	2017-03	1	4.0	0.0
4	2017-06	3	7.0	2.0
5	2017-08	1	8.0	-2.0
6	2017-09	1	9.0	0.0
7	2018-02	1	10.0	0.0
8	2018-05	1	11.0	0.0
9	2018-06	2	13.0	1.0

That's the data, now the graphs.

```
target_count_plt = (
    ggplot(target_count_cumsum_df, aes(x = 'month_year', y = 'rd_count',
    group = 1)) +
    geom_line(alpha = 0.6, colour = 'blue', size = 2) +
    labs(y = 'Count of Referring Domains', x = 'Month Year') +
    scale_y_continuous() +
    scale_x_date() +
    theme(legend_position = 'right',
```

```
      axis_text_x=element_text(rotation=90, hjust=1)
    ))

target_count_plt.save(filename = 'images/3_target_count_plt.png',
                      height=5, width=10, units = 'in', dpi=1000)

target_count_plt
```

This is a noncumulative view of the amount of referring domains. Again, this is useful for evaluating how effective a team is at acquiring links (Figure 5-6).

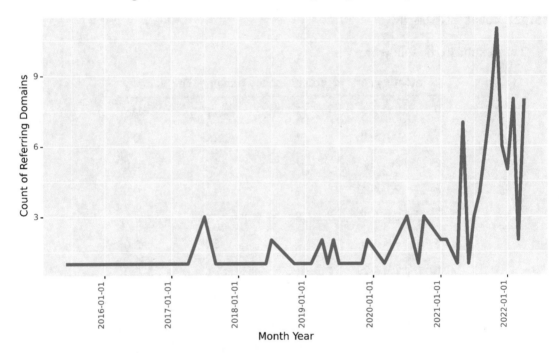

Figure 5-6. *Count of referring domains over time*

But perhaps it is not as useful for how a search engine would view the overall number of referring domains a site has.

```
target_count_cumsum_plt = (
    ggplot(target_count_cumsum_df, aes(x = 'month_year', y = 'count_
    runsum', group = 1)) +
    geom_line(alpha = 0.6, colour = 'blue', size = 2) +
    scale_y_continuous() +
    scale_x_date() +
```

```
theme(legend_position = 'right',
    axis_text_x=element_text(rotation=90, hjust=1)
    ))
```

```
target_count_cumsum_plt
```

The cumulative view shows us the total number of referring domains (Figure 5-7). Naturally, this isn't the entirely accurate picture as some referring domains may have been lost, but it's good enough to get the gist of where the site is at.

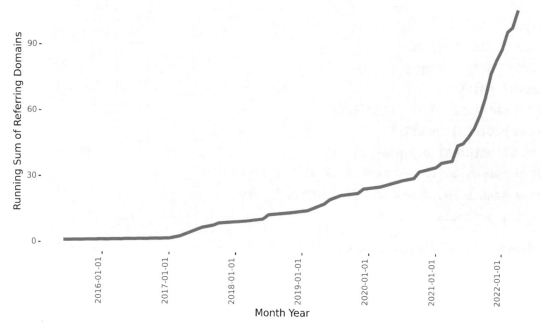

Figure 5-7. *Cumulative sum of referring domains over time*

We see that links were steadily added from 2017 for the next four years before accelerating again around March 2021. This is consistent with what we have seen with domain rating over time.

A useful extension to correlate that with performance may be to layer in

- Referring domain site traffic

- Average ranking over time

Analyzing Your Competitor's Links

Like last time, we defined the value of a site's backlinks for SEO as a product of quality and quantity – quality being the domain authority (or AHREF's equivalent domain rating) and quantity as the number of referring domains.

Again, we'll start by evaluating the link quality with the available data before evaluating the quantity. Time to code.

```
import re
import time
import random
import pandas as pd
import numpy as np
import datetime
from datetime import timedelta
from plotnine import *
import matplotlib.pyplot as plt
from pandas.api.types import is_string_dtype
from pandas.api.types import is_numeric_dtype
import uritools

pd.set_option('display.max_colwidth', None)
%matplotlib inline

root_domain = 'johnsankey.co.uk'
hostdomain = 'www.johnsankey.co.uk'
hostname = 'johnsankey'
full_domain = 'https://www.johnsankey.co.uk'
target_name = 'John Sankey'
```

Data Importing and Cleaning

We set up the file directories so we can read multiple AHREF exported data files in one folder, which is much faster, less boring, and more efficient than reading each file individually, especially when you have over ten of them:

```
ahrefs_path = 'data/'
```

The listdir() function from the OS module allows us to list all of the files in a subdirectory:

```
ahrefs_filenames = os.listdir(ahrefs_path)

ahrefs_filenames
```

This results in the following:

```
['www.davidsonlondon.com--refdomains-subdomain__2022-03-13_23-37-29.csv',
 'www.stephenclasper.co.uk--refdomains-subdoma__2022-03-13_23-47-28.csv',
 'www.touchedinteriors.co.uk--refdomains-subdo__2022-03-13_23-42-05.csv',
 'www.lushinteriors.co--refdomains-subdomains__2022-03-13_23-44-34.csv',
 'www.kassavello.com--refdomains-subdomains__2022-03-13_23-43-19.csv',
 'www.tulipinterior.co.uk--refdomains-subdomai__2022-03-13_23-41-04.csv',
 'www.tgosling.com--refdomains-subdomains__2022-03-13_23-38-44.csv',
 'www.onlybespoke.com--refdomains-subdomains__2022-03-13_23-45-28.csv',
 'www.williamgarvey.co.uk--refdomains-subdomai__2022-03-13_23-43-45.csv',
 'www.hadleyrose.co.uk--refdomains-subdomains__2022-03-13_23-39-31.csv',
 'www.davidlinley.com--refdomains-subdomains__2022-03-13_23-40-25.csv',
 'johnsankey.co.uk-refdomains-subdomains__2022-03-18_15-15-47.csv']
```

With the files listed, we'll now read each one individually using a for loop and add these to a dataframe. While reading in the file, we'll use some string manipulation to create a new column with the site name of the data we're importing:

```
ahrefs_df_lst = list()
ahrefs_colnames = list()

for filename in ahrefs_filenames:
    df = pd.read_csv(ahrefs_path + filename)
    df['site'] = filename
    df['site'] = df['site'].str.replace('www.', '', regex = False)
    df['site'] = df['site'].str.replace('.csv', '', regex = False)
    df['site'] = df['site'].str.replace('-.+', '', regex = True)
    ahrefs_colnames.append(df.columns)
    ahrefs_df_lst.append(df)

comp_ahrefs_df_raw = pd.concat(ahrefs_df_lst)

comp_ahrefs_df_raw
```

This results in the following:

	Domain	DR	Dofollow ref. domains	Dofollow linked domains	Traffic	Links to target	New links	Lost links	Dofollow links	First seen	Lost	site
0	pinterest.co.uk	92.0	189919	46703	14142143	2	0	0	2	18/08/2020 02:04	NaN	davidsonlondon.com
1	standard.co.uk	89.0	403937	7582	9931955	2	0	0	2	10/03/2021 11:53	NaN	davidsonlondon.com
2	myminifactory.com	78.0	40446	18726	293812	12	0	0	12	29/03/2017 16:29	NaN	davidsonlondon.com
3	idealhome.co.uk	77.0	18278	3133	2005577	2	0	0	0	15/02/2020 17:39	NaN	davidsonlondon.com
4	thomsonlocal.com	77.0	4895	24160	383648	22	0	6	0	18/06/2020 06:31	NaN	davidsonlondon.com
...	...	...	...	...	...	...	...	...	...	...	...	...
102	ikhatotherescue.blogspot.com	0.0	73	2314	0	1	0	0	1	18/11/2021 05:37	NaN	johnsankey.co.uk
103	peakedgehotel.blogspot.com	0.0	0	0	0	8	0	0	8	17/06/2020 23:30	NaN	johnsankey.co.uk
104	thelibertyscale.blogspot.com	0.0	0	0	1	1	0	0	1	11/04/2021 06:21	NaN	johnsankey.co.uk
105	plums-rhombus-lrg3.squarespace.com	0.0	0	0	0	1	1	1	1	14/03/2022 11:18	16/03/2022 16:51	johnsankey.co.uk
106	upholsterycleaningdozaowa.blogspot.com	0.0	0	0	0	1	0	0	1	04/04/2019 13:29	NaN	johnsankey.co.uk

5409 rows × 12 columns

Now we have the raw data from each site in a single dataframe, the next step is to tidy up the column names and make them a bit more friendlier to work with. A custom function could be used, but we'll just chain the function calls with a list comprehension:

```
competitor_ahrefs_cleancols = comp_ahrefs_df_raw.copy()
competitor_ahrefs_cleancols.columns = [col.lower().replace(' ','_').
replace('.','_').replace('__','_').replace('(','')
 .replace(')','').replace('%','')
 for col in competitor_ahrefs_cleancols.columns]
```

Having a count column and a single value column ("project") is useful for groupby and aggregation operations:

```
competitor_ahrefs_cleancols['rd_count'] = 1
competitor_ahrefs_cleancols['project'] = target_name

competitor_ahrefs_cleancols
```

This results in the following:

	domain	dr	dofollow_ref_domains	dofollow_linked_domains	traffic_	links_to_target	new_links	lost_links	dofollow_links
0	pinterest.co.uk	92.0	189919	46703	14142143	2	0	0	2
1	standard.co.uk	89.0	403937	7582	9931955	2	0	0	2
2	myminifactory.com	78.0	40446	18726	293812	12	0	0	12
3	idealhome.co.uk	77.0	18278	3133	2005577	2	0	0	0
4	thomsonlocal.com	77.0	4895	24160	383648	22	0	6	0
...	...	...	...	...	...	...	...	...	...
102	ikhatotherescue.blogspot.com	0.0	73	2314	0	1	0	0	1
103	peakedgehotel.blogspot.com	0.0	0	0	0	8	0	0	8
104	thelibertyscale.blogspot.com	0.0	0	0	1	1	0	0	1
105	plums-rhombus-lrg3.squarespace.com	0.0	0	0	0	1	1	1	1
106	upholsterycleaningdozaowa.blogspot.com	0.0	0	0	0	1	0	0	1

5409 rows × 14 columns

The columns are now cleaned up, so we'll now clean up the row data:

```
competitor_ahrefs_clean_dtypes = competitor_ahrefs_cleancols
```

For referring domains, we're replacing hyphens with zero and setting the data type as an integer (i.e., whole number). This will be repeated for linked domains, also:

```
competitor_ahrefs_clean_dtypes['dofollow_ref_domains'] =
np.where(competitor_ahrefs_clean_dtypes['dofollow_ref_domains'] == '-',
                0, competitor_ahrefs_clean_dtypes['dofollow_ref_
                domains'])
competitor_ahrefs_clean_dtypes['dofollow_ref_domains'] = competitor_ahrefs_
clean_dtypes['dofollow_ref_domains'].astype(int)

# linked_domains
competitor_ahrefs_clean_dtypes['dofollow_linked_domains'] =
np.where(competitor_ahrefs_clean_dtypes['dofollow_linked_domains'] == '-',
                0, competitor_ahrefs_clean_dtypes['dofollow_linked_
                domains'])
competitor_ahrefs_clean_dtypes['dofollow_linked_domains'] = competitor_
ahrefs_clean_dtypes['dofollow_linked_domains'].astype(int)
```

First seen gives us a date point at which links were found, which we can use for time series plotting and deriving the link age. We'll convert to date format using the to_datetime function:

```
competitor_ahrefs_clean_dtypes['first_seen'] = pd.to_datetime(competitor_
ahrefs_clean_dtypes['first_seen'],
                          format='%d/%m/%Y %H:%M')
competitor_ahrefs_clean_dtypes['first_seen'] = competitor_ahrefs_clean_
dtypes['first_seen'].dt.normalize()
competitor_ahrefs_clean_dtypes['month_year'] = competitor_ahrefs_clean_
dtypes['first_seen'].dt.to_period('M')
```

To calculate the link_age, we'll simply deduct the first seen date from today's date and convert the difference into a number:

```
competitor_ahrefs_clean_dtypes['link_age'] = dt.datetime.now() -
competitor_ahrefs_clean_dtypes['first_seen']
competitor_ahrefs_clean_dtypes['link_age'] = competitor_ahrefs_clean_
dtypes['link_age']
competitor_ahrefs_clean_dtypes['link_age'] = competitor_ahrefs_clean_
dtypes['link_age'].astype(int)
competitor_ahrefs_clean_dtypes['link_age'] = (competitor_ahrefs_clean_
dtypes['link_age']/(3600 * 24 * 1000000000)).round(0)
```

The target column helps us distinguish the "client" site vs. competitors, which is useful for visualization later:

```
competitor_ahrefs_clean_dtypes['target'] = np.where(competitor_ahrefs_
clean_dtypes['site'].str.contains('johns'),
                          1, 0)
competitor_ahrefs_clean_dtypes['target'] = competitor_ahrefs_clean_
dtypes['target'].astype('category')

competitor_ahrefs_clean_dtypes
```

This results in the following:

›mains	traffic_	links_to_target	new_links	lost_links	dofollow_links	first_seen	lost	site	rd_count	project	month_year	link_age	target
46703	14142143	2	0	0	2	2020-08-18	NaN	davidsonlondon.com	1	John Sankey	2020-08	630.0	0
7582	9931955	2	0	0	2	2021-03-10	NaN	davidsonlondon.com	1	John Sankey	2021-03	426.0	0
18726	293812	12	0	0	12	2017-03-29	NaN	davidsonlondon.com	1	John Sankey	2017-03	1868.0	0
3133	2005577	2	0	0	0	2020-02-15	NaN	davidsonlondon.com	1	John Sankey	2020-02	815.0	0
24160	383648	22	0	6	0	2020-06-18	NaN	davidsonlondon.com	1	John Sankey	2020-06	691.0	0
...	...	...	...	...	...	...	...	...	...	...	...	...	...
2314	0	1	0	0	1	2021-11-18	NaN	johnsankey.co.uk	1	John Sankey	2021-11	173.0	1
0	0	8	0	0	8	2020-06-17	NaN	johnsankey.co.uk	1	John Sankey	2020-06	692.0	1
0	1	1	0	0	1	2021-04-11	NaN	johnsankey.co.uk	1	John Sankey	2021-04	394.0	1
0	0	1	1	1	1	2022-03-14	16/03/2022 16:51	johnsankey.co.uk	1	John Sankey	2022-03	57.0	1
0	0	1	0	0	1	2019-04-04	NaN	johnsankey.co.uk	1	John Sankey	2019-04	1132.0	1

Now that the data is cleaned up both in terms of column titles and row values, we're ready to set forth and start analyzing.

Anatomy of a Good Link

When we analyzed the one target website earlier ("John Sankey"), we assumed (like the rest of the SEO industry the world over) that domain rating (DR) was the best and most reliable measure of the link quality.

But should we? Let's do a quick and dirty analysis to see if that is indeed the case or whether we can find something better. We'll start by aggregating the link features at the site level:

```
competitor_ahrefs_aggs = competitor_ahrefs_analysis.groupby('site').
agg({'link_age': 'mean',
         'dofollow_links': 'mean',    'domain': 'count', 'dr': 'mean',
'dofollow_ref_domains': 'mean',  'traffic_': 'mean', 'dofollow_
linked_domains': 'mean',    'links_to_target': 'mean',  'new_links':
'mean',    'lost_links': 'mean'}).reset_index()

competitor_ahrefs_aggs
```

This results in the following:

	site	link_age	dofollow_links	domain	dr	dofollow_ref_domains	traffic_	dofollow_linked_domains	links_to_target	new_links
0	davidlinley.com	794.675000	3.002000	1000	32.662200	22648.514000	6.456371e+06	83373.709000	4.031000	0.188000
1	davidsonlondon.com	329.118598	2.029650	371	11.003774	3515.447439	1.351592e+05	116973.943396	3.048518	0.328841
2	hadleyrose.co.uk	226.657316	1.326763	1319	1.944882	1297.971190	1.022266e+05	59388.066717	1.563306	0.146323
3	johnsankey.co.uk	540.121495	2.411215	107	26.593458	22557.794393	1.334358e+06	366136.514019	3.383178	0.644860
4	kassavello.com	301.018913	1.621749	423	6.077778	1279.536643	8.683198e+04	117990.945626	1.829787	0.108747
5	lushinteriors.co	227.137255	1.039216	306	4.706863	1202.562092	8.210043e+04	31187.594771	1.258170	0.156863
6	onlybespoke.com	267.407407	0.944444	108	16.330556	2934.129630	1.887193e+05	144268.425926	1.833333	0.296296
7	stephenclasper.co.uk	581.516129	1.516129	31	30.548387	8565.096774	7.907171e+04	302009.161290	3.612903	0.225806
8	tgosling.com	480.582474	1.783505	194	17.840722	14784.639175	1.181992e+06	197045.087629	2.221649	0.190722
9	touchedinteriors.co.uk	741.674627	5.549254	335	18.501194	8577.185075	4.250500e+05	222064.211940	7.188060	0.388060
10	tulipinterior.co.uk	273.664198	1.562963	810	7.666667	1884.629630	4.224160e+04	97459.551852	1.990123	0.149383
11	williamgarvey.co.uk	529.738272	4.024691	405	17.768889	8033.827160	1.909090e+05	325999.076543	5.728395	0.395062

The resulting table shows us aggregated statistics for each of the link features. Next, read in the list of SEMRush domain level data (which by way of manual data entry was literally typed in since it's only 11 sites):

```
semrush_viz = [10100, 2300, 931, 2400, 911, 2100, 1800, 136, 838, 428,
1100, 1700]

competitor_ahrefs_aggs['semrush_viz'] = semrush_viz

competitor_ahrefs_aggs
```

This results in the following:

	link_age	dofollow_links	domain	dr	dofollow_ref_domains	traffic_	dofollow_linked_domains	links_to_target	new_links	lost_links	semrush_viz
	794.675000	3.002000	1000	32.662200	22648.514000	6.456371e+06	83373.709000	4.031000	0.188000	0.193000	10100
	329.118598	2.029650	371	11.003774	3515.447439	1.351592e+05	116973.943396	3.048518	0.328841	0.121294	2300
	226.657316	1.326763	1319	1.944882	1297.971190	1.022266e+05	59388.066717	1.563306	0.146323	0.152388	931
	540.121495	2.411215	107	26.593458	22557.794393	1.334358e+06	366136.514019	3.383178	0.644860	0.093458	2400
	301.018913	1.621749	423	6.077778	1279.536643	8.683198e+04	117990.945626	1.829787	0.108747	0.120567	911
	227.137255	1.039216	306	4.706863	1202.562092	8.210043e+04	31187.594771	1.258170	0.156863	0.101307	2100
	267.407407	0.944444	108	16.330556	2934.129630	1.887193e+05	144268.425926	1.833333	0.296296	0.185185	1800
	581.516129	1.516129	31	30.548387	8565.096774	7.907171e+04	302009.161290	3.612903	0.225806	0.161290	136
	480.582474	1.783505	194	17.840722	14784.639175	1.181992e+06	197045.087629	2.221649	0.190722	0.164948	838
	741.674627	5.549254	335	18.501194	8577.185075	4.250500e+05	222064.211940	7.188060	0.388060	0.092537	428
	273.664198	1.562963	810	7.666667	1884.629630	4.224160e+04	97459.551852	1.990123	0.149383	0.139506	1100
	529.738272	4.024691	405	17.768889	8033.827160	1.909090e+05	325999.076543	5.728395	0.395062	0.708642	1700

The SEMRush visibility data has now been appended, so we're ready to find some r-squared, known as the coefficient of determination, which will tell which link feature can best explain the variation in SEMRush visibility:

```
competitor_ahrefs_r2 = competitor_ahrefs_aggs.corr() ** 2
competitor_ahrefs_r2 = competitor_ahrefs_r2[['semrush_viz']].reset_index()
competitor_ahrefs_r2 = competitor_ahrefs_r2.sort_values('semrush_viz',
ascending = False)

competitor_ahrefs_r2
```

This results in the following:

	index	semrush_viz
10	semrush_viz	1.000000
5	traffic_	0.890900
4	dofollow_ref_domains	0.336989
3	dr	0.214275
0	link_age	0.204189
2	domain	0.148347
6	dofollow_linked_domains	0.064904
1	dofollow_links	0.014366
7	links_to_target	0.007580
9	lost_links	0.001712
8	new_links	0.001055

Naturally, we'd expect the semrush_viz to correlate perfectly with itself. DR (domain rating) surprisingly doesn't explain the difference in SEMRush very well with an r_ squared of 21%.

On the other hand, "traffic_" which is the referring domain's traffic value correlates better. From this alone, we're prepared to disregard "dr." Let's inspect this visually:

```
comp_correl_trafficviz_plt = (
    ggplot(competitor_ahrefs_aggs,
            aes(x = 'traffic_', y = 'semrush_viz')) +
    geom_point(alpha = 0.4, colour = 'blue', size = 2) +
```

```
    geom_smooth(method = 'lm', se = False, colour = 'red', size = 3,
    alpha = 0.4)
)

comp_correl_trafficviz_plt.save(filename = 'images/2_comp_correl_
trafficviz_plt.png',
                        height=5, width=10, units = 'in', dpi=1000)

comp_correl_trafficviz_plt
```

This is not terribly convincing (Figure 5-8), due to the lack of referring domains beyond 2,000,000. Does this mean we should disregard traffic_ as a measure?

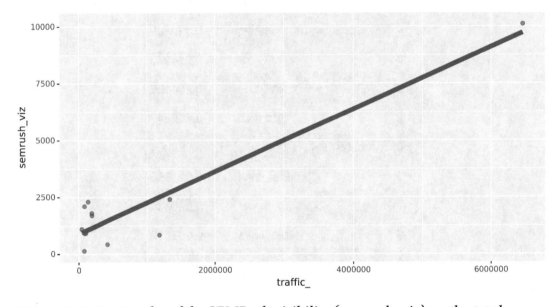

Figure 5-8. *Scatterplot of the SEMRush visibility (semrush_viz) vs. the total AHREFs backlink traffic (traffic_) of the site's backlinks*

Not necessarily. The outlier data point with 10,000 visibility isn't necessarily incorrect. The site does have superior visibility and more referring traffic in the real world, so it doesn't mean the site's data should be removed.

If anything, more data should be gathered with more domains in the same sector. Alternatively, pursuing a more thorough treatment would involve obtaining SEMRush visibility data at the page level and correlating this with page-level link feature metrics.

Going forward, we will use traffic_ as our measure of quality.

Link Quality

We start with link quality, which we've very recently discovered should be measured by "traffic_" as opposed to the industry accepted.

Let's start by inspecting the distributive properties of each link feature using the describe() function:

```
competitor_ahrefs_analysis = competitor_ahrefs_clean_dtypes
competitor_ahrefs_analysis[['traffic_']].describe()
```

The resulting table shows some basic statistics including the mean, standard deviation (std), and interquartile metrics (25th, 50th, and 75th percentiles), which give you a good idea of where most referring domains fall in terms of referring domain traffic.

	traffic_
count	5.409000e+03
mean	1.359225e+06
std	4.572404e+07
min	0.000000e+00
25%	0.000000e+00
50%	0.000000e+00
75%	8.400000e+01
max	3.191808e+09

So unsurprisingly, if we look at the median, then most of the competitors' referring domains have zero (estimated) traffic. Only domains in the 75th percentile or above have traffic.

We can also plot (and confirm visually) their distribution using the geom_boxplot function to compare sites side by side:

```
comp_dr_dist_box_plt = (
    ggplot(competitor_ahrefs_analysis, #.loc[competitor_ahrefs_
    analysis['dr'] > 0],
            aes(x = 'reorder(site, traffic_)', y = 'traffic_',
            colour = 'target')) +
```

```
geom_boxplot(alpha = 0.6) +
scale_y_log10() +
theme(legend_position = 'none',
      axis_text_x=element_text(rotation=90, hjust=1)
    ))
```

comp_dr_dist_box_plt.save(filename = 'images/4_comp_traffic_dist_box_plt. png', height=5, width=10, units = 'in', dpi=1000)

comp_dr_dist_box_plt

comp_dr_dist_box_plt compares a site's distribution of referring domain traffic side by side (Figure 5-9) and most notably the interquartile range (IQR). The competitors are in red, and the client is in blue.

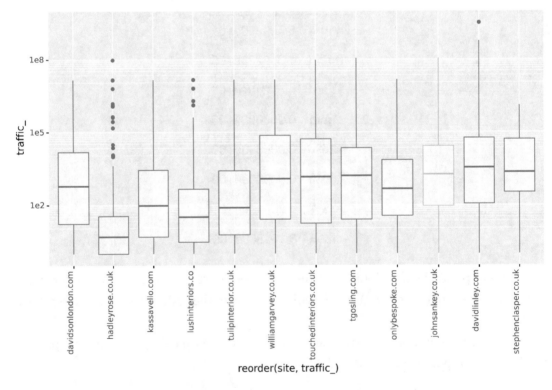

Figure 5-9. *Box plot of each website's backlink traffic (traffic_)*

The interquartile range is the range of data between its 25th percentile and 75th percentile. The purpose is to tell us

- Where most of the data is

- How much of the data is away from the median (the center)

In this case, the IQR is quantifying how much traffic each site's referring domains get and its variability.

We also see that "John Sankey" has the third highest median referring domain traffic which compares well in terms of link quality against their competitors. The size of the box (its IQR) is not the longest (quite consistent around its median) but not as short as Stephen Clasper (more consistent, with a higher median and more backlinks from referring domain sites higher than the median).

"Touched Interiors" has the most diverse range of DR compared with other domains, which could indicate an ever so slightly more relaxed criteria for link acquisition. Or is it the case that as your brand becomes more well known and visible online, this brand has naturally attracted more links from zero traffic referring domains? Maybe both.

Let's plot the domain quality over time for each competitor:

```
comp_traf_timeseries_plt = (
    ggplot(competitor_ahrefs_analysis,
          aes(x = 'first_seen', y = 'traffic_',
              group = 'site', colour = 'site')) +
    geom_smooth(alpha = 0.4, size = 2, se = False,
                method='loess'
               ) +
    scale_x_date() +
    theme(legend_position = 'right',
          axis_text_x=element_text(rotation=90, hjust=1)
         )
)

comp_traf_timeseries_plt.save(filename = 'images/4_comp_traffic_timeseries_
plt.png', height=5, width=10, units = 'in', dpi=1000)

comp_traf_timeseries_plt
```

We deliberately avoided using scale_y_log10() which would have transformed the vertical axis using logarithmic scales. Why? Because it would look very noisy and difficult to see any standout competitors.

Figure 5-10 shows the quality of links acquired over time of which the standout sites are David Linley, T Gosling, and John Sankey.

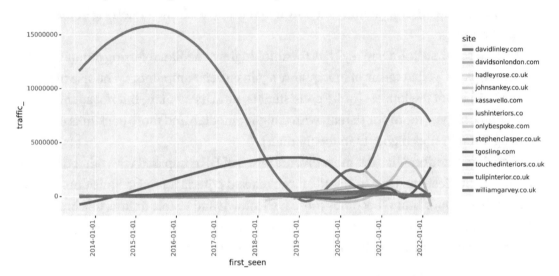

Figure 5-10. *Time series plot showing the amount of traffic each referring domain has over time for each website*

The remaining sites are more or less flat in terms of their link acquisition performance. David Linley started big, then dive-bombed in terms of link quality before improving again in 2020 and 2021.

Now that we have some concept of how the different sites perform, what we really want is a cumulative link quality by month_year as this is likely to be additive in the way search engines evaluate the authority of websites.

We'll use our trusted groupby() and expanding().mean() functions to compute the cumulative stats we want:

```
competitor_traffic_cummean_df = competitor_ahrefs_analysis.copy()
```

```
competitor_traffic_cummean_df = competitor_traffic_cummean_
df.groupby(['site', 'month_year'])['traffic_'].sum().reset_index()
```

```
competitor_traffic_cummean_df['traffic_runavg'] = competitor_traffic_
cummean_df['traffic_'].expanding().mean()
```

```
competitor_traffic_cummean_df
```

This results in the following:

	site	month_year	traffic_	traffic_runavg
0	davidlinley.com	2013-08	5770	5.770000e+03
1	davidlinley.com	2013-09	92	2.931000e+03
2	davidlinley.com	2013-10	32	1.964667e+03
3	davidlinley.com	2013-12	2	1.474000e+03
4	davidlinley.com	2014-02	0	1.179200e+03
...	...	...	...	...
502	williamgarvey.co.uk	2021-11	1940163	1.458292e+07
503	williamgarvey.co.uk	2021-12	14357281	1.458247e+07
504	williamgarvey.co.uk	2022-01	846774	1.455527e+07
505	williamgarvey.co.uk	2022-02	1628704	1.452973e+07
506	williamgarvey.co.uk	2022-03	3234	1.450108e+07

507 rows × 4 columns

Scientific formatted numbers aren't terribly helpful, nor is a table for that matter, but at least the dataframe is in a ready format to power the following chart:

```
competitor_traffic_cummean_plt = (
    ggplot(competitor_traffic_cummean_df, aes(x = 'month_year', y =
'traffic_runavg', group = 'site', colour = 'site')) +
    geom_line(alpha = 0.6, size = 2) +
    labs(y = 'Cumu Avg of traffic_', x = 'Month Year') +
    scale_y_continuous() +
    scale_x_date() +
    theme(legend_position = 'right',
        axis_text_x=element_text(rotation=90, hjust=1)
        ))

competitor_traffic_cummean_plt.save(filename = 'images/4_competitor_
traffic_cummean_plt.png', height=5, width=10, units = 'in', dpi=1000)

competitor_traffic_cummean_plt
```

The code is color coding the sites to make it easier to see which site is which.

So as we might expect, David Linley's link acquisition team has done well as their authority has made leaps and bounds over all of the competitors over time (Figure 5-11).

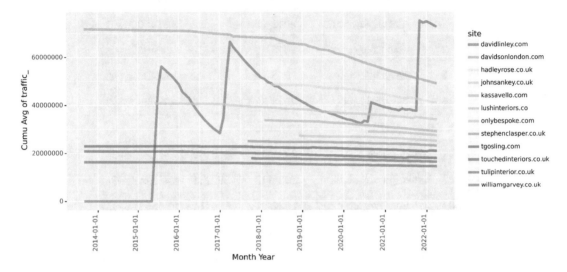

Figure 5-11. *Time series plot of the cumulative average backlink traffic for each website*

All of the other competitors have pretty much flatlined. This is reflected in David Linley's superior SEMRush visibility (Figure 5-12).

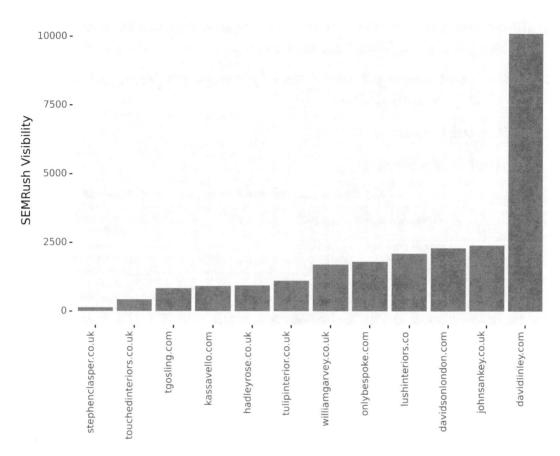

Figure 5-12. *Column chart showing the SEMRush visibility for each website*

What can we learn? So far in our limited data research, we can see that slow and steady does not win the day. By contrast, sites need to be going after links from high traffic sites in a big way.

Link Volumes

That's quality analyzed; what about the volume of links from referring domains?

Our approach will be to compute a cumulative sum of referring domains using the groupby() function:

```
competitor_count_cumsum_df = competitor_ahrefs_analysis
```

```
competitor_count_cumsum_df = competitor_count_cumsum_df.groupby(['site',
'month_year'])['rd_count'].sum().reset_index()
```

The expanding function allows the calculation window to grow with the number of rows, which is how we achieve our cumulative sum:

```
competitor_count_cumsum_df['count_runsum'] = competitor_count_cumsum_
df['rd_count'].expanding().sum()

competitor_count_cumsum_df
```

This results in the following:

	site	month_year	rd_count	count_runsum	link_velocity
0	davidlinley.com	2013-08	11	11.0	NaN
1	davidlinley.com	2013-09	1	12.0	-10.0
2	davidlinley.com	2013-10	1	13.0	0.0
3	davidlinley.com	2013-12	1	14.0	0.0
4	davidlinley.com	2014-02	1	15.0	0.0
...	...	...	...	...	...
502	williamgarvey.co.uk	2021-11	24	5324.0	-15.0
503	williamgarvey.co.uk	2021-12	36	5360.0	12.0
504	williamgarvey.co.uk	2022-01	22	5382.0	-14.0
505	williamgarvey.co.uk	2022-02	19	5401.0	-3.0
506	williamgarvey.co.uk	2022-03	8	5409.0	-11.0

507 rows × 5 columns

The result is a dataframe with the site, month_year, and count_runsum (the running sum), which is in the perfect format to feed the graph – which we will now run as follows:

```
competitor_count_cumsum_plt = (
    ggplot(competitor_count_cumsum_df, aes(x = 'month_year', y =
    'count_runsum',
     group = 'site', colour = 'site')) +
    geom_line(alpha = 0.6, size = 2) +
    labs(y = 'Running Sum of Referring Domains', x = 'Month Year') +
    scale_y_continuous() +
    scale_x_date() +
    theme(legend_position = 'right',
```

```
    axis_text_x=element_text(rotation=90, hjust=1)
  ))
```

```
competitor_count_cumsum_plt.save(filename = 'images/5_count_cumsum_smooth_
plt.png', height=5, width=10, units = 'in', dpi=1000)
```

```
competitor_count_cumsum_plt
```

The competitor_count_cumsum_plt plot (Figure 5-13) shows the number of referring domains for each site since 2014. What is quite interesting are the different starting positions for each site when they start acquiring links.

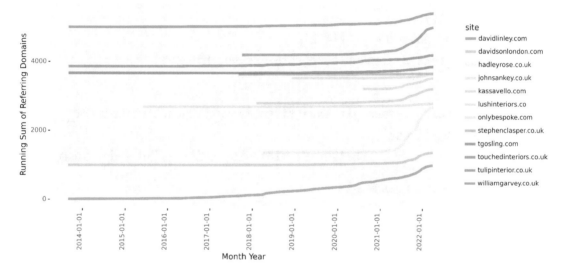

Figure 5-13. *Time series plot of the running sum of referring domains for each website*

For example, William Garvey started with over 5000 domains. I'd love to know who their digital PR team is.

We can also see the rate of growth, for example, although Hadley Rose started link acquisition in 2018, things really took off around mid-2021.

Link Velocity

Let's take a look at link velocity:

```
competitor_velocity_cumsum_plt = (
    ggplot(competitor_count_cumsum_df, aes(x = 'month_year', y = 'link_
    velocity',
     group = 'site', colour = 'site')) +
    geom_line(alpha = 0.6, size = 2) +
    labs(y = 'Running Sum of Referring Domains', x = 'Month Year') +
    scale_y_log10() +
    scale_x_date() +
    theme(legend_position = 'right',
          axis_text_x=element_text(rotation=90, hjust=1)
        ))

competitor_velocity_cumsum_plt.save(filename = 'images/5_competitor_
velocity_cumsum_plt.png',
                         height=5, width=10, units = 'in', dpi=1000)

competitor_velocity_cumsum_plt
```

The view shows the relative speed at which the sites are acquiring links (Figure 5-14). This is an unusual but useful view as for any given month you can see which site is acquiring the most links by virtue of the height of their lines.

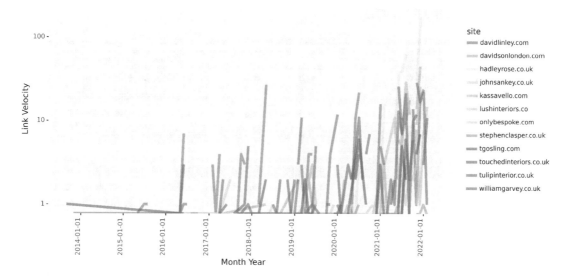

Figure 5-14. *Time series plot showing the link velocity of each website*

David Linley was winning the contest throughout the years until Hadley Rose came along.

Link Capital

Like most things that are measured in life, the ultimate value is determined by the product of their rate and volume. So we will apply the same principle to determine the overall value of a site's authority and call it "link capital."

We'll start by merging the running average stats for both link volume and average traffic (as our measure of authority):

```
competitor_capital_cumu_df = competitor_count_cumsum_df.merge(competitor_
traffic_cummean_df,
                    on = ['site', 'month_year'], how = 'left'
                    )
```

```
competitor_capital_cumu_df['auth_cap'] = (competitor_capital_cumu_
df['count_runsum'] * competitor_capital_cumu_df['traffic_runavg']).
round(1)*0.001
```

```
competitor_capital_cumu_df['auth_velocity'] = competitor_capital_cumu_
df['auth_cap'].diff()
```

```
competitor_capital_cumu_df
```

This results in the following:

	site	month_year	rd_count	count_runsum	link_velocity	traffic_	traffic_runavg	auth_cap	auth_velocity
0	davidlinley.com	2013-08	11	11.0	NaN	5770	5.770000e+03	6.347000e+01	NaN
1	davidlinley.com	2013-09	1	12.0	-10.0	92	2.931000e+03	3.517200e+01	-28.2980
2	davidlinley.com	2013-10	1	13.0	0.0	32	1.964667e+03	2.554070e+01	-9.6313
3	davidlinley.com	2013-12	1	14.0	0.0	2	1.474000e+03	2.063600e+01	-4.9047
4	davidlinley.com	2014-02	1	15.0	0.0	0	1.179200e+03	1.768800e+01	-2.9480
...	...	...	...	...	...	...	...	...	...
502	williamgarvey.co.uk	2021-11	24	5324.0	-15.0	1940163	1.458292e+07	7.763947e+07	216510.7948
503	williamgarvey.co.uk	2021-12	36	5360.0	12.0	14357281	1.458247e+07	7.816206e+07	522585.5084
504	williamgarvey.co.uk	2022-01	22	5382.0	-14.0	846774	1.455527e+07	7.833649e+07	174427.2243
505	williamgarvey.co.uk	2022-02	19	5401.0	-3.0	1628704	1.452973e+07	7.847506e+07	138573.1262
506	williamgarvey.co.uk	2022-03	8	5409.0	-11.0	3234	1.450108e+07	7.843632e+07	-38740.0970

507 rows × 9 columns

The merged table is produced with new columns auth_cap (measuring overall authority) and auth_velocity (the rate at which authority is being added).

Let's see how the competitors compare in terms of total authority over time in Figure 5-15.

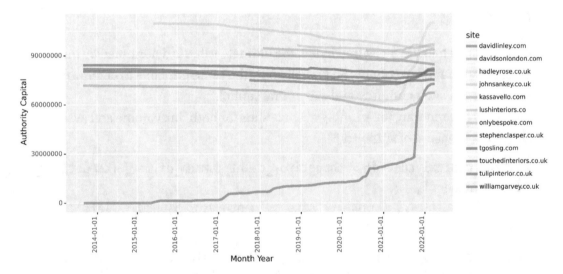

Figure 5-15. *Time series plot of authority capital over time by website*

The plot shows the link capital of several sites over time. What's quite interesting is how Hadley Rose emerged as the most authoritative with the third most consistently highest trafficked backlinking sites with a ramp-up in volume in less than a year. This has allowed them to overtake all of their competitors in the same time period (based on volume while maintaining quality).

What about the velocity in which authority has been added? In the following, we'll plot the authority velocity over time for each website:

```
competitor_capital_veloc_plt = (
    ggplot(competitor_capital_cumu_df, aes(x = 'month_year', y =
    'auth_velocity',
     group = 'site', colour = 'site')) +
    geom_line(alpha = 0.6, size = 2) +
    labs(y = 'Authority Capital', x = 'Month Year') +
    scale_y_continuous() +
    scale_x_date() +
    theme(legend_position = 'right',
          axis_text_x=element_text(rotation=90, hjust=1)
        ))

competitor_capital_veloc_plt.save(filename = 'images/6_auth_veloc_smooth_
plt.png',
                            height=5, width=10, units = 'in', dpi=1000)

competitor_capital_veloc_plt
```

The only standouts are David Linley and Hadley Rose (Figure 5-16). Should David Linley maintain the quality and the velocity of its link acquisition program?

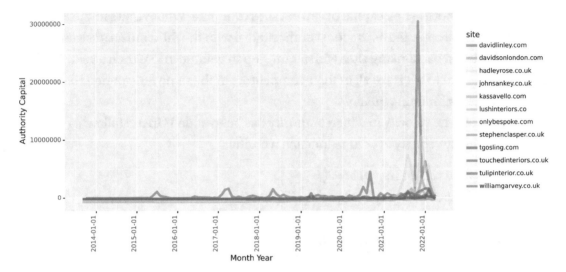

Figure 5-16. *Link capital velocity over time by website*

We're in no doubt that it will catch up and even surpass Hadley Rose, all other things being equal.

Finding Power Networks

A power network in SEO parlance is a group of websites that link to the top ranking sites for your desired keyword(s). So, getting a backlink from these websites to your website will improve your authority and thereby improve your site's ranking potential.

Does it work? From our experience, yes.

Before we go into the code, let's discuss the theory. In 1996, the quality of web search was in its infancy and highly dependent on the keyword(s) used on the page.

In response, Jon Kleinberg, a computer scientist, invented the Hyperlink-Induced Topic Search (HITS) algorithm which later formed the core algorithm for the Ask search engine.

The idea, as described in his paper "Authoritative sources in a hyperlinked environment" (1999), is a link analysis algorithm that ranks web pages for their authority and hub values. Authorities estimate the content value of the page, while hubs estimate the value of its links to other pages.

From a data-driven SEO perspective, we're not only interested in acquiring these links, we're also interested in finding out (in a data-driven manner) what these hubs are.

To achieve this, we'll group the referring domains and their traffic levels to calculate the number of sites:

```
power_doms_strata = competitor_ahrefs_analysis.groupby(['domain',
'traffic_']).agg({'rd_count': 'count'})
power_doms_strata = power_doms_strata.reset_index().sort_values('traffic_',
ascending = False)
```

A referring domain can only be considered a hub or power domain if it links to more than two domains, so we'll filter out those that don't meet the criteria. Why three or more? Because one is random, two is a coincidence, and three is directed.

```
power_doms_strata = power_doms_strata.loc[power_doms_strata['rd_
count'] > 2]

power_doms_strata
```

This results in the following:

	domain	traffic_	rd_count
3763	sitelike.org	14357011	11
1827	idcrawl.com	23024	10
3766	siteprice.org	1291812	9
1	1-2-3-4-5.com	25	9
1337	firmania.co.uk	2485	8
...	...	...	...
479	bizify.co.uk	4523	3
3891	storyandtoy.com	0	3
1748	homify.es	113111	3
1518	goldsir.ru	3	3
673	cgmood.com	6167	3

156 rows × 3 columns

The table shows referring domains, their traffic, and the number of (our furniture) sites that these backlinking domains are linking to.

Being data driven, we're not satisfied with a list, so we'll use statistics to help understand the distribution of power before filtering the list further:

```
pd.set_option('display.float_format', str)
power_doms_stats = power_doms_strata.describe()

power_doms_stats
```

This results in the following:

	traffic_	rd_count
count	156.0	156.0
mean	1015822.032051282	3.9038461538461537
std	7774169.411800991	1.4669645911413565
min	0.0	3.0
25%	0.75	3.0
50%	405.0	3.0
75%	51093.75	4.0
max	94579570.0	11.0

We see the distribution is heavily positively skewed where most of the highly trafficked referring domains are in the 75th percentile or higher. Those are the ones we want. Let's visualize:

```
power_doms_stats_plt = (
    ggplot(power_doms_strata, aes(x = 'traffic_')) +
    geom_histogram(alpha = 0.6, binwidth = 10) +
    labs(y = 'Power Domains Count', x = 'traffic_') +
    scale_y_continuous() +
    theme(legend_position = 'right',
          axis_text_x=element_text(rotation = 90, hjust=1)
          ))
```

```
power_doms_stats_plt.save(filename = 'images/7_power_doms_stats_plt.png',
                          height=5, width=10, units = 'in', dpi=1000)
```

```
power_doms_stats_plt
```

As mentioned, the distribution is massively skewed, which is more apparent from the histogram. Finally, we'll filter the domain list for the most powerful:

```
power_doms = power_doms_strata.loc[power_doms_strata['traffic_'] > power_
doms_stats['traffic_'][-2]]
```

Although we're interested in hubs, we're sorting the dataframe by traffic as these have the most authority:

```
power_doms = power_doms.sort_values('traffic_', ascending = False)
```

```
power_doms
```

This results in the following:

	domain	traffic_	rd_count
885	dailymail.co.uk	94579570	3
3763	sitelike.org	14357011	11
3135	pinterest.co.uk	14142143	6
3865	standard.co.uk	9931955	4
4614	yell.com	6335301	4
4106	thetimes.co.uk	4977287	3
1829	idealhome.co.uk	2005577	3
1726	homesandgardens.com	1975639	3
3766	siteprice.org	1291812	9
2619	minimalis.co.id	1187798	6
1743	homify.com.mx	867528	3
2389	livingetc.com	554980	3
1740	homify.com.br	485936	3
1781	houseandgarden.co.uk	462721	4
3767	sitesimilar.net	460603	3
2613	milesia.id	440193	4
1321	find-open.co.uk	411864	8
4139	thomsonlocal.com	383648	5
1751	homify.in	341049	3
1747	homify.de	287989	3
817	countryandtownhouse.co.uk	276936	5

By far, the most powerful is the daily mail, so in this case start budgeting for a good digital PR consultant or full-time employee. There are also other publisher sites like the Evening Standard (standard.co.uk) and The Times.

Some links are easier and quicker to get such as the yell.com and Thomson local directories.

Then there are more market-specific publishers such as the Ideal Home, Homes and Gardens, Livingetc, and House and Garden.

This should probably be your first port of call.

This analysis could be improved further in a number of ways, for example:

- Going more granular by looking for power pages (single backlink URLs that power your competitors)

- Checking the relevance of the backlink page (or home page) to see if it impacts visibility and filtering for relevance

- Combining relevance with traffic for a combined score for hub filtering

Taking It Further

Of course, the preceding discussion is just the tip of the iceberg, as it's a simple exploration of one site so it's very difficult to infer anything useful for improving rankings in competitive search spaces.

The following are some areas for further data exploration and analysis:

- Adding social media share data to destination URLs, referring domains, and referring pages

- Correlating overall site visibility with the running average referring domain traffic over time

- Plotting the distribution of referring domain traffic over time

- Adding search volume data on the hostnames to see how many brand searches the referring domains receive as an alternative measure of authority

- Joining with crawl data to the destination URLs to test for

  - Content relevance

  - Whether the page is indexable by confirming the HTTP response (i.e., 200)

Naturally, the preceding ideas aren't exhaustive. Some modeling extensions would require an application of the machine learning techniques outlined in Chapter 6.

Summary

Backlinks, the expression of website authority for search engines, are incredibly influential to search result positions for any website. In this chapter, you have learned about

- What site authority is and how it impacts SEO

- How brand searches could impact search visibility

- Single site analysis

- Competitor authority analysis

  - *Link anatomy*: How R^2 showed referring domain traffic was more of a predictor than domain rating for explaining visibility

  - How analyzing multiple sites adds richness and context to authority insights

- In both single and multiple site analyses

  - Authority – distribution and over time

  - Link volumes and velocity

In the next chapter, we will use data science to analyze keyword search result competitors.

CHAPTER 6

Competitors

What self-respecting SEO doesn't do competitor analysis to find out what they're missing? Back in 2007, Andreas recalls using spreadsheets collecting data on SERPs with columns representing aspects of the competition, such as the number of links to the home page, number of pages, word counts, etc. In hindsight, the idea was right, but the execution was near hopeless because of the difficulty of Excel to perform a statistically robust analysis in the short time required – something you will now learn shortly using machine learning.

And Algorithm Recovery Too!

The applications are not only useful for general competitor SEO analysis but also recovering from Google updates, especially when you don't have copies of SERPs data preceding the update to contrast what worked before to what works now.

If you did have the SERPs data leading up to the update, then you'd simply repeat the analysis for the before SERPs and compare the analysis results to the after SERPs.

Defining the Problem

Before we rush in, let's think about the challenge. With over 10,000 ranking factors, there isn't enough time nor budget to learn and optimize for the high-priority SEO items.

We propose to find the ranking factors that will make the decisive difference to your SEO campaign by cutting through the noise and using machine learning on competitor data to discover

- Which ranking factors can best explain the differences in rankings between sites

- What the winning benchmarks are

- How much a unit change in the factor is worth in terms of rank

A. Voniatis, *Data-Driven SEO with Python*, https://doi.org/10.1007/978-1-4842-9175-7_6

Outcome Metric

Let it be written that the outcome variable should be search engine ranking in Google. This approach can be adapted for any other search engine (be it Bing, Yandex, Baidu, etc.) as long as you can get the data from a reliable rank checking tool.

Why Ranking?

Because unlike user sessions, it doesn't vary according to the weather, the time of year, and so on – Query Freshness excepted. It's probably the cleanest metric. In any case, the ranking represents the order in which content of the ranking URL best satisfies the search query – the point of RankBrain, come to think of it. So in effect, we are working out how to optimize for any Google update informed by RankBrain.

From a data perspective, the ranking position must be a floating numeric data type known as a "float" in Python ("double" in R).

Features

Now that we have established the outcome metric, we must now determine the independent variables, the model inputs also known as features. The data types for the feature will vary, for example:

> first_paint_ms would be numeric.

> flesch_kincaid_reading_ease would be a character.

Naturally, you want to cover as many meaningful features as possible, including technical, content/UX, and offsite, for the most comprehensive competitor research.

Data Strategy

Now that we know the outcome and features, what to do? Given that rankings are numeric and that we want to explain the difference in rank which is a continuous variable (i.e., one flows into two, then into three, etc.), then competitor analysis in this instance is a regression problem. This means in mathematical terms

```
rank ~ w_1*feature_1 + w_2*feature_2 + … + w_n*feature_n
```

~ means explained by.

n is the nth feature.

w is the weighting of the feature.

To be clear, this is not always a linear regression exercise. Linear regression assumes all features will behave in a linear fashion – data points will all fit along a straight line. While this may be true in most cases for some features like PageSpeed, this will not be true for other features.

For example, a lot of ranking factors behave nonlinearly for some sectors. For example, the number of characters for a title tag is usually nonlinear such that there is a sweet spot.

As shown in Figure 6-1, we can see that the line of best fit is n-shaped showing the rank to get higher as we approach 28% of title characters featuring the site title and lower the further a website deviates from that winning benchmark for title tag branding proportion.

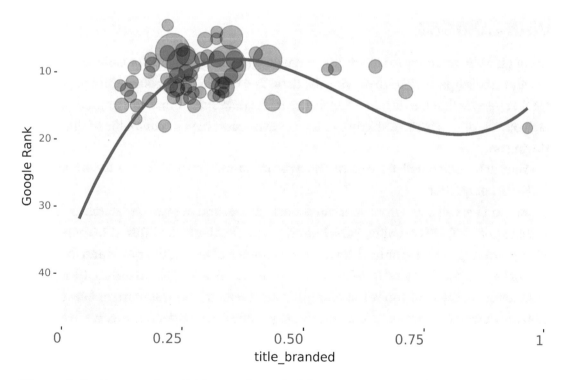

Figure 6-1. *Scatterplot of title tags branded (as a proportion of total characters) and average Google rank position*

In terms of what to collect features on, search engines rank URLs, not domains, and therefore we will focus on the former. This will save you money and time in terms of data collection costs as well as putting the actual data together. However, the downside is that you won't be able to include domain-wide features such as ranking URL distance from the home page.

To that end, we will be using a decision tree–based algorithm known as "random forest." There are other algorithms such as decision tree, better ones like AdaBoost, and XGBoost. A data scientist will typically experiment with a number of models and then pick the best one in terms of model stability and predictive accuracy.

However, we're here to get you started; the models are likely to produce similar results, and so we're saving you time while delivering you, most importantly, the intuition behind the machine learning technique for analyzing your SERP competitors for SEO.

Data Sources

Although we're not privy to Google's internal data (unless you work for Google), we rely heavily on third-party tools to provide the data. The reason the tools are third party and not first party is that the data for all websites in the data study must be comparable to each other – unless in the unlikely event you have access to your competitors' data. No? Moving on.

Your data sources will depend on the type of features you wish to test for modeling the SERP competitors.

Rank: This will come from your rank checking tool and not Google Search Console. So that's getSTAT, SEO Monitor, Rank Ranger or the DataForSEO SERPs API. There are others, although we have or have no direct experience of testing their APIs and thus cannot be mentioned. Why those three? Because they all allow you to export the top 100 URLs for every keyword you're including in your research. This is important because from the outset we don't want to assume who your SERPs competitors are. We just want to extract the data and interrogate it.

For the features:

Onsite: To test whether onsite factors can explain the differences in rank for your keyword set, use features like title tag length, page speed, number of words, reading ease, and anything your rank checking tool can tell you about a URL. You can also derive your own features such as title relevance by calculating the string distance between the title tag and the target keyword. Rest assured, we'll show you how later.

For less competitive industries and small datasets (less than 250,000 URLs per site), a tool like Screaming Frog or Sitebulb will do. For large datasets and competitive industries, it's most likely that your competitors will block desktop site auditors, so you will have to resort to an enterprise-grade tool that crawls from the cloud and has an API. We have personally found Botify, not only to have both but also to work well because most enterprise brands use them so they won't get blocked, when it comes to crawling!

Offsite: To test the impact of offsite factors, choose a reliable source with a good API. In our experience, AHREFs and BuzzSumo work well, yielding metrics such as the domain rating, number of social shares by platform, and number of internal links on the backlinking URLs. Both have APIs which allow you to automate the collection of offsite data into your R workspace.

Explore, Clean, and Transform

Now that you have the data, data science practice dictates that you explore the data to

- *Understand the distribution*: Is it skewed or multimodal (i.e., multiple peaks in the distribution)?

- *Examine the quality of data*: Are there too many NAs? Single-level factors?

- Discover new features for derivation

The idea is to improve the quality of the data you're going to feed into your model by discarding features and rows as not all of them will be informative or useful. Exploring the data will also help you understand the limits of your model for explaining the ranking factors in your search query space.

Before joining onto the SERPs data, let's explore.

To summarize the overall approach

1. Import data – both rankings and features

2. Focus on the competitors

3. Join the data

4.   Derive new features

5.   Single-level factors (SLFs)

6.   Rescale your data

7.   Near Zero Variance (NZVs)

8.   Median impute

9.   One hot encoding (OHE)

10.   Eliminate NAs

11.   Model the SERPs

12.   Evaluate the SERPs ML model

13.   The most predictive drivers of rank

14.   How much rank a ranking factor is worth

15.   The winning benchmark for a ranking factor

Naturally, there is a lot to cover, so we will explain each of these briefly and go into more detail over the more interesting secrets ML can uncover on your competitors.

Import Data – Both SERPs and Features

This can be done by importing CSV downloads from the relevant SEO tools or, for a more automated experience, using your tool provider's API into dataframes (Python's version of a spreadsheet). Some starter code is shown as follows for importing CSV data:

For regular expressions (regex), although string methods include regex by default:

```
import re
import time
import random
import pandas as pd
import numpy as np
import datetime
import re
import time
import requests
```

```
import json
from datetime import timedelta
```

For importing multiple files:

```
from glob import glob
import os
```

String methods used to compute the overlap between two text strings:

```
from textdistance import sorensen_dice
```

To extract parts of a URL:

```
import uritools
from tldextract import extract
```

For visualizing data:

```
from plotnine import *
import matplotlib.pyplot as plt
pd.set_option('display.max_colwidth', None)
%matplotlib inline
```

Some variables are initiated at the start, so that when you reuse the script on another client or site, you simply overwrite the following variable values:

```
root_url = 'https://www.johnsankey.co.uk'
target_site = 'www.johnsankey.co.uk'
root_domain = 'johnsankey.co.uk'
hostname = 'johnsankey'
target_name = 'sankey'
geo_market = 'uk'
```

A list of social sites used for subsetting or filtering the data:

```
social_sites = ['facebook.com', 'instagram.com', 'linkedin.com', 'twitter.
com', 'pinterest.com', 'tiktok.com', 'foursquare.com', 'reddit.com']
```

Start with the Keywords

As with virtually all things in SEO, start with the end in mind. That usually means the target keywords you want your site to rank for and therefore their SERPs which we will load into a Pandas dataframe:

```
serps_raw = pd.read_csv('data/keywords_serps.csv')
```

To make the dataframe easier to handle, we'll use a list comprehension to turn the column names into lowercase and replace punctuation marks and spaces with underscores:

```
serps_raw.columns = [x.lower().replace(' ', '_').replace(',', '_').
replace('__', '_') for x in serps_raw.columns]
```

The rank_absolute column is replaced by the more simplified and familiar "rank":

```
serps_raw = serps_raw.rename(columns = {'rank_absolute': 'rank'})

serps_raw
```

This results in the following:

	keyword	rank	url	location_code	language_code	se_results_count
0	msofas uk	1	https://www.msofas.co.uk/	2826	en	98
1	msofas uk	2	https://uk.trustpilot.com/review/www.msofas.co.uk	2826	en	98
2	msofas uk	3	https://www.facebook.com/Msofas/	2826	en	98
3	msofas uk	4	https://www.reviews.co.uk/company-reviews/store/msofas-co-uk	2826	en	98
4	msofas uk	5	https://www.safebuy.org.uk/business/companies/msofas	2826	en	98
...	...	...	...	...	...	...
25730	luxury fabrics	99	https://www.grahamsandersoninteriors.com/fabrics	2826	en	71000000
25731	luxury fabrics	100	https://www.kitepackaging.co.uk/blog/how-to-pack-luxury-fabrics-and-home-accessories-the-packaging-doctor/	2826	en	71000000
25732	luxury fabrics	101	NaN	2826	en	71000000
25733	luxury fabrics	1	https://www.luxuryfabricsltd.co.uk/	2826	en	71000000
25734	luxury fabrics	102	NaN	2826	en	71000000

25735 rows × 6 columns

The serps_raw dataframe has over 25,000 rows of SERPs data with 6 columns, covering all of the keywords:

```
serps_df = serps_raw.copy()
```

The URL column is set as a string for easier data manipulation:

```
serps_df['url'] = serps_df['url'].astype(str)
```

The first manipulation is to extract the domain for the "site" column. The site column will apply the uritools API function to strip of the slug and then split the URL into a list of its components using a list comprehension:

```
serps_df['site'] = [uritools.urisplit(x).authority if uritools.isuri(x)
else x for x in serps_df['url']]
```

Once split, we will extract everything in the list, taking the last three components:

```
serps_df['site'] = ['.'.join(s.split('.')[-3:]) for s in serps_df['site']]
```

Next, we want to profile the rank into strata, so that we can have rank categories. While this may not be used in this particular exercise, it's standard practice when working with SERPs data.

```
serps_df['rank_profile'] = np.where(serps_df['rank'] < 11, 'page_1',
'page_2')
serps_df['rank_profile'] = np.where(serps_df['rank'] < 3, 'top_3',
                                    serps_df['rank_profile'])
```

Rather than have zero search_volumes, we'll set these to one to avoid divide by zero errors using np.where():

```
serps_df['se_results_count'] = np.where(serps_df['se_results_count'] == 0,
1, serps_df['se_results_count'])
```

We'll set a "count" column for quicker aggregation:

```
serps_df['count'] = 1
```

We'll also count the number of keywords in a search string, known in the data science world as tokens:

```
serps_df['token_count'] = serps_df['keyword'].str.count(' ') + 1
```

These will then be categorized into head, middle, and tail, based on the token length:

```
before_length_conds = [
    serps_df['token_count'] == 1,
```

```
    serps_df['token_count'] == 2,
    serps_df['token_count'] > 2]

length_vals = ['head', 'middle', 'long']

serps_df['token_size'] = np.select(before_length_conds, length_vals)

serps_df
```

This results in the following:

	keyword	rank	url	location_code	language_code	se_results_count	site	rar
0	msofas uk	1	https://www.msofas.co.uk/	2826	en	98	msofas.co.uk	
1	msofas uk	2	https://uk.trustpilot.com/review/www.msofas.co.uk	2826	en	98	uk.trustpilot.com	
2	msofas uk	3	https://www.facebook.com/Msofas/	2826	en	98	www.facebook.com	
3	msofas uk	4	https://www.reviews.co.uk/company-reviews/store/msofas-co-uk	2826	en	98	reviews.co.uk	
4	msofas uk	5	https://www.safebuy.org.uk/business/companies/msofas	2826	en	98	safebuy.org.uk	
...	...	...	...	...	...	...	...	
25730	luxury fabrics	99	https://www.grahamsandersoninteriors.com/fabrics	2826	en	71000000	www.grahamsandersoninteriors.com	
25731	luxury fabrics	100	https://www.kitepackaging.co.uk/blog/how-to-pack-luxury-fabrics-and-home-accessories-the-packaging-doctor/	2826	en	71000000	kitepackaging.co.uk	
25732	luxury fabrics	101	nan	2826	en	71000000	nan	
25733	luxury fabrics	1	https://www.luxuryfabricsltd.co.uk/	2826	en	71000000	luxuryfabricsltd.co.uk	
25734	luxury fabrics	102	nan	2826	en	71000000	nan	

25735 rows × 12 columns

Focus on the Competitors

The SERPs data effectively tells us what content is being rewarded by Google where the rank is the outcome metric. However, much of this data is likely to be noisy, and a few of the columns are likely to have ranking factors that explain the difference in ranking between content.

The content is noisy because SERPs are likely to contain content from sites (such as reviews and references) which will prove very difficult for the commercial sites to learn from. Ultimately, when conducting this exercise, SEO is primarily interested in outranking competitor sites before these other sites become a consideration.

So, you'll want to select your competitors to make your study more meaningful. For example, if your client or your brand is in the webinar technology space, it won't make sense to include Wikipedia.com or Amazon.com in your dataset as they don't directly compete with your brand.

What you really want are near-direct to direct competitors, that is, doppelgangers, so that you can compare what it is they do or don't do to rank higher or lower than you.

The downside of this approach is that you don't get to appreciate what Google wants from the SERPs by stripping out noncompetitors. That's because the SERPs need to be analyzed as a whole, which is covered to an extent in Chapter 10. However, this chapter is about competitor analysis, so we shall proceed.

To find the competitors, we'll have to perform some aggregations starting with calculating reach (i.e., the number of content with positions in the top 10):

```
major_players_reach = serps_df.loc[serps_df['rank'] < 10]
```

With the SERPs filtered or limited to the top 10, we'll aggregate the total number of top 10s by site, using groupby site and summing the count column:

```
major_players_reach = major_players_reach.groupby('site').agg({'count': sum}).reset_index()
```

Then we'll sort the sites in descending order of reach:

```
major_players_reach = major_players_reach.sort_values('count', ascending = False)
```

The reach metric is most of the story by giving us the volume, but we also want the rank which is calculated by taking the median. This will help order sites with comparable levels of reach.

```
major_players_rank = serps_df.groupby('site').agg({'rank': 'median'}).reset_index()
major_players_rank = major_players_rank.sort_values('rank')
```

Aggregating by search engine result count helps to give a measure of the competitiveness of the keyword overall, which is aggregated by the mean value. This data is uniquely provided by DataForSEO's SERPs API. However, you could easily substitute this with the more usual search volume metric provided by other SERPs trackers such as SEO Monitor.

```
major_players_searches = serps_df.groupby('site').agg({'se_results_count':
'mean'}).reset_index()
major_players_searches = major_players_searches.sort_values('se_
results_count')
```

The rank and search result aggregations are joined onto the reach data to form one table using the merge() function. This is equivalent to a vlookup using the site column as the basis of the merge:

```
major_players_stats = major_players_reach.merge(major_players_rank, on =
'site', how = 'left')
major_players_stats = major_players_stats.merge(major_players_searches, on
= 'site', how = 'left')
```

Using all the data, we'll compute an overall visibility metric which divides the reach squared by the rank. The reach is squared to avoid sites with a few top 10s and very high rankings appearing at the top of the list.

The rank is the divisor because the higher the rank, the lower the number; therefore, dividing by a lower number will increase the value of the site's visibility should it rank higher:

```
major_players_stats['visibility'] = ((major_players_stats['reach'] ** 2) /
major_players_stats['rank']).round()
```

The social media sites are excluded to focus on commercial competitors:

```
major_players_stats = major_players_stats.loc[~major_players_stats['site'].
str.contains('|'.join(social_sites))]
```

Rename count to reach:

```
major_players_stats = major_players_stats.rename(columns = {'count':
'reach'})
```

Remove sites with nan values:

```
major_players_stats = major_players_stats.loc[major_players_stats['site']
!= 'nan']
```

```
major_players_stats.head(10)
```

This results in the following:

	site	reach	rank	se_results_count	visibility
1	furniturevillage.co.uk	60	12.0	3.279325e+08	300.0
2	dfs.co.uk	55	17.0	3.266742e+08	178.0
3	thesofaandchair.co.uk	54	20.0	2.708906e+08	146.0
4	wayfair.co.uk	49	12.0	5.283340e+08	200.0
5	darlingsofchelsea.co.uk	49	12.0	2.108414e+08	200.0
6	ebay.co.uk	45	21.0	2.998662e+08	96.0
7	www.sofasandstuff.com	42	27.0	2.922718e+08	65.0
8	sofology.co.uk	41	17.0	3.231102e+08	99.0
9	www.made.com	36	30.0	4.384029e+08	43.0
10	www.sofa.com	30	26.0	2.927668e+08	35.0

The dataframe shows the top 20 feature sites we would expect to see dominating the SERPs. A few of the top sites are not direct competitors (will probably be uncrawlable!), so these will be removed, as we're interested in learning from the most direct competitors to see their most effective SEO.

As a result, we will select the most direct competitors and store these in a list "player_sites_lst":

```
player_sites_lst = ['sofology.co.uk', 'www.designersofas4u.co.uk', 'www.heals.com', 'darlingsofchelsea.co.uk', 'www.made.com', 'www.sofasandstuff.com', 'www.arloandjacob.com', 'loaf.com', 'www.made.com', 'theenglishsofacompany.co.uk', 'willowandhall.co.uk', root_domain]
```

The list will be used to filter the SERPs to contain only content from these direct competitors:

```
direct_players_stats = major_players_stats.loc[major_players_stats['site'].isin(player_sites_lst)]

direct_players_stats
```

This results in the following:

	site	reach	rank	se_results_count	visibility
5	darlingsofchelsea.co.uk	49	12.0	2.108414e+08	200.0
7	www.sofasandstuff.com	42	27.0	2.922718e+08	65.0
8	sofology.co.uk	41	17.0	3.231102e+08	99.0
9	www.made.com	36	30.0	4.384029e+08	43.0
11	willowandhall.co.uk	30	27.0	1.465901e+08	33.0
21	theenglishsofacompany.co.uk	20	22.0	1.492079e+08	18.0
24	www.heals.com	19	30.0	2.629755e+08	12.0
38	www.arloandjacob.com	11	31.0	3.179770e+08	4.0
52	johnsankey.co.uk	7	60.0	2.253633e+08	1.0
53	loaf.com	7	50.0	3.558439e+08	1.0

The dataframe shows that Darlings of Chelsea is the leading site to "beat" with the most reach and the highest rank on average.

Let's visualize this:

```
major_players_stats_plt = (
    ggplot(direct_players_stats,
           aes(x = 'reach', y = 'rank', fill = 'site', colour = 'site',
                       size = 'se_results_count')) +
    geom_point(alpha = 0.8) +
    geom_text(direct_players_stats, aes(label = 'site'), position=position_
    stack(vjust=-0.08)) +
    labs(y = 'Google Rank', x = 'Google Reach') +
    scale_y_reverse() +
  scale_size_continuous(range = [5, 20]) +
    theme(legend_position = 'none', axis_text_x=element_text(rotation=0,
    hjust=1, size = 12))
)

major_players_stats_plt.save(filename = 'images/1_major_players_stats_
plt.png',
                        height=5, width=8, units = 'in', dpi=1000)
major_players_stats_plt
```

Although Darlings of Chelsea leads the luxury sector, Made.com has the most presence on the more highly competitive keywords as signified by the size of their data point (se_results_count) (Figure 6-2).

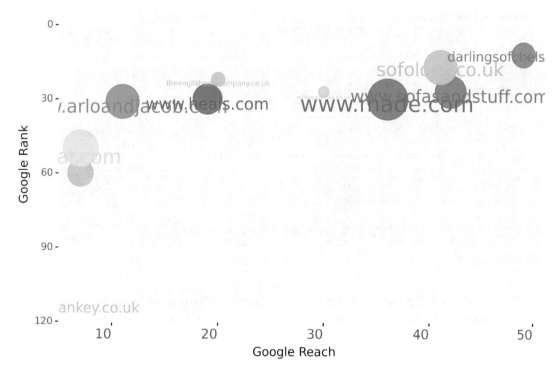

Figure 6-2. *Bubble chart comparing Google's top 10s (reach) of each website and their average Google ranking position. Circle size represents the number of search results of the queries each site appears for*

John Sankey on the other hand is the lowest ranking and has the least reach.

Filtering the SERPs data for just the direct competitors, this will make the data less noisy:

```
player_serps = serps_df[serps_df['site'].isin(player_sites_lst)]

player_serps
```

This results in the following:

	keyword	rank	url	se_results_count	site	count	token_count	token_size	compression	connecti
0	msofas uk	8	https://www.sofasandstuff.com/	98	www.sofasandstuff.com	1	2	middle	Brotli	No D:
1	msofas uk	55	https://www.sofology.co.uk/stores/cheltenham	98	sofology.co.uk	1	2	middle	Gzipped	No D:
2	msofas uk	77	https://www.made.com/sofas-and-armchairs/velvet-sofas	98	www.made.com	1	2	middle	NaN	N
3	cheap sofas derby	63	https://www.sofology.co.uk/	1070000	sofology.co.uk	1	3	long	Gzipped	No D:
4	cheap sofas derby	83	https://loaf.com/meet-the-makers/the-long-eaton-sofa-story	1070000	loaf.com	1	3	long	Brotli	No D:
...	...	...	...	...	...	...	...	...	...	...
1289	sofa with low arms	28	https://www.sofology.co.uk/leather-sofas	40300000	sofology.co.uk	1	4	long	Gzipped	No D:
1290	sofa with low arms	29	https://www.arloandjacob.com/sofas	40300000	www.arloandjacob.com	1	4	long	Gzipped	No D:
1291	sofa with low arms	30	https://www.made.com/sofas-and-armchairs/sofas	40300000	www.made.com	1	4	long	NaN	N
1292	luxury fabrics	42	https://www.johnsankey.co.uk/order-a-fabric-sample/	71000000	johnsankey.co.uk	1	2	middle	Gzipped	No D:
1293	luxury fabrics	91	https://www.sofasandstuff.com/house-and-designer-fabrics	71000000	www.sofasandstuff.com	1	2	middle	Brotli	No D:

1294 rows × 175 columns

We end with a dataframe with far less rows from 25,000 to 1,294. Although machine learning algorithms would generally work best with 10,000 rows or more, the methods are still superior (in terms of insight speed and consistency) to working manually in a spreadsheet.

We're analyzing the most relevant sites, and we can proceed to collect data on those sites. These will form our hypotheses which will form the possible ranking factors that explain the differences in ranking between sites.

Site crawls provide a rich source of data as they contain information about the content and technical SEO characteristics of the ranking pages, which will be our starting point.

We'll start by defining a function to export the URLs for site crawling to a CSV file:

```
def export_crawl(df):
    dom_name = df.domain.iloc[0]
    df = df[['url']]
    df = df[['url']].drop_duplicates()
    df.to_csv('data/1_to_crawl/' + dom_name + '_crawl_urls.csv',
    index=False)
```

The function is applied to the filtered SERPs dataframe using the groupby() function:

```
direct_players_stats.groupby(site).apply(export_crawl)
```

This results in the following:

	site	reach	rank	se_results_count	visibility
5	darlingsofchelsea.co.uk	49	12.0	2.108414e+08	200.0
7	www.sofasandstuff.com	42	27.0	2.922718e+08	65.0
8	sofology.co.uk	41	17.0	3.231102e+08	99.0
9	www.made.com	36	30.0	4.384029e+08	43.0
11	willowandhall.co.uk	30	27.0	1.465901e+08	33.0
21	theenglishsofacompany.co.uk	20	22.0	1.492079e+08	18.0
24	www.heals.com	19	30.0	2.629755e+08	12.0
38	www.arloandjacob.com	11	31.0	3.179770e+08	4.0
52	johnsankey.co.uk	7	60.0	2.253633e+08	1.0
53	loaf.com	7	50.0	3.558439e+08	1.0

Once the data is crawled, we can store the exports in a folder and read them in, one by one.

In this instance, we set the file path as a variable named "crawl_path":

```
crawl_path = 'data/2_crawled/'
crawl_filenames = os.listdir(crawl_path)
crawl_filenames

['www_johnsankey_co_uk.csv',
 'www_sofasandstuff_com.csv',
 'loaf_com.csv',
 'www_designersofas4u_co_uk.csv',
 'www_theenglishsofacompany_co_uk.csv',
 'www_willowandhall_co_uk.csv',
 'www_darlingsofchelsea_co_uk.csv',
 'www_sofology_co_uk.csv',
 'www_arloandjacob_com.csv']

crawl_df_lst = list()
crawl_colnames = list()
```

A for loop is used to go through the list of website auditor CSV exports and read the data into a list:

```
for filename in crawl_filenames:
    df = pd.read_csv(crawl_path + filename)
    df['sitefile'] = filename
    df['sitefile'] = df['sitefile'].str.replace('_', '.', regex = False)
    df['sitefile'] = df['sitefile'].str.replace('.csv', '', regex = False)
    crawl_colnames.append(df.columns)
    crawl_df_lst.append(df)
```

This list "crawl_df_lst" is then combined into a dataframe:

```
crawl_raw = pd.concat(crawl_df_lst)
```

The column names are made more data-friendly by removing formatting and converting the column names to lowercase:

```
crawl_raw.columns = [col.lower().replace(' ', '_').replace(')', '_').
replace('(', '_').replace(',', '_').replace(':', '_').replace('.', '_').
replace('__', '_') for col in crawl_raw.columns]

crawl_raw
```

This results in the following:

	url	base_url	crawl_dep
0	https://www.johnsankey.co.uk/product/fender-stool-95cm-wide-35cm-high-70cm-deep/	No Data	Not S
1	https://www.johnsankey.co.uk/collections/cushion/	No Data	Not S
2	https://www.johnsankey.co.uk/product-groups/beckett/	No Data	Not S

```
crawl_df = crawl_raw.copy()
```

Getting the site name will help us aggregate the data by site. Using a list comprehension, we'll loop through the dataframe URL column and apply the urisplit() function:

```
crawl_df['site'] = [uritools.urisplit(uri).authority if uritools.isuri(uri)
else uri for uri in crawl_df['url']]
```

Filter for HTML URLs only:

```
crawl_df = crawl_df.loc[crawl_df['content_type'] == 'HTML']

crawl_df
```

This results in the following:

	url	base_url	crawl_depth	crawl_status		host	http_protoco
0	https://www.johnsankey.co.uk/product/fender-stool-95cm-wide-35cm-high-70cm-deep/		No Data	Not Set	Success	www.johnsankey.co.uk	h:
1		https://www.johnsankey.co.uk/collections/cushion/	No Data	Not Set	Success	www.johnsankey.co.uk	h:
2		https://www.johnsankey.co.uk/product-groups/beckett/	No Data	Not Set	Success	www.johnsankey.co.uk	h:

Printing the data types using the .dtypes property helps us see which columns require potential conversion into more usable data types:

```
print(crawl_df.dtypes)
```

We can see from the following printed list highlighted in blue that there are numeric variables that are in object format but should be a float64 and will therefore require conversion.

This results in the following:

```
stylesheets_with_unused_css                  int64
total_wasted_css_kib_                        float64
total_wasted_css_percentage                  float64
total_wasted_js_kib_                         float64
total_wasted_js_percentage                   float64
cumulative_layout_shift                      object
first_contentful_paint                       object
largest_contentful_paint                     object
performance_score                            object
time_to_interactive                          object
total_blocking_time                          object
perf_budget_document                         object
perf_budget_fonts                            object
perf_budget_images                           object
perf_budget_media                            object
perf_budget_other                            object
```

We'll create a copy of the dataframe to create a new one that will have converted columns:

```
cleaner_crawl = crawl_df.copy()
```

Starting with reading time, we'll replace no data with the current timing format using np.where():

```
cleaner_crawl['reading_time_mm_ss_'] = np.where(cleaner_crawl['reading_
time_mm_ss_'] == 'No Data', '00:00',cleaner_crawl['reading_time_mm_ss_'])
cleaner_crawl['reading_time_mm_ss_'] = '00:' +cleaner_crawl['reading_time_
mm_ss_']
```

And convert it to a timing format:

```
cleaner_crawl['reading_time_mm_ss_'] = pd.to_timedelta(cleaner_
crawl['reading_time_mm_ss_']).dt.total_seconds()
```

We'll convert other string format columns to float, by first defining a list of columns to be converted:

```
float_cols = ['cumulative_layout_shift', 'first_contentful_paint',
'largest_contentful_paint', 'performance_score',
        'time_to_interactive', 'total_blocking_time']
```

Using the list, we'll use the apply column and the to_numeric() function to convert the columns:

```
cleaner_crawl[float_cols] = cleaner_crawl[float_cols].apply(pd.to_numeric,
errors='coerce')
```

We'll now view the recently converted columns:

```
cleaner_crawl[['url', 'reading_time_mm_ss_'] + float_cols]
```

This results in the following:

	url	reading_time_mm_ss_	cumulative_layout_shift	first_contentful_paint	larg
0	https://www.johnsankey.co.uk/product/fender-stool-95cm-wide-35cm-high-70cm-deep/	120.0	0.030	2913.0	
1	https://www.johnsankey.co.uk/collections/cushion/	59.0	0.010	2365.0	
2	https://www.johnsankey.co.uk/product-groups/beckett/	109.0	0.000	2360.0	
3	https://www.johnsankey.co.uk/collections/chairs/	64.0	0.011	2559.0	
4	https://www.johnsankey.co.uk/product-groups/milliner/	111.0	0.000	2724.0	
...	...	...	...	...	
505	https://www.arloandjacob.com/know-how-how-to-pick-the-perfect-snuggler/	180.0	0.205	7267.0	
508	https://www.arloandjacob.com/corner-sofas-v-normal-sofas/	204.0	0.205	6098.0	
580	https://www.arloandjacob.com/skin/frontend/rwd/arloandjacob/images/favicons/192x192.png	0.0	NaN	NaN	
582	https://www.arloandjacob.com/skin/frontend/rwd/arloandjacob/images/favicons/128x128.png	0.0	NaN	NaN	
583	https://www.arloandjacob.com/skin/frontend/rwd/arloandjacob/images/favicons/96x96.png	0.0	NaN	NaN	

242 rows × 8 columns

The columns are correctly formatted. For more advanced features, you may want to try segmenting the different types of content according to format, such as blogs, guides, categories, subcategories, items, etc.

For further features, we shall import backlink authority data. First, we'll import the data by reading all of the AHREFs data in the folder:

```
# read loop files in folder
authority_path = 'data/4_ahrefs/'
authority_filenames = os.listdir(authority_path)

authority_filenames
```

The list of AHREFs files is set ready to iterate through:

```
['darlingsofchelsea.co.uk-best-pages-by-links-subdomains-12-
Sep-2022_19-17-56.csv',
 'sofology.co.uk-best-pages-by-links-subdomains-12-Sep-2022_19-24-00.csv',
 'sofasandstuff.com-best-pages-by-links-subdomains-12-
Sep-2022_19-23-42.csv',
 'willowandhall.co.uk-best-pages-by-links-subdomains-12-
Sep-2022_19-17-03.csv',
 'theenglishsofacompany.co.uk-best-pages-by-links-subdomains-12-
Sep-2022_19-24-53.csv',
 'arloandjacob.com-best-pages-by-links-subdomains-12-
Sep-2022_19-23-19.csv',
 'designersofas4u.co.uk-best-pages-by-links-subdomains-12-
Sep-2022_19-16-38.csv',
```

```
'johnsankey.co.uk-best-pages-by-links-subdomains-12-
Sep-2022_19-16-04.csv',
'heals.com-best-pages-by-links-subdomains-12-Sep-2022_19-17-32.csv',
'loaf.com-best-pages-by-links-subdomains-12-Sep-2022_19-25-17.csv']
```

Initialize lists to contain the outputs of the for loop:

```
auth_df_lst = list()
auth_colnames = list()
```

The for loop reads in the data using the read_csv function, stores the filename as a column (so we know which file the data comes from), cleans up the column names, and adds the data to the lists created earlier:

```
for filename in authority_filenames:
    df = pd.read_csv(authority_path + filename, encoding = 'UTF-16',
    sep = '\t')
    df['sitefile'] = filename
    df['sitefile'] = df['sitefile'].str.replace('.csv', '', regex = False)
    df.columns = [x.lower().replace(' ', '_').replace('(', '_').
    replace(')', '_').replace(' ', '_').
                    replace('__', '_') for x in df.columns]
    df['sitefile'] = df['sitefile'].str.extract('(.*?)\-')
    auth_colnames.append(df.columns)
    print(df['sitefile'][0])
    auth_df_lst.append(df)
```

Once the loop has run, the lists are combined into a single dataframe using the concat() function:

```
auth_df_raw = pd.concat(auth_df_lst)
auth_df_raw = auth_df_raw.rename(columns = {'page_url': 'url'})
auth_df_raw.drop(['#', 'size', 'code', 'crawl_date', 'language',
'page_title',
                    'first_seen'], axis = 1, inplace = True)
auth_df_raw
```

The resulting auth_df_raw dataframe is shown as follows with the site pages and their backlink metrics.

	url_rating	url	referring_domains_desc_	dofollow	nofollow	redirects	
0	48	https://www.darlingsofchelsea.co.uk/	522	1663	993	13	darlingsofch
1	42	http://www.darlingsofchelsea.co.uk/	278	1210	476	1	darlingsofch
2	0	http://blog.darlingsofchelsea.co.uk/wp-content/uploads/2013/04/John-Lewis-6-Burner-Cabinet.jpg	85	121	1	0	darlingsofch
3	30	https://darlingsofchelsea.co.uk/	71	88	166	0	darlingsofch
4	0	https://www.darlingsofchelsea.co.uk/blogdata/wp-content/uploads/2017/04/francesca_large_corner_sofa_with_chaise_in_20jh_leather.jpg	59	74	3	0	darlingsofch
...	...	...	...	...	...	...	
995	0	https://assets.loaf.com/images/product_400/4071794-pantry-kitchen-table-in-pale-grey.jpg	3	5	0	0	
996	0	https://assets.loaf.com/images/product_400/4180301-oat-brushed-cotton-easy-squeeze-armchair.jpg	3	4	0	0	
997	0	https://assets.loaf.com/images/product_400/4370604-be-amt-9842.jpg	3	4	0	0	
998	0	https://assets.loaf.com/images/product_400/4373178-st-mim-0091.jpg	3	6	0	0	
999	0	https://assets.loaf.com/images/product_400/4436536-412151-sugar-bum-loveseat.jpg	3	4	0	0	

9325 rows × 7 columns

Join the Data

Now that the data from their respective tool sources are imported, they are now ready to be joined. Usually, the common column (known as the "primary key") between datasets is the URL as that is what search engines rank.

We'll start by joining the SERPs data to the crawl data. Before we do, we only require the SERPs containing the competitor sites.

```
player_serps = serps_df[serps_df['site'].isin(player_sites_lst)]
```

The vlookup to join competitor SERPs and the crawl data of the ranking URLs is achieved by using the .merge() function:

```
player_serps_crawl = player_serps.merge(cleaner_crawl, on = ['url'], how = 'left')
player_serps_crawl = player_serps_crawl.rename(columns = {'site_x': 'site'})
```

Drop unnecessary columns using the .drop() function:

```
player_serps_crawl.drop(['site_y', 'sitefile'], axis = 1, inplace = True)
player_serps_crawl
```

This results in the following:

	keyword	rank	url	location_code	language_code	se_results_count	site	rank_profile	branded	c
0	msofas uk	8	https://www.sofasandstuff.com/	2826	en	98	www.sofasandstuff.com	page_1	generic	
1	msofas uk	55	https://www.sofology.co.uk/stores/cheltenham	2826	en	98	sofology.co.uk	page_2	generic	
2	msofas uk	77	https://www.made.com/sofas-and-armchairs/velvet-sofas	2826	en	98	www.made.com	page_2	generic	

The next step is to join the backlink authority data to the dataset containing SERPs and crawl metrics, again using the merge() function:

```
player_serps_crawl_auth = player_serps_crawl.copy()
player_serps_crawl_auth = player_serps_crawl_auth.merge(auth_df_raw, on =
['url'], how = 'left')
player_serps_crawl_auth.drop(['sitefile'], axis = 1, inplace = True)
player_serps_crawl_auth
```

The data has now been joined such that each SERP URL has its onsite and offsite SEO data in a single dataframe:

	keyword	rank	url	location_code	language_code	se_results_count	site	rank_profile	branded	cour
0	msofas uk	8	https://www.sofasandstuff.com/	2826	en	98	www.sofasandstuff.com	page_1	generic	
1	msofas uk	55	https://www.sofology.co.uk/stores/cheltenham	2826	en	98	sofology.co.uk	page_2	generic	
2	msofas uk	77	https://www.made.com/sofas-and-armchairs/velvet-sofas	2826	en	98	www.made.com	page_2	generic	

Derive New Features

The great thing about combining the data is that you can derive new features that you wouldn't perhaps get from the individual datasets. For example, a highly useful feature would be to compare the similarity (or dissimilarity) of the title tag to the target search phrase. This uses the title tag of the ranking URL from the crawl data and the SERPs keyword. The new features give us additional hypotheses to test using the machine learning processes later on.

We'll start by making a new dataframe and derive a number of new data features which will be stored as additional columns:

```
hypo_serps_features = player_serps_crawl_auth.copy()
```

Add regional_tld which denotes whether the ranking URL is regionalized or not:

```
regional_tlds = ['.uk']
hypo_serps_features['regional_tld'] = np.where(hypo_serps_features['site'].
str.contains('|'.join(regional_tlds)), 1, 0)
```

Add a metric for measuring how much of the target keyword is used in the title tag using the sorensen_dice() function:

```
hypo_serps_features['title'] = hypo_serps_features['title'].astype(str)
```

```
hypo_serps_features['title_relevance'] = hypo_serps_features.loc[:,
['title', 'keyword']].apply(
    lambda x: sorensen_dice(*x), axis=1)
```

We're also interested in measuring the extent to which title tags and H1 heading consistency are influential:

```
hypo_serps_features['h1'] = hypo_serps_features['h1'].astype(str)
hypo_serps_features['title_h1'] = hypo_serps_features.loc[:, ['title',
'h1']].apply(
    lambda x: sorensen_dice(*x), axis=1)
```

Does having a brand in your title tag matter? Let's find out:

```
hypo_serps_features['site'] = hypo_serps_features['site'].astype(str)
hypo_serps_features['hostname'] = hypo_serps_features['site'].apply(lambda
x: extract(x))
hypo_serps_features['hostname'] = hypo_serps_features['hostname'].
str.get(1)
hypo_serps_features['title'] = hypo_serps_features['title'].str.lower()
hypo_serps_features['title_branded'] = hypo_serps_features.loc[:,
                                                      ['title',
'hostname']].apply(
    lambda x: sorensen_dice(*x), axis=1)
```

Another useful feature is URL parameters, that is, question marks in the ranking URL:

```
hypo_serps_features['url_params'] = np.where(hypo_serps_features['url'].
str.contains('\?'), '1', '0')
hypo_serps_features['url_params'] = hypo_serps_features['url_params'].
astype('category')
```

Another test is whether the ranking URL has Google Analytics code. It's unlikely to amount to anything, but if the data is available, why not?

```
hypo_serps_features['google_analytics_code'] = np.where(hypo_serps_
features[
    'google_analytics_code'].str.contains('UA'), '1', '0')
```

```
hypo_serps_features['google_analytics_code'] = hypo_serps_features['google_
analytics_code'].astype('category')
```

The same goes for Google Tag Manager code:

```
hypo_serps_features['google_tag_manager_code'] = np.where(
    hypo_serps_features['google_tag_manager_code'].str.contains('GTM'),
    '1', '0')
hypo_serps_features['google_tag_manager_code'] = hypo_serps_
features['google_tag_manager_code'].astype('category')
```

While tracking code in itself is unlikely to explain differences in rank, having a duplicate instance of the same tracking code might:

```
hypo_serps_features['google_tag_manager_code_second_'] = np.where(
    hypo_serps_features['google_tag_manager_code_second_'].str.
    contains('GTM'), '1', '0')
hypo_serps_features['google_tag_manager_code_second_'] = hypo_serps_
features[
    'google_tag_manager_code_second_'].astype('category')
```

A test for cache control is added to check for whether it's private, public, or other and converted to a category:

```
hypo_serps_features['cache_privacy'] = np.where(
    hypo_serps_features['cache-control'].str.contains('private'),
    'private', '0')
hypo_serps_features['cache_privacy'] = np.where(
    hypo_serps_features['cache-control'].str.contains('public'), 'public',
    hypo_serps_features['cache_privacy'])
hypo_serps_features['cache_privacy'] = np.where(
    hypo_serps_features['cache-control'].str.contains('0'), 'other', hypo_
    serps_features['cache_privacy'])
hypo_serps_features['cache_privacy'] = hypo_serps_features['cache_
privacy'].astype('category')
```

A cache age has also been added by extracting the numerical component of the cache-control string. This is achieved by splitting the string on the "=" sign and then using the .get() function, before converting to a numerical float data type:

```
hypo_serps_features['cache_age'] = hypo_serps_features['cache-control'].
str.split('\=')
hypo_serps_features['cache_age'] = hypo_serps_features['cache_age'].
str.get(-1)
hypo_serps_features['cache_age'] = np.where(hypo_serps_features['cache_
age'].isnull(), 0,
                                      hypo_serps_
                                      features['cache_age'])
hypo_serps_features['cache_age'] = np.where(hypo_serps_features['cache_
age'].str.contains('[a-z]'),
                                      0, hypo_serps_
                                      features['cache_age'])
hypo_serps_features['cache_age'] = hypo_serps_features['cache_age'].
astype(float)
```

Here's a test for whether the ranking URL is canonical or not:

```
hypo_serps_features['self_canonicalised'] = np.where(hypo_serps_features[
    'canonical_url'] == hypo_serps_features['url'], 1, 0)
```

We drop identifiers such as the canonical URL as these are individual records that identify a single row which will add nothing to the analysis.

We're only interested in the characteristics or trend of this unique data value in itself.

We also drop hypotheses which are likely to be redundant or not interested in testing, such as the HTTP protocol. This relies on your own SEO experience and judgment.

```
hypo_serps_features.drop(['cache-control', 'canonical_url', 'base_url',
'crawl_status', 'host', 'encoding', 'indexable_status', 'meta_robots_
response_', 'title', 'title_response_', 'title_second_', 'title_render_
status', 'meta_description', 'meta_description_response_', 'h1', 'h1_
second_', 'h2', 'h2_second_', 'open_graph_audio', 'twitter_card_site',
'twitter_card_creator', 'twitter_card_description', 'twitter_card_image_
url', 'twitter_card_title', 'content-security-policy', 'referrer-policy',
'hostname', 'open_graph_description', 'open_graph_image_url', 'open_graph_
locale', 'open_graph_site_name', 'open_graph_title' , 'open_graph_url',
```

```
'meta_robots_rendered_', 'twitter_card_description', 'twitter_card_title',
'http_protocol', 'http_status_code' ], axis = 1, inplace = True)
```

Once done, we'll create another copy of the dataframe and export as CSV in preparation for machine learning, starting with single-level factors:

```
hypo_serps_pre_slf = hypo_serps_features.copy()
```

```
hypo_serps_pre_slf.to_csv('data/'+ geo_market +'_hypo_serps_pre_slf.csv',
index = False)
```

```
hypo_serps_pre_slf
```

This results in the following:

	keyword	rank	url	location_code	language_code	se_results_count	site	rank_profile	branded	c
0	msofas uk	8	https://www.sofasandstuff.com/	2826	en	98	www.sofasandstuff.com	page_1	generic	
1	msofas uk	55	https://www.sofology.co.uk/stores/cheltenham	2826	en	98	sofology.co.uk	page_2	generic	
2	msofas uk	77	https://www.made.com/sofas-and-armchairs/velvet-sofas	2826	en	98	www.made.com	page_2	generic	
3	cheap sofas derby	63	https://www.sofology.co.uk/	2826	en	1070000	sofology.co.uk	page_2	generic	
4	cheap sofas derby	83	https://loaf.com/meet-the-makers/the-long-eaton-sofa-story	2826	en	1070000	loaf.com	page_2	generic	
...	...	...	...	...	...	...	...	...	...	
1289	sofa with low arms	28	https://www.sofology.co.uk/leather-sofas	2826	en	40300000	sofology.co.uk	page_2	generic	
1290	sofa with low arms	29	https://www.arloandjacob.com/sofas	2826	en	40300000	www.arloandjacob.com	page_2	generic	
1291	sofa with low arms	30	https://www.made.com/sofas-and-armchairs/sofas	2826	en	40300000	www.made.com	page_2	generic	
1292	luxury fabrics	42	https://www.johnsankey.co.uk/order-a-fabric-sample/	2826	en	71000000	johnsankey.co.uk	page_2	generic	
1293	luxury fabrics	91	https://www.sofasandstuff.com/house-and-designer-fabrics	2826	en	71000000	www.sofasandstuff.com	page_2	generic	

1294 rows × 340 columns

Single-Level Factors (SLFs)

A single-level factor is any column which has the same value throughout, which would not only be redundant, the machine learning code would fail. For example, all of the ranking URL titles might be branded, in which case, these should be removed.

To remove SLFs, we'll iterate through the dataframe column by column to identify any column that has data containing 70% or more of the same value and store the column names in a list. 70% is an arbitrary threshold; you could choose 80% or 90%, for example; however, that comes with a risk of removing some insightful ranking factors – even if it only applies to a smaller number of URLs which might ironically be the top ranking URLs.

```
slf_cols = []
slf_limit = .7

for col in hypo_serps_pre_slf.columns:
    if hypo_serps_pre_slf[col].value_counts().iat[0] >= (hypo_serps_pre_
    slf.shape[0] * slf_limit):
        slf_cols.append(col)

slf_cols
```

The columns with 70% identical data are printed as follows and will be removed from the dataset:

```
['location_code',
 'language_code',
 'rank_profile',
 'branded',
 'count',
 'crawl_depth',
 'is_subdomain',
 'no_query_string_keys',
 'query_string_contains_filtered_parameters',
 'query_string_contains_more_than_three_keys',
 'query_string_contains_paginated_parameters',
 'query_string_contains_repetitive_parameters',
 'query_string_contains_sorted_parameters',
 'scheme',...
```

Let's examine a few of these SLF ranking factors using the groupby() function. Starting with branded, we can see all of the ranking URL titles are branded, so these can be removed:

```
hypo_serps_pre_slf.groupby('branded')['count'].sum().sort_values()
```

```
branded
generic    1294
Name: count, dtype: int64
```

```
hypo_serps_pre_slf.groupby('url_params')['count'].sum().sort_values
```

Parameterized URLs also appear to be redundant with only 17 URLs that are parameterized. However, these may still provide insight in unexpected ways.

```
<bound method Series.sort_values of url_params
0     1277
1       17
Name: count, dtype: int64>
```

Having identified the SLFs, we'll process these in a new dataframe where these will be removed using a list comprehension:

```
hypo_serps_pre_mlfs = hypo_serps_pre_slf.copy()
```

The list of columns to be removed is nuanced further as we'd like to keep url_params as mentioned earlier and the count column for further aggregation in future processes:

```
slf_cols = [elem for elem in slf_cols if not elem in ['count', 'url_
params']]
```

Drop the SLF columns:

```
hypo_serps_pre_mlfs.drop(slf_cols, axis = 1, inplace = True)
```

```
hypo_serps_pre_mlfs
```

The resulting hypo_serps_pre_mlfs dataframe has the SLF columns removed:

	keyword	rank	url	se_results_count	site	count	token_count	token_size	compression	connecti
0	msofas uk	8	https://www.sofasandstuff.com/	98	www.sofasandstuff.com	1	2	middle	Brotli	No Da
1	msofas uk	55	https://www.sofology.co.uk/stores/cheltenham	98	sofology.co.uk	1	2	middle	Gzipped	No Da
2	msofas uk	77	https://www.made.com/sofas-and-armchairs/velvet-sofas	98	www.made.com	1	2	middle	NaN	N
3	cheap sofas derby	63	https://www.sofology.co.uk/	1070000	sofology.co.uk	1	3	long	Gzipped	No Da
4	cheap sofas derby	83	https://loaf.com/meet-the-makers/the-long-eaton-sofa-story	1070000	loaf.com	1	3	long	Brotli	No Da
...	...	...	...	...	...	...	...	...	...	
1289	sofa with low arms	28	https://www.sofology.co.uk/leather-sofas	40300000	sofology.co.uk	1	4	long	Gzipped	No Da
1290	sofa with low arms	29	https://www.arloandjacob.com/sofas	40300000	www.arloandjacob.com	1	4	long	Gzipped	No Da
1291	sofa with low arms	30	https://www.made.com/sofas-and-armchairs/sofas	40300000	www.made.com	1	4	long	NaN	N
1292	luxury fabrics	42	https://www.johnsankey.co.uk/order-a-fabric-sample/	71000000	johnsankey.co.uk	1	2	middle	Gzipped	No Da
1293	luxury fabrics	91	https://www.sofasandstuff.com/house-and-designer-fabrics	71000000	www.sofasandstuff.com	1	2	middle	Brotli	No Da

1294 rows × 175 columns

Rescale Your Data

Whether you're using linear models like linear regression or decision trees, both benefit from rescaling the data, because the data becomes "normalized," making it easier for the ML to detect variation when comparing rank with page speed, for example:

```
hypo_serps_preml_prescale = hypo_serps_pre_mlfs.copy()
```

Separate columns into numeric and nonnumeric so we can rescale the numeric columns using .dtypes:

```
hypo_serps_preml_num = hypo_serps_preml_prescale.select_
dtypes(include=np.number)
hypo_serps_preml_num_colnames = hypo_serps_preml_num.columns
```

Nonnumeric columns are saved into a separate dataframe, which will be used for joining later:

```
hypo_serps_preml_nonnum = hypo_serps_preml_prescale.select_
dtypes(exclude=np.number)
```

277

hypo_serps_preml_num

The resulting hypo_serps_preml_num is shown as follows, which includes the numeric columns only, ready for rescaling:

	rank	se_results_count	count	token_count	expires_date	no_cookies	file_size_kib_	total_page_size_kib_	no_canonical_links	total_canonicals	no_internal_
0	8	98	1	2	0.0	3.0	154.51	6777.26	0.0	0.0	
1	55	98	1	2	0.0	7.0	798.63	5168.04	1.0	1.0	
2	77	98	1	2	NaN	NaN	NaN	NaN	NaN	NaN	
3	63	1070000	1	3	1200.0	32.0	993.53	4273.68	1.0	1.0	
4	83	1070000	1	3	0.0	8.0	124.01	1024.95	1.0	1.0	
...	...	...	...	...	...	...	...	...	...	...	
1289	28	40300000	1	4	1200.0	34.0	940.58	5459.99	1.0	1.0	
1290	29	40300000	1	4	0.0	16.0	428.77	1733.56	1.0	1.0	
1291	30	40300000	1	4	NaN	NaN	NaN	NaN	NaN	NaN	
1292	42	71000000	1	2	0.0	8.0	514.73	1841.09	1.0	1.0	
1293	91	71000000	1	2	0.0	14.0	116.04	4453.60	1.0	1.0	

1294 rows × 136 columns

We'll make use of the MinMaxScaler() from the preprocessing functions of the sklearn API:

```
from sklearn import preprocessing
```

Convert the column values into a numpy array and then use the MinMaxScaler() function to rescale the data:

```
x = hypo_serps_preml_num.values
min_max_scaler = preprocessing.MinMaxScaler()
x_scaled = min_max_scaler.fit_transform(x)
hypo_serps_preml_num_scaled = pd.DataFrame(x_scaled, index=hypo_serps_
preml_num.index, columns = hypo_serps_preml_num_colnames)
```

hypo_serps_preml_num_scaled

This results in the following:

	rank	se_results_count	count	token_count	expires_date	no_cookies	file_size_kib_	total_page_size_kib_	no_canonical_links	total_canonicals	no_inte
0	0.069307	1.186552e-09	0.0	0.2	0.000000	0.00000	0.090331	0.154479	0.0	0.0	
1	0.534653	1.186552e-09	0.0	0.2	0.000000	0.12500	0.481670	0.114133	1.0	1.0	
2	0.752475	1.186552e-09	0.0	0.2	NaN	NaN	NaN	NaN	NaN	NaN	
3	0.613861	7.052868e-05	0.0	0.4	0.001984	0.90625	0.600083	0.091710	1.0	1.0	
4	0.811881	7.052868e-05	0.0	0.4	0.000000	0.15625	0.071801	0.010257	1.0	1.0	
...	...	...	...	...	...	...	...	...	...	...	
1289	0.267327	2.656554e-03	0.0	0.6	0.001984	0.96875	0.567913	0.121453	1.0	1.0	
1290	0.277228	2.656554e-03	0.0	0.6	0.000000	0.40625	0.256960	0.028024	1.0	1.0	
1291	0.287129	2.656554e-03	0.0	0.6	NaN	NaN	NaN	NaN	NaN	NaN	
1292	0.405941	4.680285e-03	0.0	0.2	0.000000	0.15625	0.309185	0.030720	1.0	1.0	
1293	0.891089	4.680285e-03	0.0	0.2	0.000000	0.34375	0.066959	0.096220	1.0	1.0	

1294 rows × 136 columns

Near Zero Variance (NZVs)

The next stage is to eliminate redundant numerical data columns which are similar to SLFs known as Near Zero Variance (NZVs). While the values are different, there may not be much variation that can reliably explain the differences in ranking positions, and we will therefore want these removed.

To identify NZVs, we'll use the VarianceThreshold function from the SK Learn API:

```
from sklearn.feature_selection import VarianceThreshold
```

```
variance = hypo_serps_preml_num_scaled.var()
columns = hypo_serps_preml_num_scaled.columns
```

Save the names of variables having variance more than a threshold value:

```
highvar_variables = [ ]
nz_variables = [ ]
```

We'll iterate through the numeric columns setting the threshold at 7% such that there must be at least 7% variation in the data to remain in the dataset. Again, 7% is an arbitrary choice. The high variation columns are stored in the list we created earlier called highvar_variables:

```
for i in range(0,len(variance)):
    if variance[i]>=0.07:
        highvar_variables.append(columns[i])
    else:
```

```
        nz_variables.append(columns[i])
```

```
highvar_variables
```

The highvar_variables are shown as follows:

```
['rank',
 'no_canonical_links',
 'total_canonicals',
 'no_internal_followed_linking_urls',
 'no_internal_followed_links',
 'no_internal_linking_urls',
 'no_internal_links_to_url',
 'url_rank', ...]
```

```
nz_variables
```

This results in the following:

```
['se_results_count',
 'count',
 'token_count',
 'expires_date',
 'no_cookies',
 'file_size_kib_',
 'total_page_size_kib_', ...]
```

The NZVs identified and stored in nz_variables are shown earlier. We can see that more web pages, for example, have highly similar numbers of keywords in the search query ("token count") and HTML page sizes ("total_page_size_kib_"), so we'll be happy to remove these.

Here's a quick sanity check to ensure there are no columns that are listed as both high variation and NZV:

```
[x for x in highvar_variables if x in nz_variables]
```

An empty list is returned, so thankfully there is no crossover:

```
[]
```

Let's examine a couple of the NZV columns identified. Although identified as an NZV, the title relevance has some variation as shown in the following using the describe() function. We can see the data ranges from 0 to 1 and has an interquartile range of 0.32 to 0.62, which is of course after rescaling. We'll keep "title_relevance" as from SEO experience, it is an important ranking factor:

```
hypo_serps_preml_num_scaled['title_relevance'].describe()
```

This results in the following:

```
count    1294.000000
mean        0.478477
std         0.199106
min         0.000000
25%         0.323529
50%         0.512712
75%         0.622487
max         1.000000
Name: title_relevance, dtype: float64
```

The scaled_images column on the other hand is NZV, as shown in the following, where most values are zero until the 75th percentile of 0.17 showing very little variation and should therefore be excluded:

```
hypo_serps_preml_num_scaled['scaled_images'].describe()
```

This results in the following:

```
count     977.000000
mean        0.114348
std         0.163530
min         0.000000
25%         0.000000
50%         0.000000
75%         0.179487
max         1.000000
Name: scaled_images, dtype: float64
```

We'll redefine the highvar_variables list to include some NZVs we think should remain in the dataset:

```
highvar_variables = highvar_variables + ['title_relevance', 'title_
branded', 'no_content_words', 'first_contentful_paint', 'scaled_images',
'no_outgoing_links']
```

Save a new dataframe to include only columns listed in highvar_variables:

```
hypo_serps_preml_num_highvar = hypo_serps_preml_num_scaled[highvar_
variables]
hypo_serps_preml_num_highvar
```

The hypo_serps_preml_num_highvar df is shown as follows and has gone from 136 to 38 columns, removing 98 columns.

	rank	no_canonical_links	total_canonicals	no_internal_followed_linking_urls	no_internal_followed_links	no_internal_linking_urls	no_internal_links_to_url
0	0.069307	0.0	0.0	0.966667	0.496454	0.966667	0.496454
1	0.534653	1.0	1.0	0.000000	0.000000	0.000000	0.000000
2	0.752475	NaN	NaN	NaN	NaN	NaN	NaN
3	0.613861	1.0	1.0	0.533333	0.113475	0.533333	0.113475
4	0.811881	1.0	1.0	0.000000	0.000000	0.000000	0.000000
...	...	...	...	...	...	...	...
1289	0.267327	1.0	1.0	0.533333	0.397163	0.533333	0.397163
1290	0.277228	1.0	1.0	0.766667	1.000000	0.766667	1.000000
1291	0.287129	NaN	NaN	NaN	NaN	NaN	NaN
1292	0.405941	1.0	1.0	0.733333	0.177305	0.733333	0.177305
1293	0.891089	1.0	1.0	0.033333	0.007092	0.033333	0.007092

1294 rows × 38 columns

Next, we'll also remove ranking factors that are highly correlated to each other (known as multicollinearity), using the variance_inflation_factor() function from the statsmodels API to detect large variance inflation factors (VIF).

Multicollinearity is an issue because it reduces the statistical significance of the ranking features used to model the search result rankings.

A large variance inflation factor (VIF) on a ranking feature or any modeling variable hints at a highly correlated relationship to other ranking factors. Removing those variables will improve the model's predictive consistency, that is, more stable and less degree of error when making forecasts.

```
from statsmodels.stats.outliers_influence import variance_inflation_factor
```

Remove rows with missing values (np.nan) and infinite values (np.inf, -np.inf):

```
vif_input = hypo_serps_preml_num_highvar[~hypo_serps_preml_num_highvar.
isin([np.nan, np.inf, -np.inf]).any(1)]
```

Store in X_variables.

```
X_variables = vif_input
```

Determine columns that are highly correlated by applying the variance_inflation_factor() function:

```
vif_data = pd.DataFrame()
vif_data["feature"] = X_variables.columns
vif_data["vif"] = [variance_inflation_factor(X_variables.values, i) for i
in range(len(X_variables.columns))]
vif_data.sort_values('vif')
```

The VIF data distribution using the describe() function is printed to get an idea of what level of intercolumn correlation is and act as our threshold for rejecting columns:

```
vif_data['vif'].describe()
```

This results in the following:

```
count       38.000000
mean              inf
std               NaN
min          3.254763
25%         26.605281
50%         76.669063
75%       3504.833113
max               inf
Name: vif, dtype: float64
```

Having determined the VIF range, we'll discard any ranking factor with a VIF above the median. Technically, best practice is that a VIF of five or above is highly correlated; however, in this case, we're just looking to remove excessively correlated ranking factors, which is still an improvement:

```
hypo_serps_preml_lowvif = hypo_serps_preml_num_highvar.copy()

vif_exclude_df = vif_data.loc[vif_data['vif'] > vif_data['vif'].median()]
vif_exclude_cols = vif_exclude_df['feature'].tolist()

hypo_serps_preml_lowvif.drop(vif_exclude_cols, inplace = True, axis = 1)

hypo_serps_preml_lowvif
```

We've now gone from 38 to 19 columns. As you may come to appreciate by now, machine learning is not simply a case of plugging in the numbers to get a result as much work must be done to get the numbers into a usable format.

	rank	no_anchors_with_no_text	no_outgoing_navigation_links	documents_files_	scripts_with_unused_js	total_wasted_js_kib_	performance_score	no_go
0	0.069307	0.000000	0.696884	0.000000	0.023256	0.029986	0.191919	
1	0.534653	0.007692	0.453258	0.833333	0.581395	0.837929	0.101010	
2	0.752475	NaN	NaN	NaN	NaN	NaN	NaN	
3	0.613861	0.007692	0.657224	0.833333	0.000000	0.000000	1.000000	
4	0.811881	0.030769	0.000000	0.000000	0.186047	0.222888	0.313131	
...	...	...	...	...	...	...	...	
1289	0.267327	0.007692	0.487252	0.833333	0.651163	0.824791	0.101010	
1290	0.277228	0.000000	0.971671	0.833333	0.139535	0.805868	0.151515	
1291	0.287129	NaN	NaN	NaN	NaN	NaN	NaN	
1292	0.405941	0.015385	0.133144	0.166667	0.534884	0.424763	0.181818	
1293	0.891089	0.000000	0.643059	0.333333	0.023256	0.057513	0.101010	

1294 rows × 19 columns

Median Impute

We want to retain as many rows of data as possible as any rows with missing values in any column will have to be removed.

One technique is to use median impute where the median value for a given column of data will be estimated to replace the missing value.

Of course, the median is likely to be more meaningful if it is calculated at the domain level rather than an entire column, as we're pitting sites against each other. So where possible, we will use median impute at the domain level, otherwise at the column level.

Import libraries to detect data types used in the for loop to detect columns that are not numeric for median imputation:

```
from pandas.api.types import is_string_dtype
from pandas.api.types import is_numeric_dtype

hypo_serps_preml_median = hypo_serps_preml_lowvif.copy()
```

Variables are set so that the for loop can groupby() the entire column and at the domain level ("site"):

```
hypo_serps_preml_median['site'] = hypo_serps_preml_prescale['site']
hypo_serps_preml_median['project'] = 'competitors'

for col, col_val in hypo_serps_preml_median.iteritems():
    if col in ['http_status_code'] or not is_numeric_dtype(hypo_serps_
    preml_median[col]):
        continue
    hypo_serps_preml_median[col].fillna(hypo_serps_preml_median.
    groupby('site')[col].transform('median'), inplace=True)
    hypo_serps_preml_median[col].fillna(hypo_serps_preml_median.
    groupby('project')[col].transform('median'), inplace=True)

hypo_serps_preml_median.drop(['site', 'project'], axis = 1, inplace = True)

hypo_serps_preml_median
```

The result is a dataframe with less missing values, improving data retention.

	rank	no_anchors_with_no_text	no_outgoing_navigation_links	documents_files_	scripts_with_unused_js	total_wasted_js_kib_	performance_score	no_go
0	0.069307	0.000000	0.696884	0.000000	0.023256	0.029986	0.191919	
1	0.534653	0.007692	0.453258	0.833333	0.581395	0.837929	0.101010	
2	0.752475	0.007692	0.453258	0.166667	0.139535	0.297817	0.262626	
3	0.613861	0.007692	0.657224	0.833333	0.000000	0.000000	1.000000	
4	0.811881	0.030769	0.000000	0.000000	0.186047	0.222888	0.313131	
...	...	...	...	...	...	...	...	
1289	0.267327	0.007692	0.487252	0.833333	0.651163	0.824791	0.101010	
1290	0.277228	0.000000	0.971671	0.833333	0.139535	0.805868	0.151515	
1291	0.287129	0.007692	0.453258	0.166667	0.139535	0.297817	0.262626	
1292	0.405941	0.015385	0.133144	0.166667	0.534884	0.424763	0.181818	
1293	0.891089	0.000000	0.643059	0.333333	0.023256	0.057513	0.101010	

1294 rows × 19 columns

One Hot Encoding (OHE)

One hot encoding (OHE) is a technique to help statistical models convert categorical data into binary format (1s and 0s) that they can interpret more easily. It achieves this by creating additional columns for each value of a given categorical data column. Then depending on the data point, they will have a value of one or zero assigned to the appropriate column. Rather than give an example here, we'll run the code, which will hopefully be obvious in the resulting dataframe.

We don't want to create OHEs out of columns such as keywords and URLs as these are not ranking factors, so we'll drop these from the dataframe:

```
stop_cols = ['keyword', 'url', 'site']

hypo_serps_preml_cat = hypo_serps_preml_nonnum.drop(stop_cols, axis = 1)
```

Store the categorical data columns in a list:

```
categorical_cols = hypo_serps_preml_cat.columns.tolist()
```

Use a list comprehension to update the categorical_cols list and ensure the stop columns are not in there:

```
categorical_cols = [feat for feat in categorical_cols if feat not in stop_cols]

categorical_cols
```

The following are the categorical columns that will now be one hot encoded:

```
['token_size',
 'compression',
 'connection',
 'charset',
 'canonical_status',
 'canonical_url_render_status',
 'flesch_kincaid_reading_ease',
 'sentiment',
 'contains_paginated_html', ... ]

hypo_serps_preml_cat
```

The following is the dataframe with only the OHE columns selected.

	token_size	http_status_code	compression	connection	charset	canonical_status	canonical_url_render_status	flesch_kincaid_reading_ease	sentiment	cor
0	middle	200	Brotli	No Data	utf-8	Missing	No Change	Very Easy	Neutral	
1	middle	200	Gzipped	No Data	utf-8	To Self	Created	Fairly Easy	Positive	
2	middle	NaN	NaN	NaN	NaN	NaN	NaN	NaN	NaN	
3	long	200	Gzipped	No Data	utf-8	To Self	Created	Standard	Positive	
4	long	200	Brotli	No Data	utf-8	To Self	No Change	Very Easy	Positive	
...	...	...	...	...	...	...	...	...	...	
1289	long	200	Gzipped	No Data	utf-8	To Self	Created	Fairly Easy	Positive	
1290	long	200	Gzipped	No Data	UTF-8	To Self	No Change	Fairly Easy	Positive	
1291	long	NaN	NaN	NaN	NaN	NaN	NaN	NaN	NaN	
1292	middle	200	Gzipped	No Data	UTF-8	To Self	No Change	Fairly Easy	Neutral	
1293	middle	200	Brotli	No Data	utf-8	To Self	No Change	Standard	Positive	

1294 rows × 38 columns

The get_dummies() will be used to create the OHE columns for each categorical rank factor:

```
hypo_serps_preml_ohe = pd.get_dummies(hypo_serps_preml_cat, columns = categorical_cols)

hypo_serps_preml_ohe
```

This results in the following:

	token_size_head	token_size_long	token_size_middle	compression_Brotli	compression_Gzipped	connection_No Data	connection_keep-alive, Keep-Alive	charset_UTF-8	charset_u
0	0	0	1	1	0	1	0	0	
1	0	0	1	0	1	1	0	0	
2	0	0	1	0	0	0	0	0	
3	0	1	0	0	1	1	0	0	
4	0	1	0	1	0	1	0	0	
...	...	...	...	...	...	...	...	...	
1289	0	1	0	0	1	1	0	0	
1290	0	1	0	0	1	1	0	1	
1291	0	1	0	0	0	0	0	0	
1292	0	0	1	0	1	1	0	1	
1293	0	0	1	1	0	1	0	0	

1294 rows × 95 columns

With OHE, the category columns have now expanded from 38 to 95 columns. For example, the compression column has been replaced by two new columns compression_Brotli and compression_Gzipped, as there were only two values for that ranking factor.

Eliminate NAs

With the numeric and category data columns cleaned and transformed, we're now ready to combine the data and eliminate the missing values.

Combine the dataframes using concat():

```
hypo_serps_preml_ready = pd.concat([hypo_serps_preml_ohe, hypo_serps_preml_
median], axis = 1)
```

```
hypo_serps_preml_ready
```

The dataframes are now combined into a single dataframe "hypo_serps_preml_ready."

	token_size_head	token_size_long	token_size_middle	compression_Brotli	compression_Gzipped	connection_No Data	connection_keep-alive, Keep-Alive	charset_UTF-8	charset_
0	0	0	1	1	0	1	0	0	
1	0	0	1	0	1	1	0	0	
2	0	0	1	0	0	0	0	0	
3	0	1	0	0	1	1	0	0	
4	0	1	0	1	0	1	0	0	
...	...	...	...	...	...	...	...	...	...
1289	0	1	0	0	1	1	0	0	
1290	0	1	0	0	1	1	0	1	
1291	0	1	0	0	0	0	0	0	
1292	0	0	1	0	1	1	0	1	
1293	0	0	1	1	0	1	0	0	

1294 rows × 114 columns

The next preparation step is to eliminate "NA" values as ML algorithms don't cope very well with cell values that have "not available" as a value.

First of all, check which columns have a proportion of NAs, by taking the sum of null values in each column and dividing by the total number of rows:

```
percent_missing = hypo_serps_preml_ready.isnull().sum() * 100 / len(hypo_
serps_preml_ready)
```

We put our calculations of missing data into a separate dataframe and then sort values:

```
missing_value_df = pd.DataFrame({'column_name': hypo_serps_preml_ready.
columns,
                                'percent_missing': percent_missing})
missing_value_df.sort_values('percent_missing')
```

We can see that there are no columns with missing values, which is great news, onto the next stage.

If there were missing values, the columns would be removed as we've done what we can to improve the data to get to this point.

	column_name	percent_missing
token_size_head	token_size_head	0.0
x-frame-options_Invalid	x-frame-options_Invalid	0.0
x-content-type-options_nosniff	x-content-type-options_nosniff	0.0
x-content-type-options_Not Found	x-content-type-options_Not Found	0.0
x-content-type-options_None	x-content-type-options_None	0.0
...	...	...
perf_budget_images_No	perf_budget_images_No	0.0
perf_budget_fonts_Yes	perf_budget_fonts_Yes	0.0
perf_budget_fonts_No	perf_budget_fonts_No	0.0
is_paginated_No	is_paginated_No	0.0
no_outgoing_links	no_outgoing_links	0.0

114 rows × 2 columns

Modeling the SERPs

A quick reminder, modeling the SERPs is a formula that will predict rank based on the features of SEO, that is

```
rank ~ w_1*feature_1 + w_2*feature_2 + … + w_n*feature_n
```

~ means explained by.

n is the nth feature.

w is the weighting of the feature.

Here are some points worth mentioning:

- Split the dataset into test (20%) and train (80%). The model will learn from the most of the dataset (train) and will be applied to the test dataset (data the model has not seen before). We do this to see how the model really performs in a real-world situation.

- We will use a decision tree–based model, that is, random forest. A random forest uses a number of decision trees and takes the average of all the decision trees to arrive at the final model that best generalizes over the dataset and is therefore likely to perform well on unseen data. A random forest can also handle nonlinearities, which linear models can't.

- Set Seed is there to control the randomness of the model to make the results reproducible should another SEO/data scientist wish to evaluate the research with the same data and get the same results you were getting.

- Cross-validation will be used to make the model as robust as possible with no hyperparameter tuning.

- Typically, a random forest model (or any machine learning model) performs best with 10,000 rows or more, but it can still deliver useful insight with much less.

Import the relevant APIs and libraries which are mostly from scikit-learn, a free machine learning software library for Python:

```
from sklearn.model_selection import cross_val_score
from sklearn.model_selection import RepeatedKFold
from sklearn.ensemble import RandomForestRegressor
from sklearn.model_selection import train_test_split
from sklearn.preprocessing import StandardScaler
import category_encoders as ce
from sklearn import metrics

hypo_serps_ml = hypo_serps_preml_ready.copy()
```

Encode the data:

```
encoder = ce.HashingEncoder()
serps_features_ml_encoded = encoder.fit_transform(hypo_serps_ml)
serps_features_ml_encoded
```

This results in the following:

	token_size_head	token_size_long	token_size_middle	compression_Brotli	compression_Gzipped	connection_No Data	connection_keep-alive, Keep-Alive	charset_UTF-8	charset_u
0	0	0	1	1	0	1	0	0	
1	0	0	1	0	1	1	0	0	
2	0	0	1	0	0	0	0	0	
3	0	1	0	0	1	1	0	0	
4	0	1	0	1	0	1	0	0	
...	...	...	...	...	...	...	...	...	
1289	0	1	0	0	1	1	0	0	
1290	0	1	0	0	1	1	0	1	
1291	0	1	0	0	0	0	0	0	
1292	0	0	1	0	1	1	0	1	
1293	0	0	1	1	0	1	0	0	

1294 rows × 114 columns

Set the target variable as rank, which is the outcome we're looking to explain and guide our SEO recommendations:

```
target_var = 'rank'
```

Assign the ranking factor data to X and rank to y:

```
X, y = serps_features_ml_encoded.drop(target_var, axis=1), serps_features_
ml_encoded[target_var]
```

To train our model, we're using RandomForestRegressor because it tends to deliver better results than linear regression models. Alternatives you may wish to trial in parallel are XGBoost, LightGBM (especially for much larger datasets), and AdaBoost.

Instantiate the model:

```
regressor = RandomForestRegressor(n_estimators=20, random_state=1231)
```

Cross-validate the model. When a model is cross-validated, what is happening is that the model is being evaluated by splitting the train dataset further to see how well the model generalizes across all of the training data and hopefully the real world too.

In our case, we're splitting the model five times and storing the result in n_scores:

```
n_scores = cross_val_score(regressor, X, y, scoring='neg_mean_absolute_
error', cv=5, n_jobs=-1)
n_scores
```

Split the data randomly into train and test:

```
X_train, X_test, y_train, y_test = train_test_split(X, y, test_size=0.2,
random_state=1231)
```

Fit the machine learning model based on the training set:

```
regressor.fit(X_train, y_train)
```

Test the model on the test dataset and store the forecasts into y_pred:

```
y_pred = regressor.predict(X_test)
```

Evaluate the SERPs ML Model

Now that we have our model, we can now use it to test its efficacy. The general principles are

- Feeding the predict command, the test dataset, and the model

- Calculating the Root Mean Squared Error (RMSE) and r-squared

Given the modeling and prediction of rank is a regression problem, we use RMSE and r-squared as evaluation metrics. So what do they tell us?

The RMSE tells us what the average margin of error is for a predicted rank. For example, an RMSE of 5 would tell us that the model will predict ranking positions + or – 5 from the true value on average.

The r-squared has the formal title of "coefficient of determination." What does that mean? In practical terms, the r-squared represents the proportion of data points in the dataset that can be explained by the model. It is computed by taking the square of the correlation coefficient (r), hence r-squared. An r-squared of 0.4 means that 40% of the data can be explained by the model.

Beware of models with an r-squared of 1 or anything remotely close, especially in SEO. The chances are there's an error in your code or it's overfitting or you work for Google. Either way, you need to debug.

The r-squared is nowhere near as useful as the RMSE, so we won't be covering it here. However, if you still wish to get an idea of what the r-squared is, then you can view the r-squared of your training model (i.e., based on the training data) by running

```
print('MAE: %.3f (%.3f)' % (np.mean(n_scores), np.std(n_scores)))
print('Mean Absolute Error:', metrics.mean_absolute_error(y_test, y_pred))
print('Mean Squared Error:', metrics.mean_squared_error(y_test, y_pred))
print('Root Mean Squared Error:', np.sqrt(metrics.mean_squared_error(y_test, y_pred)))
```

```
MAE: -0.222 (0.010)
Mean Absolute Error: 0.23340777636881724
Mean Squared Error: 0.0861631648297419
Root Mean Squared Error: 0.2935356278712039
```

You might be wondering what good or reasonable values for each of these metrics are. The truth is it depends on how good you need your model to be and how you intend to use it.

If you intend to use your model as part of an automated SEO system that will directly make changes to your content management system (CMS), then the RMSE needs to be really accurate, so perhaps no more than five ranking positions. Even then, that depends on the starting position of your rankings, as five is a significant difference for a page already ranking on page 1 compared to a ranking on page 3!

If the intended use case for the model is simply to gain insight into what is driving the rankings and what you should prioritize for split A/B testing or optimization, then an RMSE of 20 or less is acceptable.

The Most Predictive Drivers of Rank

So what secrets can machine learning model tell us? We'll extract the ranking factors and the model importance data into a single dataframe:

```
df_imp = pd.DataFrame({"feature": X.columns.values,
                       "importance": regressor.feature_importances_,
                      })
df_imp = df_imp.sort_values('importance', ascending = False)
```

```
df_imp = df_imp
```

```
df_imp
```

The following dataframe result shows the most influential SERP features or ranking factors in descending order of importance.

	feature	importance
107	title_relevance	466.821502
96	no_outgoing_navigation_links	100.065025
108	title_branded	50.479885
100	performance_score	49.559890
110	first_contentful_paint	41.219836
...	...	...
105	sitelinks_search_box	0.021771
16	flesch_kincaid_reading_ease_Fairly Difficult	0.003895
48	viewport_error_Initial Scale Missing	0.002879
64	open_graph_type_object	0.000000
82	x-frame-options_Invalid	0.000000

113 rows × 2 columns

Plot the importance data in a bar chart using the plotnine library:

```
RankFactor_plt = (ggplot(df_imp.head(7), aes(x = 'reorder(feature,
importance)', y = 'importance')) +
 geom_bar(stat = 'identity', fill = 'blue', alpha = 0.6) +
 labs(y = 'Google Influence', x = '') +
 theme_classic() +
 coord_flip() +
 theme(legend_position = 'none')
)
```

```
RankFactor_plt.save(filename = 'images/1_RankFactor_plt.png', height=5,
width=5, units = 'in', dpi=1000)
RankFactor_plt
```

In this particular case, Figure 6-3 shows that the most important factor was "title_relevance" which measures the string distance between the title tag and the target keyword. This is measured by the string overlap, that is, how much of the title tag string is taken up by the target keyword.

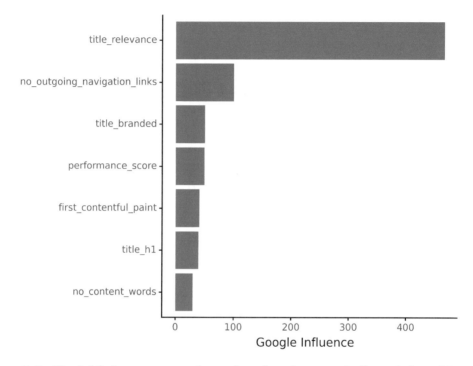

Figure 6-3. *Variable importance chart showing the most influential ranking factors identified by the machine learning algorithm*

No surprise there for the SEO practitioner; however, the value here is providing empirical evidence to the nonexpert business audience that doesn't understand the need to optimize the title tags. Data like this can also be used to secure buy-in from non-SEO colleagues such as developers to prioritize SEO change requests.

Other factors of note in this industry are as follows:

- no_cookies: The number of cookies

- dom_ready_time_ms: A measure of page speed

- no_template_words: The number of words outside the main body content section

- link_root_domains_links: Count of links to root domains
- no_scaled_images: Count of images scaled that need scaling by the browser to render

Every market or industry is different, so the preceding text is not a general result for the whole of SEO!

How Much Rank a Ranking Factor Is Worth

Now that you have your model, you'll probably want to communicate your findings to colleagues and clients alike. We'll examine one of the ranking factors as an example of how to communicate the findings of the machine learning model.

Store the most influential ranking factors in a list:

```
influencers = ['title_relevance', 'no_outgoing_navigation_links', 'title_
branded', 'performance_score', 'first_contentful_paint', 'title_h1', 'no_
content_words']
```

Select performance_score as the ranking factor we want to examine. According to Python's zero indexing, that would be three for the fourth item in the list:

```
i = 3
```

Calculate the stats to average the site CWV performance and Google rank:

```
num_factor_agg = hypo_serps_features.groupby(['site']).
agg({str(influencers[i]): 'mean', 'rank': 'mean', 'se_results_count':
'sum', 'count': 'sum'}).reset_index()
num_factor_agg = num_factor_agg.sort_values(str(influencers[i]))
```

To show the client in a different color to the competitors, we'll create a new column "target," such that if the website is the client, then it's 1, otherwise 0:

```
num_factor_agg['target'] = np.where(num_factor_agg['site'].str.
contains(hostname), 1, 0)
```

```
num_factor_agg
```

The following is the dataframe that will be used to power the chart in Figure 6-4.

	site	performance_score	rank	se_results_count	count	target
5	willowandhall.co.uk	7.350746	35.859259	19789670000	135	0
6	www.arloandjacob.com	12.436975	38.907563	37839268000	119	0
9	www.sofasandstuff.com	25.044776	34.067164	39164423098	134	0
1	johnsankey.co.uk	26.751880	57.052632	29973316080	133	1
2	loaf.com	36.016807	51.537815	42345429000	119	0
3	sofology.co.uk	47.314286	29.142857	45235434098	140	0
4	theenglishsofacompany.co.uk	50.446602	31.211538	15517618000	104	0
0	darlingsofchelsea.co.uk	68.821053	27.354545	23192559000	110	0
7	www.heals.com	NaN	34.837398	32345983000	123	0
8	www.made.com	NaN	35.836158	77597312098	177	0

This function returns a polynomial line of best fit according to whether you'd like it straight (degree 1) or curved (2 or more degrees):

```
def poly(x, degree=1):
    """

    Fit Polynomial
    These are non orthogonal factors, but it may not matter if
    we only need this for smoothing and not extrapolated
    predictions.
    """

    d = {}
    for i in range(degree+1):
        if i == 1:
            d['x'] = x
        else:
            d[f'x**{i}'] = np.power(x, i)
    return pd.DataFrame(d)
```

Plot the chart:

```
num_factor_viz_plt = (
    ggplot(num_factor_agg,
        aes(x = str(influencers[i]), y = 'rank', fill = 'target', colour
        = 'target', #shape = 'cat_item',
```

```
                                    size = 'se_results_count')) +
    geom_point(alpha = 0.3) +
    geom_smooth(method = 'lm', se = False, formula = 'y ~ poly(x,
    degree=1)', colour = 'blue', size = 1.5) +
    labs(y = 'Google Rank', x = str(influencers[i])) +
    scale_y_reverse() +
  scale_size_continuous(range = [5, 20]) +
    theme(legend_position = 'none', axis_text_x=element_text(rotation=0,
    hjust=1, size = 12))
)

num_factor_viz_plt
```

Plotting the average Core Web Vitals (CWV) vs. average Google rank by website which also includes a line of best fit (Figure 6-4), we can estimate the ranking impact per unit improvement in CWV. In this case, is about 0.5 rank position gain per 1 unit improvement in CWV.

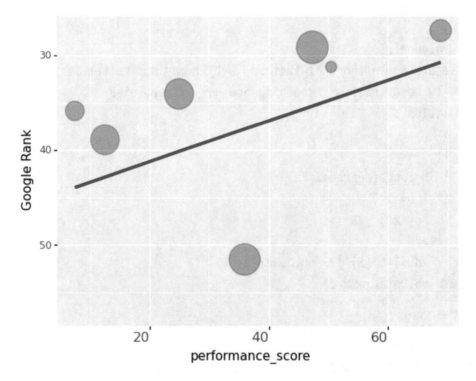

Figure 6-4. *Bubble chart of websites comparing Google rank and CWV performance score*

The Winning Benchmark for a Ranking Factor

The winning benchmark also appears to be 70, which may come as a relief to developers as achieving a score of 90 or above may be incredibly resource intensive to attain.

Thanks to machine learning, we're not only able to surface the most important factors, when taking a deep dive, we can also see the winning benchmark.

Tips to Make Your Model More Robust

Naturally, no model is perfect and never will be. The usefulness of the model is down to

- Your imagination, inventiveness, SEO knowledge, and ability to form hypotheses that are informative for model inclusion

- Your ability to translate these hypotheses into measurable metrics that can be gathered as data on your site and your competitors

- The way you structure the data that is meaningful for the model to spot patterns

Activation

With your model outputs, you're now ready to make some decisions on your SEO strategy in terms of

- Changes you'd like to make sitewide because they're a "no-brainer," such as site speed or increasing brand searches (either through programmatic advertising, content marketing, or both)

- Split A/B testing of factors included in your model

- Further research into the ranking factor itself to guide your recommendations

Automating This Analysis

The preceding application of ML is great for getting some ideas to split A/B test and improve the SEO program with evidence-driven change requests. It's also important to recognize that this analysis is made all the more powerful when it is ongoing.

Why? Because the ML analysis is just a snapshot of the SERPs for a single point in time. Having a continuous stream of data collection and analysis means you get a more true picture of what is really happening with the SERPs for your industry.

This is where SEO purpose–built data warehouse and dashboard systems come in, and these products are available today. What these systems do are

- Ingest your data from your favorite SEO tools daily

- Combine the data

- Use ML to surface insights like before in a front end of your choice like Google Data Studio

To build your own automated system, you would deploy into a cloud infrastructure like Amazon Web Services (AWS) or Google Cloud Platform (GCP) what is called ETL, that is, extract, transform, and load, so that your data collection, analysis, and visualization are automated in one place. This is explained more fully in Chapter 8.

Summary

In this chapter, you learned

- The data science principles behind understanding the ranking factors for competitor analysis

- How to combine data sources

- How to prepare data for machine learning

- How to train a machine learning model

- How to use the model outputs to generate SEO recommendations

Competitor research and analysis in SEO is hard because there are so many ranking factors that are available and so many to control for. Spreadsheet tools are not up to the task due to the amounts of data involved, let alone the statistical capabilities that data science languages like Python offer.

When conducting SEO competitor analysis using machine learning (ML), it's important to understand that this is a regression problem, the target variable is Google rank, and the hypotheses are the ranking factors.

In Chapter 7, we will cover experiments which are something that would naturally follow the outputs of competitor statistical analysis.

## CHAPTER 7

Experiments

It's quite exciting to unearth insights from data or your own thought experiments that could be implemented on your site and drive real, significant organic improvements. With the rise of split testing platforms such as Distilled ODN and RankScience, it's of no surprise that experimentation is playing an ever-increasing role in SEO.

If you're running a small site where the change leading to a negative impact is inconsequential or the change is seemingly an obvious SEO best practice, then you may forgo formal experimentation and simply focus on shipping the changes you believe are required.

On the other hand, if you're working on a large enterprise website, be it in-house or as part of an agency, then any changes will be highly consequential, and you'll want to make sure you test these changes in order to both understand the impact (both positive and negative) as well as help shape your understanding to help inform new hypotheses to test.

How Experiments Fit into the SEO Process

To run experiments and split A/B tests successfully, you'll need a process which starts from idea generation (otherwise referred to as a hypothesis) all the way to implementation. We have outlined the steps as follows:

1. Hypothesis generation

2. Experiment design

3. Running the experiment

4. Evaluation

5. Implementation

We will cover these steps in the following sections.

A. Voniatis, *Data-Driven SEO with Python*, https://doi.org/10.1007/978-1-4842-9175-7_7

Generating Hypotheses

Before any experiment starts, you need to base it around your hypothesis, that is, a belief in what you believe will significantly change your target variable or outcome metric, for example, organic impressions or URLs crawled.

This step is crucially important because without clear hypotheses, you won't begin to know what it is you will be testing to influence your outcome metric. So think about what it is you want to learn from that could help you improve your SEO performance.

There's a number of areas to source hypotheses from:

- Competitor analysis

- Website articles and social media

- You/your team's ideas

- Recent website updates

- Conference events and industry peers

- Past experiment failures

I usually like to use the format "*We believe that* Google will give a greater weighting to URLs linked from by other prominent pages of a website." This statement is then expanded to consider what you're proposing to test and how you'll measure (i.e., "We'll know if the hypothesis is valid when…").

Competitor Analysis

The competitor analysis that you carried out (explained in Chapter 6) will be a natural source of ideas to test because they have some statistical foundation to them, surfacing things that your competitors are doing or not doing to benefit from superior organic performance. These hypotheses have the added advantage of knowing what the metric is that you'll be testing from the outset. After all, you had to get the data into your analysis in the first place.

Website Articles and Social Media

Often, we read studies, articles, and social media memes that claim to have driven or decreased organic performance. That's not to say these claims are untrue or not substantiated. However, these claims are not in the context of your situation, and if they made the news, they most probably merit testing. As an aside, in the early days of our SEO careers before data science was actually a thing, the best way to really know your data was to test everything we read about SEO online, such as Webmaster World, BlackHatWorld forums, etc., and see what worked and what didn't work. If you didn't have sites banned from the index in Google, you weren't respected as an SEO or perhaps you were just not being bold with your experiments. The very essence was "optimizing" for search engines.

Naturally, things have moved on, and most of us in SEO are working for brands and established businesses. So some of the creative wild experiments would be inappropriate or rather career limiting, which we're not advocating to do.

You/Your Team's Ideas

The test hyoptheses are not limited to your immediate SEO team (in-house or agency), not even your digital marketing team. This could be colleagues that have any exposure to the website with any ideas. Most of them might be unviable to devote resources to an experiment. However, their ideas, since they (should) care about the business, are worthy of some consideration. Naturally, your immediate SEO team may have the better ideas of things to test from an SEO perspective.

Recent Website Updates

Usually, it's better to test things before a large website update impacts your organic traffic. However, you may not get such luxuries with other competing priorities or tight timelines. Nevertheless, if a product update is expected to impact your organic traffic, good or bad, such as the launch of a stripped back top-level navigation, you'll ideally want to know why and get ahead of its full launch so that you can test it to understand the impact for SEO.

Conference Events and Industry Peers

Why limit your ideas to online and your company? Attending industry events can be a great way of not only finding new things to test for SEO but also meeting people at the events who may be wrestling with the same problems. Diversity of thought is highly valued and could lead to breakthrough experiment ideas.

Past Experiment Failures

If and when you fail, try, try, and try again. Usually, if an experiment fails, it's often due to the experiment not having the required sample size or the experiment was designed or ran incorrectly. Learn from it and reiterate. Whether it's designing, running it correctly, or reformulating your hypotheses, do it and keep iterating until you get the desired result. You may get taken in a different direction to the original failed experiment, but you will have learned so much in the process.

Experiment Design

Having decided on what hypotheses you're going to test, you're now ready to design your experiment. In the following code example, we will be designing an experiment to see the impact of a test item (it could be anything, say a paragraph of text beneath the main heading) on organic impressions at the URL level.

Let's start by importing the APIs:

```
import re
import time
import random
import pandas as pd
import numpy as np
import datetime
import requests
import json
from datetime import timedelta
from glob import glob
import os
from plotnine import *
import matplotlib.pyplot as plt
from pandas.api.types import is_string_dtype
from pandas.api.types import is_numeric_dtype
from datetime import datetime, timedelta
```

These are required for carrying out the actual split test:

```
from statsmodels.discrete.discrete_model import NegativeBinomial
from statsmodels.tools.tools import add_constant
from collections import Counter
import uritools

pd.set_option('display.max_colwidth', None)
%matplotlib inline
```

Because we're using website analytics data of which popular brands include Google, Looker, Adobe, this is easily exported from a landing report by date. If this is a large site, then you may need to use an API (see Chapter 8 for Google Analytics).

Depending on your website analytics package, the column names may vary; however, if you're looking to test the difference in impressions between URL groups over time, you will require

- Date

- Landing page URL

- Sessions

Other outcomes than sessions can also be tested, such as impressions, ctr, position, etc.

Assuming you have a CSV export, read the data from your website analytics package using pd.read_csv:

```
analytics_raw = pd.read_csv("data/expandable-content.csv")
```

Print the data types of the columns to check for numerical data that might be imported as a character string which would need changing:

```
print(analytics_raw.dtypes)
```

```
landing_page      object
date              object
sessions         float64
dtype: object
```

The session data is a numerical float which is fine, but the data is classed as "object" which means it will need converting to a date format.

We'll make a copy of the dataframe using the .copy() method. This is so that any changes we make won't affect the original imported table. That way, if we're not happy with the change, we don't have to go all the way to the top and reimport.

```
analytics_clean = analytics_raw.copy()
```

The date column uses the to_datetime() function which takes the column and the format the date is in and is then normalized to convert the string into a date format.

This will be important for filling in missing dates and plotting data over time later:

```
analytics_clean['date'] = pd.to_datetime(analytics_clean['date'],
format='%d/%m/%Y').dt.normalize()
```

```
analytics_clean
```

The Pandas dataframe below shows 'analytics_clean' which now has the data in a usable format for further manipulation such as graphing sessions over time.

	landing_page	date	sessions
1	https://www.next.com/shop/1-state-boots/	2019-08-09	2.0
3	https://www.next.com/shop/1-state-boots/	2019-08-12	1.0
5	https://www.next.com/shop/1-state-boots/	2019-08-21	1.0
7	https://www.next.com/shop/1-state-boots/	2019-08-13	1.0
9	https://www.next.com/shop/1-state-boots/	2019-08-25	1.0
...	...	...	...
315347	https://www.next.com/shop/zuhair-murad-tops/	2019-08-22	1.0
315349	https://www.next.com/shop/zuhair-murad-tops/	2019-08-25	1.0
315351	https://www.next.com/shop/zuhair-murad-tops/	2019-08-24	1.0
315353	https://www.next.com/shop/zuhair-murad-tops/	2019-08-18	1.0
315355	https://www.next.com/shop/zuhair-murad-tops/	2019-08-29	1.0

157678 rows × 3 columns

Let's explore the sessions' distribution using .describe():

```
analytics_raw.describe()
```

The following screenshot shows the distribution of sessions including count (number of data points), the average (mean), and others.

	sessions
count	157678.000000
mean	2.155976
std	4.009260
min	1.000000
25%	1.000000
50%	1.000000
75%	2.000000
max	152.000000

We can see that the average (mean) number of sessions per URL on any given date is about 2, which varies wildly shown by the standard deviation (sd) value of 4. Given the mean is 2 and a landing page can't have a session of –2 (mean of 2 less standard deviation of 4), this implies that some outlier pages are getting extremely high sessions, which explains the variation.

Now look at the dates:

```
analytics_clean['date'].describe()
```

This results in the following:

```
count                     157678
unique                        28
top          2019-08-18 00:00:00
freq                        6351
first        2019-08-06 00:00:00
last         2019-09-02 00:00:00
Name: date, dtype: object
```

There's not much to infer other than the data's date range of about a month in August.

Zero Inflation

Web analytics typically only logs data against a web page when there is an impression.

What about the days when the page doesn't receive an impression? What then?

Zero inflation is where we add null records for pages that didn't record an organic impression on a given day. If we didn't zero-inflate, then there would be a distortion of the mean of the data for a given web page, let alone for a group of pages, namely, A and B.

For example, URL X may have had 90 sessions on 10 days within a given day period logged in analytics which should suggest that average impression per day is 9 per day.

However, because they happened on the 10 days, your calculations would mislead you to think the URL is better than it is. By zero-inflating the data, that is, adding null rows to the dataset for the days in the 30-day period, when URL X didn't have any organic impressions, the average calculated would be restored to the expected 3 per day.

Zero inflation also gives us another useful property to work from, and that is the Poisson distribution.

It's beyond the scope of this book to explain the Poisson distribution. Still, what you need to know is that the Poisson distribution is common for rare events when we test for the difference between groups A and B.

Any statistically significant difference between the two groups will hopefully show that the test group B had significantly less zeros than A. Enough science, let's go.

There is a much easier (and less comprehensible) way to fill in missing dates. Both methods are given in this chapter starting with the longer yet easier to read.

Here, we use the function date_range() to set the date range from the minimum and maximum dates found in the analytics dataframe, with an interval of one day. This is saved into a variable object called "datelist":

```
datelist = pd.date_range(start=analytics_clean['date'].min(),
                         end=analytics_clean['date'].max(), freq='1d')
```

nd is the length of days in the date range, and nu is the unique list of landing pages we want the dates for:

```
nd = len(datelist)
nu = len(analytics_clean['landing_page'].unique())
```

Here, we create a dataframe with all the possible landing page and date combinations by a cross-product of the landing page and the dates:

```
analytics_expanded = pd.DataFrame({'landing_page': analytics_
clean['landing_page'].unique().tolist() * nd,
                        'date':np.repeat(datelist, nu)})
```

Then we look up which dates and landing pages have sessions logged against them:

```
analytics_expanded = analytics_expanded.merge(analytics_clean, how='left')
```

Any that are unmatched (and thus null) are filled with zeros:

```
analytics_expanded[['date','sessions']] = analytics_expanded.
groupby('landing_page')[
    ['date', 'sessions']].fillna(0)
```

Convert the sessions to numerical float for easier data handling:

```
analytics_expanded['sessions'] = analytics_expanded['sessions'].
astype('int64')

analytics_expanded
```

The resulting dataframe "analytics_expanded," shown as follows, is our zero-inflated dataframe ready for split testing. Note the original analytics data had 157,678 rows, and with zero inflation, it's now 807,212 rows.

	landing_page	date	sessions
0	https://www.next.com/shop/1-state-boots/	2019-08-06	0
1	https://www.next.com/shop/1-state-flats/	2019-08-06	0
2	https://www.next.com/shop/1-state-heels/	2019-08-06	0
3	https://www.next.com/shop/1-state-jackets/	2019-08-06	0
4	https://www.next.com/shop/1-state-jumpsuits/	2019-08-06	0
...		...	...	...
807207	https://www.next.com/shop/zuhair-murad-dresses/	2019-09-02	3
807208	https://www.next.com/shop/zuhair-murad-jackets/	2019-09-02	0
807209	https://www.next.com/shop/zuhair-murad-jumpsuits/	2019-09-02	1
807210	https://www.next.com/shop/zuhair-murad-pants/	2019-09-02	0
807211	https://www.next.com/shop/zuhair-murad-tops/	2019-09-02	0

807212 rows × 3 columns

Let's explore the data using the .describe() which will tell us how the distribution of sessions has changed having been zero-inflated:

```
analytics_expanded.describe()
```

The following screenshot shows the distribution of sessions following zero inflation.

	sessions
count	807212.000000
mean	0.421141
std	1.967350
min	0.000000
25%	0.000000
50%	0.000000
75%	0.000000
max	152.000000

And what a difference! The mean has shrunk by over 75% from 2.16 to 0.42, and hardly any pages get over one session just under a month.

Split A/A Analysis

A/A testing is the process by which we test the same items against themselves over a time period. This is the type of split test popularized by SearchPilot (formerly Distilled ODN) as illustrated in Figure 7-1.

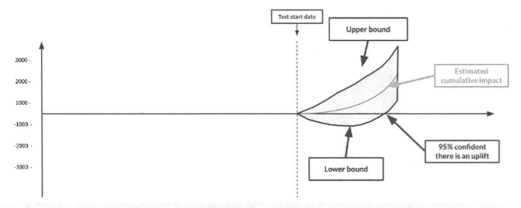

There is a 95% chance of the true impact of the test lying within the shaded area. Statistical significance is reached at any point when the shaded area is **all** above or below the x-axis

Figure 7-1. *Split A/B analysis by SearchPilot*
Source: `www.searchpilot.com/features/seo-a-b-testing/`

For example, we take a sample of URLs and benchmark the performance before implementing a test on said URL sample to see if a significant impact results or not.

The main motivation for us to conduct A/A testing is to determine whether the A/B test design is reliable enough to proceed with or not. What we're looking for are no differences between A before and A after.

We'll test a period of 13 days, assuming now changes have been made, although in a real setting, you would check nothing has changed before testing.

Why 13 days? This is an arbitrary number; however, methods are given later on determining sample size for a robust A/B test to ensure any differences detected are significant. The same methods could be applied here.

This A/A test is just an illustration of how to create the data structures and test. So if you wanted to conduct a "SearchPilot" style of split testing, then sample size and testing period determination aside, the following code would help you run it:

```
aa_test_period = 13
```

Set the cutoff date to be the latest date less the test period:

```
cutoff = analytics_expanded['date'].max() - timedelta(days = aa_
test_period)
```

Create a dataframe copy for A/A testing "analytics_phased":

```
analytics_phased = analytics_expanded.copy()
```

Set the A/A group based on the date before or after the cutoff:

```
analytics_phased['aa_group'] = np.where(analytics_expanded['date'] <
cutoff, "pre_test", "test_period")
```

```
analytics_phased
```

This should result in the following output:

	landing_page	date	sessions	aa_group
0	https://www.next.com/shop/1-state-boots/	2019-08-06	0	pre_test
1	https://www.next.com/shop/1-state-flats/	2019-08-06	0	pre_test
2	https://www.next.com/shop/1-state-heels/	2019-08-06	0	pre_test
3	https://www.next.com/shop/1-state-jackets/	2019-08-06	0	pre_test
4	https://www.next.com/shop/1-state-jumpsuits/	2019-08-06	0	pre_test
...	...	...	...	...
807207	https://www.next.com/shop/zuhair-murad-dresses/	2019-09-02	3	test_period
807208	https://www.next.com/shop/zuhair-murad-jackets/	2019-09-02	0	test_period
807209	https://www.next.com/shop/zuhair-murad-jumpsuits/	2019-09-02	1	test_period
807210	https://www.next.com/shop/zuhair-murad-pants/	2019-09-02	0	test_period
807211	https://www.next.com/shop/zuhair-murad-tops/	2019-09-02	0	test_period

807212 rows × 4 columns

Before testing, let's determine analytically the statistical properties of both A/A groups, which is indicative of what the actual split A/A test result might be.

The first function is day_range() which returns the number of days in the date range which is the latest date less the earliest date:

```
def day_range(date):
    return (max(date) - min(date)).days
```

First, we calculate the means by filtering the data for nonzero sessions and then aggregate the date range and average by A/A group:

```
aa_means = (
    analytics_phased.loc[analytics_phased["sessions"] != 0]
    .groupby(["aa_group"])
    .agg({"date": ["min", "max", day_range], "sessions": "mean"})
)
```

```
aa_means
```

The resulting aa_means dataframe shows the following output:

	date			sessions
	min	max	day_range	mean
aa_group				
pre_test	2019-08-06	2019-08-19	13	2.279396
test_period	2019-08-20	2019-09-02	13	2.007706

We can see that the day ranges and the date range are correct, and the averages per group are roughly the same when rounded to whole numbers.

```
aa_means = analytics_phased.loc[analytics_phased['sessions'] != 0]
```

Let's determine the variation between groups:

```
aa_zeros = analytics_phased.copy()
```

Create a zeros column so we can count zeros and the ratio:

```
aa_zeros['zeros'] = np.where(aa_zeros['sessions'] == 0, 1, 0)
aa_zeros['rows'] = 1
```

Aggregate the number of zeros and data points by A/A group:

```
aa_means_sigmas = aa_zeros.groupby('aa_group').agg({'zeros': sum, 'rows':
sum}).reset_index()
```

Calculate the variation "sigma" which is 99.5% of the ratio of zeros to the total possible sessions:

```
aa_means_sigmas['sigma'] = aa_means_sigmas['zeros']/aa_means_
sigmas['rows'] * 0.995
```

```
aa_means_sigmas
```

This should result in the following output:

	aa_group	zeros	rows	sigma
0	pre_test	317556	403606	0.782863
1	test_period	331978	403606	0.818417

We can see the variation is very similar before and after the cutoff, so that gives us some confidence that the URLs are stable enough for A/B testing.

Put it together using the .merge() function (the Python equivalent of Excel's vlookup):

```
aa_means_stats.merge(aa_means_sigmas, on = 'aa_group', how = 'left')
```

This should result in the following output:

	aa_group	min_date	max_date	sessions	date_range	zeros	rows	sigma
0	pre_test	2019-08-06	2019-08-19	2.279396	13 days	317556	403606	0.782863
1	test_period	2019-08-20	2019-09-02	2.007706	13 days	331978	403606	0.818417

If you were conducting an A/A test to see the effect of an optimization, then you'd want to see the test_period group report a higher session rate with the same sigma or lower. That would indicate your optimization succeeded in increasing SEO traffic.

Let's visualize the distributions using the histogram plotting capabilities of plotnine:

```
aa_test_plt = (
    ggplot(analytics_phased,
            aes(x = 'sessions', fill = 'aa_group')) +
    geom_histogram(alpha = 0.8, bins = 30) +
    labs(y = 'Count', x = '') +
    theme(legend_position = 'none',
            axis_text_y =element_text(rotation=0, hjust=1, size = 12),
            legend_title = element_blank()
            ) +
    facet_wrap('aa_group')
)

aa_test_plt.save(filename = 'images/2_aa_test_plt.png',
                                height=5, width=8, units = 'in', dpi=1000)
aa_test_plt
```

The chart shown in Figure 7-2 confirms visually there is no difference between the groups.

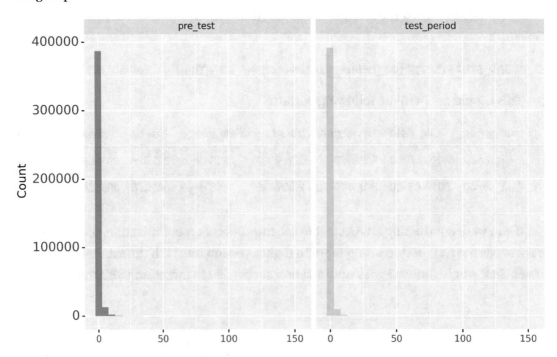

Figure 7-2. *Histogram plots of pretest and test period A/A group data*

316

The box plot gives more visual detail of the two groups' distribution which will now be used:

```
aa_test_box_plt = (
    ggplot(analytics_phased,
           aes(x = 'aa_group', y = 'sessions',
               fill = 'aa_group', colour = 'aa_group')) +
    geom_boxplot(alpha = 0.8) +
    labs(y = 'Count', x = '') +
    theme(legend_position = 'none',
          axis_text_y =element_text(rotation=0, hjust=1, size = 12),
          legend_title = element_blank()
         )
)

aa_test_box_plt.save(filename = 'images/2_aa_test_box_plt.png',
                              height=5, width=8, units = 'in', dpi=1000)
aa_test_box_plt
```

Figure 7-3 shows again in aa_test_box_plt that there is no difference between the groups other than the pretest group having a larger number of higher value outliers.

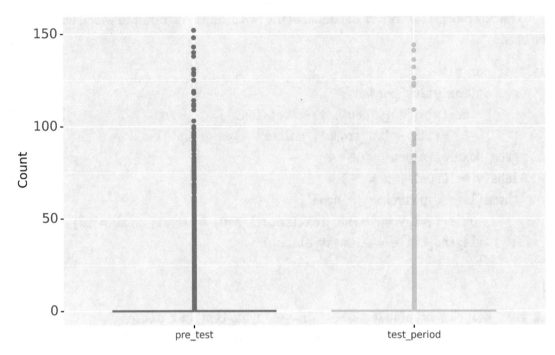

Figure 7-3. *Box plots of pretest and test period groups*

Let's perform the actual A/A test using a statistical model. We'll create an array, which is a list of numbers marking data points as either 0 (for pretest) or 1 (test_period), which will then be assigned to X:

```
X = np.where(analytics_phased['aa_group'] == 'pre_test', 0.0, 1.0)
X = add_constant(X)
X = np.asarray(X)
X
```

This results in the following:

```
array([[1., 0.],
       [1., 0.],
       [1., 0.],
       ...,
       [1., 1.],
       [1., 1.],
       [1., 1.]])
```

X is used to feed the NegativeBinomial() model which will be used to test the difference in the number of sessions between the two A/A groups.

The arguments are the outcome metric (sessions) and the independent variable (aa_group):

```
aa_model = NegativeBinomial(analytics_phased['sessions'], X).fit()
```

Then we'll see the model results using the .summary() attribute:

```
aa_model.summary()
```

The resulting aa_model.summary() is shown as follows:

```
Optimization terminated successfully.
        Current function value: 0.762045
        Iterations: 3
        Function evaluations: 8
        Gradient evaluations: 8
```

NegativeBinomial Regression Results

Dep. Variable:	sessions	No. Observations:	807212
Model:	NegativeBinomial	Df Residuals:	807210
Method:	MLE	Df Model:	1
Date:	Tue, 18 Oct 2022	Pseudo R-squ.:	0.001896
Time:	08:50:29	Log-Likelihood:	-6.1513e+05
converged:	True	LL-Null:	-6.1630e+05
Covariance Type:	nonrobust	LLR p-value:	0.000

	coef	std err	z	P>|z|	[0.025	0.975]
const	-0.7216	0.004	-163.064	0.000	-0.730	-0.713
x1	-0.3104	0.006	-48.458	0.000	-0.323	-0.298
alpha	5.8461	0.024	247.865	0.000	5.800	5.892

The printout shows that p-value (LLR p-value) is zero, which means there is significance. However, the x1 is –0.31, indicating there is a small difference between groups.

Is that enough of a difference to stop the A/B test? That's a business question. In this case, it isn't; however, this is subjective; the graphs and analytical tables would support the claim that there is no real difference – onward.

Determining the Sample Size

Getting the sample size right is absolutely key because it can really make or break the reliability of your test results. Some basic principles are to ensure you have enough to get conclusive results in a reasonable time period and to ensure you don't terminate a test early like many software do when significant changes are observed before the experiment has run its course.

We won't be going into more detail into the factors that determine the appropriate sample size for your SEO experiment, such as power, because this isn't a statistics textbook (and there are plenty of really good textbooks that teach statistics such as the *OpenIntro Statistics*).

The main factor that determines your required sample size is the required level of statistical significant difference between test and control. The typical and most conventional level of cutoff is 95%. That is, there's a 5% (or less) chance that the test results are the same as the control results due to random noise, and therefore you may reject the null hypothesis that there is no difference between test (B) and control (A).

The reality is that the 95% rule, while conventional, is not set in stone. You can decide what is good enough. For example, you may wish to go for 89%, which is absolutely fine, because 89 times out of 100, your test will beat control. That's the way to think about it.

The following is some code to do exactly that. We'll estimate some parameters which will be used to help us determine the required sample size based on the minimum number of sessions:

```
num_rows = analytics_phased["sessions"].count()
mu = analytics_phased[analytics_phased["sessions"] != 0].agg({"sessions":
"mean"})
sigma = get_sigma(analytics_phased["sessions"])

print(num_rows, mu, sigma)

807212 sessions     2.155976
dtype: float64 0.8006401416232662
```

With the parameters set, these will feed the following functions. python_rzip will generate and return a random Poisson distribution based on the parameters:

```python
def python_rzip(n, mu, sigma):
    rng = np.random.default_rng()
    P = rng.poisson(mu, n)

    return [p if random.random() > sigma else 0 for p in P]
```

simulate_sample uses the python_rzip function to return a split test between two groups of data assuming there is a difference of 20% or more:

```python
def simulate_sample(n, difference=0.2):
    control = python_rzip(n, mu, sigma)
    test = python_rzip(n, mu + difference, sigma)
    test = stats.ttest_ind(control, test)
    return test[1]
```

Finally, run_simulations uses simulate_sample to estimate the significance of a sample size of a given level of traffic:

```python
def run_simulations(n, difference=0.2, n_simulations=100):
    p_values = [simulate_sample(n, difference) for i in range(n_
    simulations)]
    significant = sum(map(lambda x: x <= 0.05, p_values))
    return significant / n_simulations
```

With the three functions defined, we can test for significance at varying levels of traffic. If you fancy a challenge to avoid repetitive code and stretch your Python skills, try implementing the run_simulations function as part of a list comprehension:

```python
print(run_simulations(n=100), ": 100")
print(run_simulations(n=1000), ": 1000")
print(run_simulations(n=10000), ": 10000")
print(run_simulations(n=15000), ": 15000")
print(run_simulations(n=18000), ": 16000")
print(run_simulations(n=18000), ": 18000")
print(run_simulations(n=20000), ": 20000")
```

```
print(run_simulations(n=25000), ": 25000")
print(run_simulations(n=30000), ": 30000")
print(run_simulations(n=50000), ": 50000")
```

This results in the following:

```
0.04 : 100
0.08 : 1000
0.74 : 10000
0.85 : 15000
0.86 : 16000
0.9 : 18000
0.96 : 20000
0.97 : 25000
1.0 : 30000
1.0 : 50000
```

The preceding output shows the levels of significance (p-value) achieved at different sample size levels, which in our case are the required number of sessions. If we would be happy with a 90% (or higher) chance that a 20% difference would be observed, then we'd require 18,000 sessions per group or more.

So we'll set the experiment sample size as appropriate:

```
exp_sample_size = 18000
```

Test and Control Assignment

Once you've set your sample size at the desired level of statistical significance, you're now ready to start assigning URLs for test and control at random.

We aggregate the average sessions and number of days by landing page and store this as a dataframe "urls_agg":

```
urls_agg = analytics_clean.groupby('landing_page').agg({'sessions': 'mean',
'date': 'count'}).reset_index()
```

The testing_days, which is the maximum number of days to run the test, will be set at 30, which is an arbitrary number set by the business. This of course can be lower.

```
testing_days = 30
```

With the max period of testing days set, we'll need the minimum URLs for the test group to hit the required number of user sessions in that time period. Dividing the sample session size of 18,000 by the number of testing days will give us that approximate number.

Bear in mind that it can take up to two weeks (and sometimes longer) for Google to register the site changes and reflect these in the search results (i.e., they have to crawl, index, and rerank their results). So to limit the risk of ending the experiment early, we'll double the minimum URLs required for testing, in order to increase the likelihood of Google, in the first instance, crawling the test URLs (those with the change(s)):

```
url_sample_size = int(exp_sample_size / testing_days) * 2

print(url_sample_size)
```

```
1200
```

1200 URLs are required for the test group. Note that it's implicitly assumed that the test URLs are much smaller than the control such that there are plenty of URLs in the control group to hit the minimum sessions during the testing period.

```
urls_agg
```

Our resulting dataframe shows each landing page and their average sessions and number of days where sessions are generated. Some URLs get more than one day of sessions as shown by the date column.

	landing_page	sessions	date
0	https://www.next.com/shop/1-state-boots/	1.125000	8
1	https://www.next.com/shop/1-state-flats/	1.000000	3
2	https://www.next.com/shop/1-state-heels/	1.000000	5
3	https://www.next.com/shop/1-state-jackets/	1.500000	8
4	https://www.next.com/shop/1-state-jumpsuits/	1.100000	10
...	...	...	...
28824	https://www.next.com/shop/zuhair-murad-dresses/	1.840000	25
28825	https://www.next.com/shop/zuhair-murad-jackets/	1.000000	1
28826	https://www.next.com/shop/zuhair-murad-jumpsuits/	1.722222	18
28827	https://www.next.com/shop/zuhair-murad-pants/	1.142857	7
28828	https://www.next.com/shop/zuhair-murad-tops/	1.000000	5

28829 rows × 3 columns

We will now sample the dataframe based on the required 1200 URLs and assign these to the "test" group:

```
urls_test = urls_agg.sample(url_sample_size).assign(ab_group="test")
```

Drop the sessions and date column as we only need the URLs to send to the web developer team for allocation:

```
urls_test.drop(['sessions', 'date'], axis = 1, inplace = True)
urls_test
```

The urls_test is shown as follows:

	landing_page	ab_group
7129	https://www.next.com/shop/henrik-vibskov-coats/	test
16561	https://www.next.com/shop/mens-le-mont-st-michel-jackets/	test
13017	https://www.next.com/shop/mens-cartier-belts/	test
16169	https://www.next.com/shop/mens-jw-anderson-t-shirts/	test
8949	https://www.next.com/shop/kensie-totes/	test
...	...	...
9813	https://www.next.com/shop/lizzie-fortunato-shoulder-bags/	test
12857	https://www.next.com/shop/mens-buscemi-sneakers/	test
18681	https://www.next.com/shop/mens-raey-jackets/	test
6000	https://www.next.com/shop/forever-21-knitwear/	test
20060	https://www.next.com/shop/mens-the-quiet-life-hats/	test

1200 rows × 2 columns

Our dataframe shows the test landing pages:

```
urls_test_list = urls_test["landing_page"]
urls_test_list
```

This results in the following:

```
7129                    https://www.next.com/shop/henrik-vibskov-coats/
16561        https://www.next.com/shop/mens-le-mont-st-michel-jackets/
13017                    https://www.next.com/shop/mens-cartier-belts/
16169            https://www.next.com/shop/mens-jw-anderson-t-shirts/
8949                            https://www.next.com/shop/kensie-totes/
                                   ...
9813        https://www.next.com/shop/lizzie-fortunato-shoulder-bags/
12857              https://www.next.com/shop/mens-buscemi-sneakers/
18681                https://www.next.com/shop/mens-raey-jackets/
6000                  https://www.next.com/shop/forever-21-knitwear/
20060          https://www.next.com/shop/mens-the-quiet-life-hats/
Name: landing_page, Length: 1200, dtype: object
```

Test landing pages are converted to a list which will be used to mark the other (non test allocated) URLs as control:

```
urls_control = urls_agg[~urls_agg["landing_page"].isin(urls_test_list.
values)].assign(
    ab_group="control"
)

urls_control.drop(['sessions', 'date'], axis = 1, inplace = True)

urls_control
```

The urls_control dataframe shows the control groups:

	landing_page	ab_group
0	https://www.next.com/shop/1-state-boots/	control
1	https://www.next.com/shop/1-state-flats/	control
2	https://www.next.com/shop/1-state-heels/	control
3	https://www.next.com/shop/1-state-jackets/	control
4	https://www.next.com/shop/1-state-jumpsuits/	control
...	...	...
28824	https://www.next.com/shop/zuhair-murad-dresses/	control
28825	https://www.next.com/shop/zuhair-murad-jackets/	control
28826	https://www.next.com/shop/zuhair-murad-jumpsuits/	control
28827	https://www.next.com/shop/zuhair-murad-pants/	control
28828	https://www.next.com/shop/zuhair-murad-tops/	control

27629 rows × 2 columns

Both test and control groups will now be combined into a single dataframe showing which URLs are test and control:

```
split_ab_dev = pd.concat([urls_control, urls_test], axis=0).sort_index()
split_ab_dev
```

The following shows the split_ab_dev dataframe:

	landing_page	ab_group
0	https://www.next.com/shop/1-state-boots/	control
1	https://www.next.com/shop/1-state-flats/	control
2	https://www.next.com/shop/1-state-heels/	control
3	https://www.next.com/shop/1-state-jackets/	control
4	https://www.next.com/shop/1-state-jumpsuits/	control
...	...	...
28824	https://www.next.com/shop/zuhair-murad-dresses/	control
28825	https://www.next.com/shop/zuhair-murad-jackets/	control
28826	https://www.next.com/shop/zuhair-murad-jumpsuits/	control
28827	https://www.next.com/shop/zuhair-murad-pants/	control
28828	https://www.next.com/shop/zuhair-murad-tops/	control

28829 rows × 2 columns

```
split_ab_dev.to_csv("data/split_ab_developers.csv")
```

The final dataframe is combined and exported into a CSV for the software development team's reference.

Running Your Experiment

So far, we have assumed near perfect lab conditions, and there are quite a number of pitfalls that could scupper your experiment. We'll deal with these in turn.

Ending A/B Tests Prematurely

Whatever you do, ensure you run your tests to the full sample size. Just because your test group might reach a statistically significant difference before the required sample size, it doesn't mean the test result is conclusive.

What can and does happen is that the test could regress back to similar levels of performance as the control group after outperforming control. However, if you end the experiment prematurely, you won't know and therefore end up wasting your time and company resources on an invalid experiment.

So if your experiment requires 20,000 pageviews, make sure your experiment reaches 20,000 pageviews for both groups.

Not Basing Tests on a Hypothesis

If you've got this far and you haven't based this on a hypothesis, start again and form a hypothesis. Having a hypothesis helps frame the outcome of your experiment, so that you can be certain of what it is you've actually learned from the experiment. For example, if you're testing whether 100 worded body copy below the H1 will increase SEO impressions, then simply state it so. Just make sure you do it from the outset. This will help you to be more precise about what it is you are testing, how you will test it, and what it is you have learned.

Simultaneous Changes to Both Test and Control

This happens more often than you might think. We were once asked by a few CTOs whether it would be acceptable to make changes to both test and control while the experiment was running.

We advised it would not because even though the change may be applied equally to both groups, one or both of them may have an interaction with the simultaneous change and not necessarily in the same direction. Of course, it sounds unlikely, so if you're ever tempted or get asked if the simultaneous change is okay, avoid or refuse it.

One thing you can do is to make the control group for your experiment be the control group for other experiments. Just ensure the test group is left intact and untouched.

So if you wanted to run another experiment, assuming you have plenty of control URLs to hit the minimum sessions for the first experiment, some could be allocated to a second test.

Non-QA of Test Implementation and Experiment Evaluation

It may be obvious, but we shall state it nonetheless: do check and QA the implementation of the test across all browsers, operating systems, and device types to avoid any kind of bias in the experiment results.

With the experiment having run, you're now ready to evaluate the experiment. How do you know it's run? When both groups have reached the required number set earlier, in this case, 18,000 pageviews.

To repeat, it can take up to two weeks to crawl the changes on your test group pages and another two weeks to recalculate the effects of those changes before reflecting them in the SERPs, be they better, worse, or no change.

With the experiment run, we'll import data from our website analytics. Just as before, the data is a CSV extract from a website analytics software:

```
test_analytics = pd.read_csv('data/sim_split_ab_data.csv')
```

Convert the date to date format:

```
test_analytics["date"] = pd.to_datetime(test_analytics["date"],
format="%Y/%m/%d")
```

```
test_analytics
```

You'll see that you now have a dataframe with all the URLs by date and outcomes labeled as test and control. As is the nature of analytics data, some dates are missing:

	landing_page	ab_group	date	sessions
0	https://www.next.com/shop/1-state-boots/	control	2019-09-05	0
1	https://www.next.com/shop/1-state-boots/	control	2019-09-10	0
2	https://www.next.com/shop/1-state-boots/	control	2019-09-12	1
3	https://www.next.com/shop/1-state-boots/	control	2019-09-13	0
4	https://www.next.com/shop/1-state-boots/	control	2019-09-16	1
...	...	...	...	...
427207	https://www.next.com/shop/zoe-morgan-earrings/	test	2019-09-18	6
427208	https://www.next.com/shop/zoe-morgan-earrings/	test	2019-09-22	5
427209	https://www.next.com/shop/zoe-morgan-earrings/	test	2019-09-26	4
427210	https://www.next.com/shop/zoe-morgan-earrings/	test	2019-09-28	0
427211	https://www.next.com/shop/zoe-morgan-earrings/	test	2019-09-29	2

427212 rows × 4 columns

Add missing dates as some URLs from either group will not have logged a pageview. We'll use the list comprehension technique to fill in the missing dates where for every unique landing page, we'll create a new date row where none exists:

```
test_analytics_expand = pd.DataFrame(
    [(x, y)
    for x in test_analytics['landing_page'].unique()
    for y in test_analytics['date'].unique()], columns=("landing_page",
    "date"),)

test_analytics_expand
```

The following is a screenshot of test_analytics_expand:

	landing_page	date
0	https://www.next.com/shop/1-state-boots/	2019-09-05
1	https://www.next.com/shop/1-state-boots/	2019-09-10
2	https://www.next.com/shop/1-state-boots/	2019-09-12
3	https://www.next.com/shop/1-state-boots/	2019-09-13
4	https://www.next.com/shop/1-state-boots/	2019-09-16
...	...	...
807207	https://www.next.com/shop/zoe-morgan-earrings/	2019-09-09
807208	https://www.next.com/shop/zoe-morgan-earrings/	2019-09-17
807209	https://www.next.com/shop/zoe-morgan-earrings/	2019-09-26
807210	https://www.next.com/shop/zoe-morgan-earrings/	2019-09-28
807211	https://www.next.com/shop/zoe-morgan-earrings/	2019-09-08

807212 rows × 2 columns

Note that there are more rows than before because of the added missing dates. These will need to have session data added, which will be achieved by merging the original analytics data:

```
test_analytics_expanded = test_analytics_expand.merge(
    split_ab_dev, how="left", on=['landing_page'])
test_analytics_expanded = test_analytics_expanded.merge(
    test_analytics, how="left", on=["date", "landing_page", 'ab_group'])
```

Post merge, any landing pages with missing dates will have "NaNs" (not a number), which is dealt with by filling those with zeros and converting the data type to an integer:

```
test_analytics_expanded['sessions'] = test_analytics_expanded['sessions'].
fillna(0).astype(int)
test_analytics_expanded
```

The following is a screenshot of test_analytics_expanded which has the landing pages labeled by their ab_group:

	landing_page	date	ab_group	sessions
0	https://www.next.com/shop/1-state-boots/	2019-09-05	control	0
1	https://www.next.com/shop/1-state-boots/	2019-09-10	control	0
2	https://www.next.com/shop/1-state-boots/	2019-09-12	control	1
3	https://www.next.com/shop/1-state-boots/	2019-09-13	control	0
4	https://www.next.com/shop/1-state-boots/	2019-09-16	control	1
...	...	...	...	...
807207	https://www.next.com/shop/zoe-morgan-earrings/	2019-09-09	test	2
807208	https://www.next.com/shop/zoe-morgan-earrings/	2019-09-17	test	0
807209	https://www.next.com/shop/zoe-morgan-earrings/	2019-09-26	test	4
807210	https://www.next.com/shop/zoe-morgan-earrings/	2019-09-28	test	0
807211	https://www.next.com/shop/zoe-morgan-earrings/	2019-09-08	test	0

807212 rows × 4 columns

Our dataset is ready for some data exploration before finally testing. We explore the data to observe the distribution of sessions, which helps with our model selection.

Split A/B Exploratory Analysis

We'll estimate the parameters including the average sessions and their variation by A/B group:

```
ab_means = (
    test_analytics_expand[test_analytics_expand["sessions"] != 0]
    .groupby(["ab_group"])
    .agg({"date": ["min", "max", day_range], "sessions": "mean"})
)
ab_sigmas = test_analytics_expand.groupby(["ab_group"]).agg({"sessions":
[get_sigma]})

pd.concat([ab_means, ab_sigmas], axis=1)
```

The pd.concat([ab_means, ab_sigmas], axis=1) dataframe is shown as follows:

	date		sessions			
	min	max	day_range	sum	mean	get_sigma
ab_group						
control	2019-09-05	2019-10-02	27	173547	2.165467	0.891922
test	2019-09-05	2019-10-02	27	72813	4.399577	0.504903

The dataframe shows that the minimum sample sessions were comfortably hit and that it looks like the test group has made a significant difference, that is, a statistically significant higher number of sessions.

Let's plot the data by test and control groups to explore it more, starting with an overall time trend. The data will be aggregated by date and ab_group with the sessions averaged:

```
simul_abgroup_trend = sim_split_ab_data.groupby(["date", 'ab_group']).
agg({"sessions": "mean"}).reset_index()
simul_abgroup_trend.head()
```

The simul_abgroup_trend.head() is shown as follows:

	date	ab_group	sessions
0	2019-09-05	control	0.505078
1	2019-09-05	test	4.068471
2	2019-09-06	control	0.500034
3	2019-09-06	test	4.046225
4	2019-09-07	control	0.466905

Once aggregated, we can now plot:

```
simul_abgroup_trend_plt = (
    ggplot(simul_abgroup_trend,
           aes(x = 'date', y = 'sessions', colour = 'ab_group',
           group = 'ab_group')) +
    geom_line(alpha = 0.6, size = 3) +
    labs(y = 'Count', x = '') +
```

```
    theme(legend_position = 'right',
          axis_text_y =element_text(rotation=0, hjust=1, size = 12),
          legend_title = element_blank()
          )
)

simul_abgroup_trend_plt.save(filename = 'images/3_simul_abgroup_trend_
plt.png',
                             height=5, width=8, units = 'in', dpi=1000)
simul_abgroup_trend_plt
```

Figure 7-4 shows the resulting time series plot of simul_abgroup_trend_plt. Both groups experienced dips during that period; however, the test group has outperformed the control group.

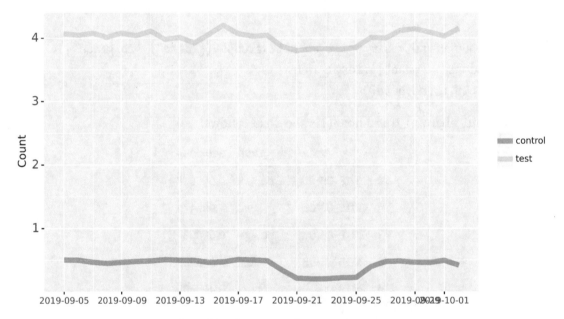

Figure 7-4. *Time series plot of both test and control group sessions over time*

Next, we'll inspect the distribution of sessions overall, starting with a histogram:

```
ab_assign_plt = (
    ggplot(test_analytics_expanded,
           aes(x = 'sessions', fill = 'ab_group')) +
    geom_histogram(alpha = 0.6, bins = 30) +
```

```
    labs(y = 'Count', x = '') +
    #scale_y_log10() +
    #coord_flip() +
    theme(legend_position = 'none',
          axis_text_y =element_text(rotation=0, hjust=1, size = 12),
          legend_title = element_blank()
        ) +
    facet_wrap('ab_group', scales = 'free')
)

ab_assign_plt.save(filename = 'images/4_ab_test_plt.png',
                              height=5, width=8, units = 'in', dpi=1000)
ab_assign_plt
```

Figure 7-5 shows ab_assign_plt, which is a side-by-side comparison of both control and test distributions of sessions. The chart shows that the distribution of the test group has much more data points above zero, which looks promising.

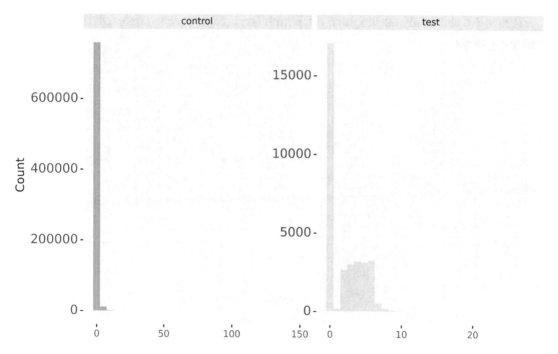

Figure 7-5. *Histogram distribution plots of both Control and Test*

The box plot method will be used to contrast the distributions further:

```
ab_assign_box_plt = (
    ggplot(test_analytics_expand,
           aes(x = 'ab_group', y = 'sessions',
               fill = 'ab_group', colour = 'ab_group')) +
    geom_boxplot(alpha = 0.8) +
    labs(y = 'Count', x = '') +
    theme(legend_position = 'none',
          axis_text_y =element_text(rotation=0, hjust=1, size = 12),
          legend_title = element_blank()
          )
)

ab_assign_box_plt.save(filename = 'images/4_ab_test_box_plt.png',
                               height=5, width=8, units = 'in', dpi=1000)
ab_assign_box_plt
```

Figure 7-6 shows ab_assign_box_plt, which is a box plot comparison of the control and test groups.

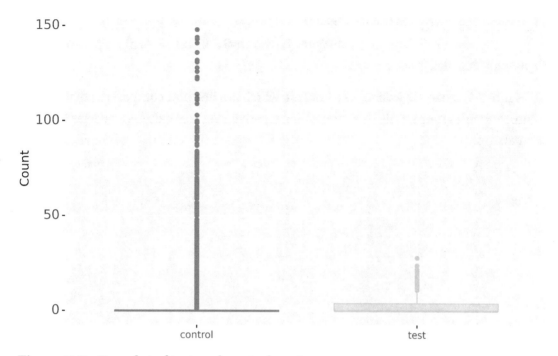

Figure 7-6. *Box plot of test and control sessions*

The control group has many more outliers, but the test group has much less zeros than the control group.

The scales make this hard to distinguish, so we'll take a logarithm of the session scale to visualize this further:

```
ab_assign_log_box_plt = (
    ggplot(test_analytics_expanded,
           aes(x = 'ab_group', y = 'sessions',
               fill = 'ab_group', colour = 'ab_group')) +
    geom_boxplot(alpha = 0.6) +
    labs(y = 'Count', x = '') +
    scale_y_log10() +
    theme(legend_position = 'none',
          axis_text_y =element_text(rotation=0, hjust=1, size = 12),
          legend_title = element_blank()
          )
)
```

```
ab_assign_log_box_plt.save(filename = 'images/4_ab_assign_log_box_plt.png',
                          height=5, width=8, units = 'in', dpi=1000)
ab_assign_log_box_plt
```

Figure 7-7 shows ab_assign_log_box_plt, which is a box plot comparison of the control and test groups, only this time with a logarized vertical axis for an easier visual comparison.

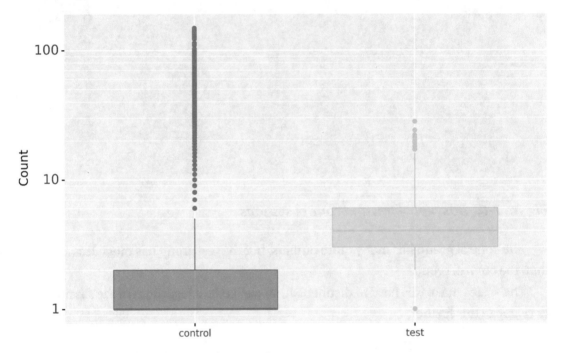

Figure 7-7. *Box plot of test and control*

In all cases, we can see that the average sessions are close to zero, and there are many landing pages on any given day with zero sessions, which indicates that sessions are a rare event. This type of distribution is known as "Poisson."

As a consequence, we'll use a negative binomial distribution to test the differences between test and control for significance.

First, we'll mark up the data as being test (1.0) or control (0.0), then convert it to an array:

```
X = np.where(test_analytics_expand['ab_group'] == 'control', 0.0, 1.0)
X = add_constant(X)
X = np.asarray(X)
X

array([[1., 0.],
       [1., 0.],
       [1., 0.],
       ...,
       [1., 1.],
       [1., 1.],
       [1., 1.]])
```

Fit a model of sessions by ab_group using negative binomial:

```
ab_model = NegativeBinomial(test_analytics_expand['sessions'],
    X).fit()
```

Print the model summary:

```
ab_model.summary()
```

The following is a screenshot of ab_model.summary():

```
Optimization terminated successfully.
        Current function value: 1.155104
        Iterations: 9
        Function evaluations: 10
        Gradient evaluations: 10
```

NegativeBinomial Regression Results

Dep. Variable:	sessions	No. Observations:	807212
Model:	NegativeBinomial	Df Residuals:	807210
Method:	MLE	Df Model:	1
Date:	Thu, 20 Oct 2022	Pseudo R-squ.:	0.02809
Time:	08:16:19	Log-Likelihood:	-4.3946e+05
converged:	True	LL-Null:	-4.5216e+05
Covariance Type:	nonrobust	LLR p-value:	0.000

	coef	std err	z	P>|z|	[0.025	0.975]
const	-1.4918	0.004	-350.822	0.000	-1.500	-1.484
x1	2.2453	0.018	126.281	0.000	2.210	2.280
alpha	9.5439	0.049	196.050	0.000	9.449	9.639

From the preceding result, we can conclude that the change was indeed significant. The test group (shown by *x1*) exhibited 2.24 more pageviews on average compared to control.

In terms of significance, the LLR p-value is zero, so the chances of the difference occurring due to random noise are incredibly slim.

Interestingly, the pseudo r-squared which measures the extent to which ab_group can explain the sessions per se is very low at 0.029, which means the model is very noisy and would require many more other factors to predict levels of traffic other than ab_group.

Inconclusive Experiment Outcomes

Experiments may not go the way you expected for a number of reasons:

- The expected difference is too high – so consider revising and rerunning the experiment.

- The hypothesis needs to be tested differently – perhaps using a different measure or a different test.

- You need a different time period – despite meeting the sample size requirements, it could be down to seasonal effects such as the time of year or the data fulfilling the sample requirement before a full week is run or Google wasn't given a chance to process the changes (see the previous discussion).

- Other external forces.

By setting your hypothesis in the first instance, regardless of the outcome, you will have learned something, and you will be able to move forward with a sensible plan, be it your next test or a sitewide implementation of your test.

Summary

Experiments have always been a part of the SEO expert's skill in determining what tactics are likely to work, even if sometimes the scientific rigor is missing. In the enterprise setting, a rigorous experiment design is essential due to the impact on revenue and the need to prove recommendations are beneficial, before rolling out changes sitewide. While there are tools that assist in this area, it is also useful to understand the data science behind SEO split tests and the considerations that must be borne in mind. In this chapter, we covered

- The importance of experiments in SEO

- Generating hypotheses

- Experiment design

- Running your experiments

- Evaluating your experiments

- And what to do if your experiment "fails"

In the next chapter, we will cover SEO reporting in the form of dashboards.

CHAPTER 8

Dashboards

Although a performance dashboard system in itself doesn't solve SEO problems directly, having the infrastructure can be a very useful repository for data to support SEO science as well as create visuals that communicate useful trends, changes, threats, and opportunities.

Even more importantly, SEO is data rich, and there are numerous data sources and a good many number of things you can possibly measure in SEO, so the picture can look very noisy and at worst can be useless if you can't clearly see and get to the signal.

Having a performance reporting system that uses well-designed and well-thought-out dashboards will help highlight the most important trends from the noisy data. It will also be easier to identify causal effects.

We will be supplying some code, written in SQL, to help you understand how to achieve some of the most valuable visuals.

Data Sources

The types of data sources you would want for your dashboard will be anything that (a) offers an API and (b) obviously adds information to understanding your SEO performance more effectively. These may include (and this is by no means exhaustive)

- *Website analytics*: Google Analytics (GA), Adobe Analytics, Looker, Segment

- *Webmaster tools*: Google Search Console (GSC), Bing Webmaster

- *Cloud web crawlers*: DeepCrawl, OnCrawl, and Botify

- *SERPs*: getSTAT, SEO Monitor, DataForSEO, AWR, AccuRanker

- *Link checkers*: AHREFs, Majestic, DataForSEO

© Andreas Voniatis 2023
A. Voniatis, *Data-Driven SEO with Python*, https://doi.org/10.1007/978-1-4842-9175-7_8

- *Social*: BuzzSumo

- *Keywords*: SEMRush, Keywords.io

- *Ad platforms*: Google Ads, DV360

Don't Plug Directly into Google Data Studio

Google provides convenient connectors to plugging in data sources, like Google Analytics (GA) and Google Search Console (GSC). This allows for data to be imported directly into Google Data Studio (GDS) which makes SEO dashboards easy.

However, this is a missed opportunity because there is no way to overlay the data between the two data sources in GDS. GDS is a front end for visualizing data.

Without a process that goes between the data source and the front end, the data is raw, undistilled, and less useful for spotting trends and uncovering insights.

Using Data Warehouses

This is where a data warehouse like Google's BigQuery or Amazon's Redshift comes in. You store the data in those data warehouses, and the front end, be it GDS, Tableau, DOMO, or others, will use custom Structured Query Language (SQL) to query the data warehouse and get the data in a format ready to drive the charts you want to show.

We will share SQL code with you and some charts to help you on your way to building your own SEO dashboards.

However, before you start building, you need to get the data into the data warehouse. So, how do we achieve that?

Extract, Transform, and Load (ETL)

Extract, transform, and load (ETL) is a method by which you literally

- *Extract* your data from your data source APIs.

- *Transform* is where you run calculations and create new calculated fields from the data extracted.

- *Load* is the part where you load the transformed data into the data warehouse.

There are numerous configurations you can pursue depending on which cloud stack you go with, your team's cloud engineering skills, and your budget.

Extracting Data

The extract process will usually be automated where your APIs get queried on a daily basis (known as "polling") using a virtual machine running the script. The data gets stored either in storage or a data warehouse.

If your cloud engineering skills are nonexistent, you can still upload data via CSV format to the data warehouse.

The following is some code to extract data from a number of APIs which will be the main Google products and some of (not all of) the more well-known SEO processes:

- Google Analytics

- Google Search Console

- DataForSEO SERPs API

- Google PageSpeed API

We'll now provide Python code for you to connect to these APIs not just for reporting purposes as this code can be adapted to support other SEO science activities covered in other chapters.

Google Analytics

Traffic remains a key lever of growth, and Google Analytics is widely used as a web analytics package. However, more organic search traffic will not always correlate directly with more revenue, but it may indicate engagement through other means.

The following will detail code to extract data from the most well-known and used website analytics APIs being Google Analytics version 4.

Import the API libraries:

```
import pandas as pd
from pathlib import Path
import os
from datetime import date, timedelta
```

Set the file path of the credential keys which is a JSON file and obtainable from your Google Cloud Platform account under API Libraries ➤ Credentials:

```
credentials_path = Path("keys/xxxxx.json")
credentials_path_str = str(credentials_path.absolute())

os.environ["GOOGLE_APPLICATION_CREDENTIALS"] = credentials_path_str
```

Import the other APIs from the other Google library:

```
from google.analytics.data_v1beta import BetaAnalyticsDataClient
from google.analytics.data_v1beta.types import DateRange
from google.analytics.data_v1beta.types import Dimension
from google.analytics.data_v1beta.types import Metric
from google.analytics.data_v1beta.types import RunReportRequest

client = BetaAnalyticsDataClient()
```

Define a function to run an aggregated report which will require a date range and the property ID of the GA4 account.

In this function, we query the API using the inputs to build the request which includes the metrics we want and store the API response.

We've set the dimension as landingPage because that's how we want the traffic numbers broken down. Other dimensions may be used which are listed here (https://developers.google.com/analytics/devguides/reporting/data/v1/api-schema).

```
def aggregated_run_report(client, property_id="[your-GA-property-id]",
date_ranges=[DateRange(start_date="2020-03-31", end_date="today")]):

    request = RunReportRequest(
        property=f"properties/{property_id}",
        dimensions=[Dimension(name="landingPage")],
        metrics=
            [
                Metric(name="activeUsers"),
                Metric(name="screenPageViewsPerSession"),
                Metric(name="bounceRate"),
                Metric(name="averageSessionDuration"),
                Metric(name="userConversionRate"),
```

```
            Metric(name="ecommercePurchases"),
        ],
    date_ranges=date_ranges,
)
response = client.run_report(request)

return response

response = aggregated_run_report(client)
print("Report result:")
for row in response.rows:
    print(row.dimension_values[0].value, row.metric_values[0].value)

Report result:
/ 11347
/blog/sell-airtime-over-charged-your-line-dont-panic 8423
/faq 4870
/blog/sell-airtime-over-charged-your-line-dont-panic 2355
/privacy 1338
```

The next function uses the API response result rows and packages it into a single dataframe:

```
def ga4_response_to_df(response):
    dim_len = len(response.dimension_headers)
    metric_len = len(response.metric_headers)
    all_data = []
    for row in response.rows:
        row_data = {}
        for i in range(dim_len):
            row_data.update({response.dimension_headers[i].name: row.
            dimension_values[i].value})
        for i in range(metric_len):
            row_data.update({response.metric_headers[i].name: row.metric_
            values[i].value})
        all_data.append(row_data)
    df = pd.DataFrame(all_data)
    return df
```

```
df = ga4_response_to_df(response)
df.info()
```

This results in the following:

```
<class 'pandas.core.frame.DataFrame'>
RangeIndex: 418 entries, 0 to 417
Data columns (total 8 columns):
 #   Column          Non-Null Count  Dtype
---  ------          --------------  -----
 0   landingPage 418 non-null    object
 1   dateRange   418 non-null    object
 2   activeUsers 418 non-null    object
 3   screenPageViewsPerSession  418 non-null    object
 4   bounceRate   418 non-null    object
 5   averageSessionDuration     418 non-null    object
 6   userConversionRate         418 non-null    object
 7   ecommercePurchases         418 non-null    object
dtypes: object(8)
memory usage: 26.2+ KB
```

Printing the dataframe's properties via df.info() tells us the data types which all appear to be strings, which is okay for the landing page but not for metrics, such as activeUsers, which should be converted to numeric before the data can be processed further.

```
df.head()
```

The following resulting dataframe shows the dimensions "landingPage" along with the metrics. The data is aggregated across the entire date range.

	landingPage	dateRange	activeUsers	screenPageViewsPerSession	bounceRate	averageSessionDuration	userConversionRate	ecommercePurchases
0	/blog/sell-airtime-over-charged-your-line-dont...	2022-11-26	305.0	1.759878	0.294833	186.317980	0.0	0.0
1	/blog/sell-airtime-over-charged-your-line-dont...	2022-11-25	297.0	1.803175	0.298413	159.541673	0.0	0.0
2	/blog/sell-airtime-over-charged-your-line-dont...	2022-11-27	279.0	1.835526	0.256579	206.459100	0.0	0.0
3	/	2022-11-25	265.0	2.515901	0.155477	90.767197	0.0	0.0
4	/	2022-11-26	261.0	2.536496	0.145985	71.557897	0.0	0.0

But suppose you wanted the data broken down by date as well.

The following function will do just that with the default number of day parameters set to two years:

```python
def dated_run_report_to_df(client, property_id="xxxxxxxxx", n_days=365*2):

    date_ranges = []
    count = 0
    df_output = pd.DataFrame()
    for i in range(n_days):
        count += 1

        current = date.today() - timedelta(days=i)
        before = date.today() - timedelta(days=i+1)
        date_ranges.append(DateRange(start_date=before.strftime
        ("%Y-%m-%d"),
    end_date=current.strftime("%Y-%m-%d"),
    name=current.strftime("%Y-%m-%d")))

        if count == 4:
            response = aggregated_run_report(client,
            property_id=property_id, date_ranges=date_ranges)
            df = ga4_response_to_df(response)
            df_output = pd.concat([df_output, df], ignore_index=True)

            # Re-initialize
            count = 0
            date_ranges = []

    return df_output
```

Run the function; in this case, we'll extract the last 90 days:

```
df = dated_run_report_to_df(client, n_days=90)
```

The following function converts the column data formats from str to their appropriate formats which are mostly numeric:

```
def format_df(df):
    df["dateRange"] = pd.to_datetime(df["dateRange"])
    df["activeUsers"] = df["activeUsers"].astype("float")
    df["screenPageViewsPerSession"] = df["screenPageViewsPerSession"].
    astype("float")
    df["bounceRate"] = df["bounceRate"].astype("float")
    df["averageSessionDuration"] = df["averageSessionDuration"].
    astype("float")
    df["userConversionRate"] = df["userConversionRate"].astype("float")
    df["ecommercePurchases"] = df["ecommercePurchases"].astype("float")

    return df

df = format_df(df)
df.info()
```

This results in the following:

```
<class 'pandas.core.frame.DataFrame'>
RangeIndex: 376 entries, 0 to 375
Data columns (total 8 columns):
 #   Column          Non-Null Count  Dtype
---  ------          --------------  -----
 0   landingPage 376 non-null    object
 1   dateRange   376 non-null    datetime64[ns]
 2   activeUsers 376 non-null    float64
 3   screenPageViewsPerSession  376 non-null    float64
 4   bounceRate  376 non-null    float64
 5   averageSessionDuration     376 non-null    float64
 6   userConversionRate         376 non-null    float64
 7   ecommercePurchases         376 non-null    float64
```

```
dtypes: datetime64[ns](1), float64(6), object(1)
memory usage: 23.6+ KB
```

df

This results in the following:

	landingPage	dateRange	activeUsers	screenPageViewsPerSession	bounceRate	averageSessionDuration	userConversionRate	ecommercePurchases
0	/blog/sell-airtime-over-charged-your-line-dont...	2022-11-26	305.0	1.759878	0.294833	186.317980	0.0	0.0
1	/blog/sell-airtime-over-charged-your-line-dont...	2022-11-25	297.0	1.803175	0.298413	159.541673	0.0	0.0
2	/blog/sell-airtime-over-charged-your-line-dont...	2022-11-27	279.0	1.835526	0.256579	206.459100	0.0	0.0
3	/	2022-11-25	265.0	2.515901	0.155477	90.767197	0.0	0.0
4	/	2022-11-26	261.0	2.536496	0.145985	71.557897	0.0	0.0

The result is a dataframe ready for transformation.

DataForSEO SERPs API

The DataForSEO SERPs API is purpose built to return the Google search results for a given keyword. This is useful for checking rankings, understanding the search intent of keywords, and other SEO research.

We start by defining our target keyword list:

keywords_lst

This results in the following:

```
['airtime app',
 'airtime to cash app',
 'airtime transfer',
 'app to sell airtime',
 'app to transfer airtime from one network to another',
 'bet with airtime and win cash',
 'buy airtime online',
 'buy airtime with discount',
 'buy recharge card online',
 'buy recharge card online with debit card',
```

```
'can i subscribe dstv with airtime?',
'can i use my airtime to buy electricity?',
'can you convert airtime to cash?', ...]
```

With this API, you'll need your DataForSEO client file which resides in the same folder as the Jupyter notebook script file running this code:

```
from client import RestClient

client = RestClient("[your-username]", "xxxxxxxxxxxxxxx")
```

The API will need to know which country you'd like to see the search results for, the device, and the language, which are defined as follows. The countries list may be found here (https://docs.dataforseo.com/v3/serp/google/locations/?bash).

```
location = 2826
language = "en"
device_input = 'mobile'
```

The following are functions to query the API. set_post_data will set the parameters for the search:

```
def set_post_data(search_query):
    post_data = dict()
    post_data[len(post_data)] = dict(
        language_code = language,
        location_code = location,
        device = device_input,
        keyword = search_query,
        calculate_rectangles = True)
    return post_data
```

The function get_api_result uses the preceding function to structure the input which will be used to request the search results. The API result is stored in a variable named "response."

There is a try loop in place so that should there be an issue with the API call, the function carries on and moves on to the next keyword, to prevent holding up the entire operation or stalling:

```
def get_api_result(search_query):
    post_data = set_post_data(search_query)
    response = client.post("/v3/serp/google/organic/live/advanced",
    post_data)

    try:
        return response
    except Exception as e:
        print(response)
        print(e)
        return None
```

With multiple keywords to be queried, we'll want to call the function multiple times, so we'll do that using a for loop.

Initialize an empty dictionary to store the individual API results:

```
desktop_serps_returned = {}
```

Add a for loop to query the API for each and every keyword in the list:

```
i = 0

for search_query in set(keywords_lst):
    print(search_query, i + 1, len(keywords_lst) - i - 1)
    i += 1
    serp_dict = get_api_result(search_query)
    desktop_serps_returned[search_query] = serp_dict
```

Printing the entire output in this book and in the Jupyter notebook would be too impractical. Instead, we'll print the keys of the dictionary where the data is stored which shows the keywords that have API data:

```
desktop_serps_returned.keys()
```

```
dict_keys(['buy recharge card online with debit card', 'can i use my
airtime to buy electricity?', 'buy recharge card online', 'can you convert
airtime to cash?', 'airtime to cash app', 'app to sell airtime', 'bet with
airtime and win cash', 'airtime sell', 'buy airtime online', 'airtime
sharing', 'can i subscribe dstv with airtime?', 'buy mtn', 'airtime buy',
'airtime transfer', 'buy airtime with discount', 'airtime app', 'airtime
application', 'airtime bills', 'airtime funding', 'app to transfer airtime
from one network to another'])
```

With the data stored, the dictionary requires unpacking into a dataframe format, which will be carried out as follows.

Initialize an empty list:

```
desktop_serps_flat_df = []
```

Using a for loop, we'll iterate through the dictionary keys which will be used to select parts of the dictionary by keyword. Then we loop through the contents of the dictionary data for that keyword and add these to the empty list initialized earlier:

```
for serp in desktop_serps_returned.keys():
    single_serp = desktop_serps_returned[serp]
    keyword = serp
    for task in single_serp['tasks']:
        cost = task['cost']
        task_id = task['id']
        se = task['data']['se']
        device = task['data']['device']
        os = task['data']['os']

        for res in task['result']:
            for idx, item in enumerate(res['items']):
```

```
desktop_serps_flat_df.append(
    (
        cost, task_id, se, device, os, res['keyword'],
        res['location_code'], res['language_code'],
        res['se_results_count'], res['type'], res['se_domain'],
        res['check_url'], item['rank_group'], item['rank_absolute'],
        item.get('url', None), item.get('domain'), item.get('is_image'),
        item.get('is_featured_snippet'),
        item.get('is_video'), item.get('is_malicious'), item.get
        ('is_web_story'),
        item.get('description'), item.get('pre_snippet'), item.get
        ('amp_version'),
        item.get('rating'), item.get('price'), item.get('highlighted'),
        item.get('links'), item.get('faq'), item.get('extended_people_
        also_search'),
        item.get('timestamp'), item.get('rectangle'),
        res['datetime'], item.get('title'), item.get('cache_url')
    )
)
```

Once the list has all the added keyword SERP data, it is converted into a dataframe:

```
desktop_full_df = pd.DataFrame(
    desktop_serps_flat_df,
    columns=[
        'cost', 'task_id', 'se', 'device', 'os', 'keyword',
        'location_code', 'language_code', 'se_results_count',
        'type', 'se_domain', 'check_url', 'rank_group',
        'rank_absolute', 'url', 'domain', 'is_image', 'is_featured_snippet',
        'is_video', 'is_malicious', 'is_web_story',
        'description', 'pre_snippet', 'amp_version',
        'rating', 'price', 'highlighted', 'links',
        'faq', 'extended_people_also_search', 'timestamp', 'breadcrumb',
        'datetime', 'title', 'cache_url'
    ]
)
desktop_full_df.head(2)
```

This results in the following:

	cost	task_id	se	device	os	keyword	location_code	language_code	se_results_count	type	...	price	highlighted	links	
0	0.004	08171635-2300-0139-0000-60754972b3ac	google	desktop	windows	buy recharge card online with debit card	2840	en	9280000	organic	...	None	None	None	N
1	0.004	08171635-2300-0139-0000-60754972b3ac	google	desktop	windows	buy recharge card online with debit card	2840	en	9280000	organic	...	None	None	[{'type': 'ad_link_element', 'title': 'myWalgr...	N

The result is the API data in a dataframe which is ready for reporting or prereporting transformation.

Google Search Console (GSC)

Google Search Console (GSC) is first-party data and the source of truth for most SEOs. Here, we will show you how to extract data from the API which will provide more rows than the standard 1000 rows available in the interface.

The API will require a Google Cloud Platform (GCP) account in which you will have to create a GCP project and, within that project, some credentials with a JSON key.

Let's start by importing some libraries:

```
from apiclient import errors
from apiclient.discovery import build

import datetime
import httplib2
import re
import pandas as pd
import numpy as np
from collections import defaultdict

from oauth2client.client import OAuth2WebServerFlow
from datetime import datetime, timedelta, date
from dateutil.relativedelta import relativedelta
import calendar
import time
```

The script is constructed to allow you to query multiple domains, which could be useful for an agency reporting system where you look after more than one client or, if you're in the client side, multiple sites:

```
site_list = ['https://www.babywishiest.com']
site = 'https://www.babywishiest.com'
client_name = babywishiest
```

The dimensions will give a breakdown of the data, while no dimensions will return summary data for the date range:

```
dimensions = ['query', 'page']
```

To filter to a device, enter MOBILE, DESKTOP, or TABLET or leave it blank for all devices:

```
device_filter = ''
```

To filter to a search type, enter WEB, IMAGE, VIDEO, or discover. This defaults to WEB if left blank:

```
search_filter = ''
```

To filter to a specific three-digit country code (e.g., FRA). A list of country codes is available here (https://en.wikipedia.org/wiki/ISO_3166-1_alpha-3). If left blank, the API will default to all:

```
country_filter = ''
```

To filter pages which contain a string, you can use operators such as "equals," "contains," "notContains," or "notEquals":

```
page_filter_string = ''
page_filter_operator = 'equals'
```

The same can be applied to queries:

```
query_filter_string = ''
query_filter_operator = ''
```

State your date range for the query, which will be converted in datetime format:

```
start_date = '2022-08-01'
end_date = '2022-11-30'

start_date_datetime = datetime.strptime(start_date, '%Y-%m-%d').date()
end_date_datetime = datetime.strptime(end_date, '%Y-%m-%d').date()
print(start_date_datetime, end_date_datetime)
```

Enter a date grouping to break down the data. Use D (day), W (week), M (month), or A (all):

```
date_grouping = 'A'
```

Enter API credentials obtainable from the APIs section of your GCP project:

```
CLIENT_ID = 'xxxxxxx'
CLIENT_SECRET = 'xxxxxx'
```

Add sleep time between requests. Increase this if you are hitting limits or getting errors:

```
sleep_time = 10

2022-08-01 2022-11-30
```

With the parameters specified, the next block deals with authentication using the OAuth method:

```
OAUTH_SCOPE = 'https://www.googleapis.com/auth/webmasters.readonly'
REDIRECT_URI = 'urn:ietf:wg:oauth:2.0:oob'
```

Run through the OAuth flow and retrieve credentials:

```
flow = OAuth2WebServerFlow(CLIENT_ID, CLIENT_SECRET, OAUTH_SCOPE, redirect_
uri=REDIRECT_URI)
authorize_url = flow.step1_get_authorize_url()
print ('Go to the following link in your browser: ' + authorize_url)
code = input('Enter verification code: ').strip()
credentials = flow.step2_exchange(code)
```

Create an httplib2.Http object and authorize it with your credentials:

```
http = httplib2.Http()
http = credentials.authorize(http)

webmasters_service = build('searchconsole', 'v1', http=http)
```

Go to the following link in your browser: https://accounts.google.com/o/ oauth2/v2/auth?client_id=xxxxx&redirect_uri=xxxxx&scope=https%3A%2F%2Fwww. googleapis.com%2Fauth%2Fwebmasters.readonly&access_type=offline&response_ type=code
Enter verification code: xxxxx

Once authenticated, set the custom number of rows to retrieve from the API per request:

```
row_limit = 25000
```

Create a dataframe to store the full output:

```
output = pd.DataFrame()

request = {
            'rowLimit': row_limit,
            'startRow': 0
        }

if dimensions:
  request['dimensions'] = dimensions

if search_filter:
  request['searchFilterGroups'] = [{'filters':[{'dimension':'search',
  'expression':search_filter}]}]
```

Build dimension filters from the settings:

```
dimension_filters = []

if device_filter:
  dimension_filters.append({'dimension':'device', 'expression':device_
  filter})
if country_filter:
```

```
dimension_filters.append({'dimension':'country', 'expression':country_
filter})
if page_filter_string:
  dimension_filters.append({'dimension':'page','expression':page_filter_
  string, 'operator': page_filter_operator})
if query_filter_string:
  dimension_filters.append({'dimension':'query','expression':query_filter_
  string, 'operator': query_filter_operator,})
request['dimensionFilterGroups'] = [{'filters':dimension_filters}]

print(f'Filter: {dimension_filters}')
```

Loop through all the dates from start to end, inclusive and populate the request start and end dates with the date from the loop:

```
for site in site_list:
  for single_date in daterange(start_date_datetime, end_date_datetime,
  date_grouping):

    request['startDate'] = f"{single_date[0].strftime('%Y')}-{single_
    date[0].strftime('%m')}-{single_date[0].strftime('%d')}"
    request['endDate'] = f"{single_date[1].strftime('%Y')}-{single_date[1].
    strftime('%m')}-{single_date[1].strftime('%d')}"

    print(site + ' - ' + request['startDate'] + ' to ' +
    request['endDate'])

    run = True
    rowstart = 0
    request['startRow'] = rowstart

    while run:

      try:
        response_page = execute_request(webmasters_service, site, request)
        scDict_results = defaultdict(list)

        try:
          for row in response_page['rows']:
```

```
        if dimensions:
          for i,dimension in enumerate(dimensions):
          scDict_results[dimension].append(row['keys'][i] or 0)

        scDict_results['clicks'].append(row['clicks'] or 0)
        scDict_results['ctr'].append(row['ctr'] or 0)
        scDict_results['impressions'].append(row['impressions'] or 0)
        scDict_results['position'].append(row['position'] or 0)

      df = pd.DataFrame(data = scDict_results)
      df['start_date'] = request['startDate']
      df['end_date'] = request['endDate']
      df['site'] = site

      frames = [output, df]
      output = pd.concat(frames)
      print(str(len(df)) + ' results')

      time.sleep(sleep_time)

      if len(df) == row_limit:
        rowstart += row_limit
        request['startRow'] = rowstart
      else:
        run=False
    except:
      print('No results found for this date range')
      run=False

  except HttpError:
    print('Got an error. Retrying in 1m.')
    time.sleep(60)

Filter: []
https://www.babywishiest.com - 2022-08-01 to 2022-11-30
1672 results

output
```

This results in the following:

	query	page	clicks	ctr	impressions	position	start_date	end_date	site
0	baby wishlist	https://www.babywishiest.com/	154	0.564103	273	1.139194	2022-08-01	2022-11-30	https://www.babywishiest.com
1	baby wish list	https://www.babywishiest.com/	23	0.718750	32	1.000000	2022-08-01	2022-11-30	https://www.babywishiest.com
2	baby wishlist	https://www.babywishiest.com/login/	12	0.044118	272	1.139706	2022-08-01	2022-11-30	https://www.babywishiest.com
3	babywishlist	https://www.babywishiest.com/	6	1.000000	6	1.000000	2022-08-01	2022-11-30	https://www.babywishiest.com
4	lulworth changing unit	https://www.babywishiest.com/products/motherca...	5	0.031447	159	8.547170	2022-08-01	2022-11-30	https://www.babywishiest.com
...	...	...	...	...	...	...	...	...	...
1667	вишлист создать онлайн	https://www.babywishiest.com/	0	0.000000	1	98.000000	2022-08-01	2022-11-30	https://www.babywishiest.com
1668	онлайн вишлист	https://www.babywishiest.com/	0	0.000000	2	46.000000	2022-08-01	2022-11-30	https://www.babywishiest.com
1669	создать вишлист онлайн	https://www.babywishiest.com/	0	0.000000	14	72.785714	2022-08-01	2022-11-30	https://www.babywishiest.com
1670	בייבי ביורן מירקל	https://www.babywishiest.com/products/baby-bjo...	0	0.000000	9	28.555556	2022-08-01	2022-11-30	https://www.babywishiest.com
1671	וישליסט	https://www.babywishiest.com/	0	0.000000	2	90.500000	2022-08-01	2022-11-30	https://www.babywishiest.com

1672 rows × 9 columns

Although it's a large block of code, the API can be used to extract 100,000 rows of data if not much more.

Google PageSpeed API

The PageSpeed API is another core metric for SEOs especially with the Core Web Vitals (CWV) initiative introduced by Google in April 2020. The API is not only useful for checking your own site's CWV scores but also those of your SERP competitors.

To make use of this API, a key will be required which is obtainable from Google Cloud Platform (GCP) in the APIs section.

Start by defining your list of URLs to check CWV scores against:

```
desktop_serps_urls = ['https://pay.jumia.com.ng/services/airtime',
 'https://pay.jumia.com.ng/',
 'https://vtpass.com/',
 'https://www.gloverapp.co/products/airtime-to-cash',
 'https://www.zoranga.com/',
 'https://airtimeflip.com/',
```

```
 'https://www.tingtel.com/blog/sell-airtime-over-charged-your-line-
 dont-panic',
 'https://vtpass.com/payment',
 'https://pay.jumia.com.ng/services/mobile-data/mtn-mobile-data', ...]
```

```
"https://www.googleapis.com/pagespeedonline/v5/runPagespeed?url=[test-url]&
key=xxxxxxxxxxxxxxxxxxxxxxxxxxxxxxxxxxxxxxxxxxxxx"
```

Set the parameters for the API:

```
base_url = 'https://www.googleapis.com/pagespeedonline/v5/
runPagespeed?url='
strategy = '&strategy=desktop'
api_url = '&key=xxxxxxxxxxxxxxxxxxxxxxxxxxxxxxxxxxxxxxxxxxxxx'
```

Initialize an empty dictionary to store the data and a counter to keep track of the number of URLs being queried:

```
desktop_cwv = {}
i = 0
```

Loop through the list of URLs to query the API:

```
for url in desktop_serps_urls:
    request_url = base_url + url + strategy + api_url
    response = json.loads(requests.get(request_url).text)
    i += 1
    print(i, " ", request_url)
    desktop_cwv[url] = response
```

The keys are printed to list the URLs queried successfully:

```
desktop_cwv.keys()
```

```
dict_keys(['https://pay.jumia.com.ng/services/airtime', 'https://pay.
jumia.com.ng/', 'https://vtpass.com/', 'https://www.tingtel.com/blog/buy-
airtime-get-discount-on-every-airtime-recharge', 'https://www.gloverapp.
co/products/airtime-to-cash', 'https://www.zoranga.com/', 'https://
airtimeflip.com/', 'https://www.tingtel.com/blog/sell-airtime-over-charged-
your-line-dont-panic', 'https://www.tingtel.com/', 'https://www.tingtel.
```

com/faq/airtime-sell', 'https://vtpass.com/payment', 'https://pay.jumia.
com.ng/services/mobile-data/mtn-mobile-data', 'https://www.tingtel.com/
blog/transfer-airtime-one-airtime-works-for-all-networks', 'https://www.
tingtel.com/blog/airtime-bills-settle-electricity-cable-tv-bills-with-
airtime', 'https://www.tingtel.com/blog/fund-wallet-with-airtime'])

Iterate through the PageSpeed API JSON Response dictionary, starting with an empty list:

```
desktop_psi_lst = []
```

Loop through the dictionary by key to extract the different CWV metrics and store them in the list:

```
for key, data in desktop_cwv.items():
    if 'lighthouseResult' in data:
        FCP = data['lighthouseResult']['audits']['first-contentful-paint']
        ['numericValue']
        LCP = data['lighthouseResult']['audits']['largest-contentful-
        paint']['numericValue']
        CLS = data['lighthouseResult']['audits']['cumulative-layout-shift']
        ['numericValue']
        FID = data['lighthouseResult']['audits']['max-potential-fid']
        ['numericValue']
        SIS = data['lighthouseResult']['audits']['speed-index']
        ['score'] * 100
        desktop_psi_lst.append([key, FCP, LCP, CLS, FID, SIS])
desktop_psi_df = pd.DataFrame(desktop_psi_lst, columns = ['url', 'FCP',
'LCP', 'CLS', 'FID', 'SIS'])
desktop_psi_df
```

This results in the following:

	url	FCP	LCP	CLS	FID	SIS
0	https://pay.jumia.com.ng/services/airtime	730.000000	1116.000000	0.010691	267.0	98.0
1	https://pay.jumia.com.ng/	770.000000	997.000000	0.535293	267.0	89.0
2	https://vtpass.com/	870.000000	2390.000000	0.003771	81.0	90.0
3	https://www.tingtel.com/blog/buy-airtime-get-discount-on-every-airtime-recharge	1010.000000	1246.000000	0.316606	23.0	97.0
4	https://www.gloverapp.co/products/airtime-to-cash	423.945510	2043.964767	1.002655	179.0	74.0
5	https://www.zoranga.com/	1752.145607	2586.145607	0.004675	27.5	12.0
6	https://airtimeflip.com/	522.000000	3822.500000	0.029568	439.0	53.0
7	https://www.tingtel.com/blog/sell-airtime-over-charged-your-line-dont-panic	1000.000000	1080.000000	0.257226	58.0	97.0
8	https://www.tingtel.com/	576.008945	1040.000000	0.011780	33.0	99.0
9	https://www.tingtel.com/faq/airtime-sell	1000.000000	1000.000000	0.000192	28.0	97.0
10	https://vtpass.com/payment	785.000000	1317.500000	0.006833	36.0	0.0
11	https://pay.jumia.com.ng/services/mobile-data/mtn-mobile-data	730.000000	1465.000000	0.103316	330.0	97.0
12	https://www.tingtel.com/blog/transfer-airtime-one-airtime-works-for-all-networks	1012.000000	1152.000000	0.257163	46.0	97.0
13	https://www.tingtel.com/blog/airtime-bills-settle-electricity-cable-tv-bills-with-airtime	1000.000000	1140.000000	0.257163	49.0	97.0
14	https://www.tingtel.com/blog/fund-wallet-with-airtime	1011.000000	1191.000000	0.263009	59.0	97.0

The result is a dataframe showing all the CWV scores for each URL.

Transforming Data

The purpose of transforming the data, which has been extracted by the API or other means from your data source, is to

- Clean it up for further calculated metrics

- Derive meaningful stats such as month-on-month (mom) variance

- Insert it (i.e., loading) into a data warehouse

The code is going to continue from the Google Analytics (GA) data extracted earlier where we will cover the preceding points.

We start by copying the GA dataframe:

```
df_clean = df.copy()
```

Reprofile "averageSessionDuration" to be the number of seconds:

```
df_clean['averageSessionDuration'] = (df_clean['averageSessionDuration']
/ 60).round(1)
```

Create new columns for easier transformation based on time and calendar date units:

```
df_clean['month'] = df_clean['dateRange'].dt.strftime('%m')
df_clean['year'] = df_clean['dateRange'].dt.strftime('%Y')
df_clean['month_year'] = df_clean['dateRange'].dt.strftime('%Y-%m')

df_clean
```

This results in the following:

	landingPage	dateRange	activeUsers	screenPageViewsPerSession	bounceRate	averageSessionDuration	userConversionRate	ecommercePurchases	month
0	/blog/sell-airtime-over-charged-your-line-dont...	2022-11-26	305.0	1.759878	0.294833	3.1	0.0	0.0	11
1	/blog/sell-airtime-over-charged-your-line-dont...	2022-11-25	297.0	1.803175	0.298413	2.7	0.0	0.0	11
2	/blog/sell-airtime-over-charged-your-line-dont...	2022-11-27	279.0	1.835526	0.256579	3.4	0.0	0.0	11
3	/	2022-11-25	265.0	2.515901	0.155477	1.5	0.0	0.0	11
4	/	2022-11-26	261.0	2.536496	0.145985	1.2	0.0	0.0	11

With the data formatted, we can start transforming to derive new columns of trend data such as

- Averages

- Standard deviations (for variation)

- Periodic changes (such as month-on-month)

Let's make a copy and rename it to reflect that we're aggregating by landing page and by month:

```
ga4_lp_agg_month = df_clean.copy()
```

We'll create some basic summary statistics which will be the average ("mean") and total ("sum") of various GA metrics using the groupby() and agg() functions:

```
ga4_lp_agg_month_basic = ga4_lp_agg_month.groupby(['landingPage',
'month_year']).agg({'activeUsers':'sum',
        'screenPageViewsPerSession':'mean',
        'bounceRate':'mean',
        'averageSessionDuration':'mean',
        'userConversionRate':'mean',
        'ecommercePurchases':'sum'
    }).reset_index()

ga4_lp_agg_month_basic
```

This results in the following:

	landingPage	month_year	activeUsers	screenPageViewsPerSession	bounceRate	averageSessionDuration	userConversionRate	ecommercePurchases
0	(not set)	2022-08	110.0	0.000000	0.977922	0.436364	0.0	0.0
1	(not set)	2022-09	195.0	0.000000	0.989683	0.183333	0.0	0.0
2	(not set)	2022-10	196.0	0.000000	0.977181	0.432258	0.0	0.0
3	(not set)	2022-11	220.0	0.000000	1.000000	0.032143	0.0	0.0
4	/	2022-08	1753.0	2.323391	0.218771	3.418182	0.0	0.0

The metrics as shown earlier are summarized by landing page and month_year which can be used to feed a basic SEO dashboard reporting system.

Usually in all cases, we track the mean and standard deviation. The average gives us a useful indicator of where a channel is at in terms of performance, as it will indicate where most data points were or will be for a given category of data for a given point of time.

Averages, as every statistician (and many others being statistically aware) will tell you, can be dangerous on their own when making inferences or decisions even. This is why we also track the standard deviation as this indicator tells us something about the variation of a given metric, that is, how consistent it is.

In practical terms, the standard deviation tells us how close the data points are to the average. And what can we deduce from this?

We can deduce which averages we're more likely to trust or rely on for comparing between months. So the standard deviation can tell us a bit about the quality of the averages for the purpose of confidence in the data and for comparisons and also how the metric we're tracking is behaving over time.

For example, you might find that the standard deviation is increasing or decreasing and should therefore try to understand what the reason behind it is. Could it be

- Changes in Google through algorithm updates?

- Changes in your user behavior, search intent of the query, your brand positioning, or the market?

- Changes in your site's UX or content or an architectural change?

Tracking the standard deviation could help you see whether something is afoot for the better or worse:

```
ga4_lp_agg_month = df_clean.copy()
```

Perform an aggregation to derive various average and standard deviation statistics:

```
ga4_lp_agg_month_mean = ga4_lp_agg_month.groupby(['landingPage',
    'month_year']).agg({'activeUsers':'mean',
        }).reset_index()

ga4_lp_agg_month_mean = ga4_lp_agg_month_mean.rename(columns =
{'activeUsers':'activeUsers_avg'})
```

While "mean" calculates the average, "std" calculates the standard deviation:

```
ga4_lp_agg_month_std = ga4_lp_agg_month.groupby(['landingPage',
    'month_year']).agg({'activeUsers':'std',
        'bounceRate':'std',
        }).reset_index()
```

Rename the columns:

```
ga4_lp_agg_month_std = ga4_lp_agg_month_std.rename(columns =
{'activeUsers':'activeUsers_std',
  'bounceRate':'bounceRate_std'
 })
```

Join the data to the basic dataframe created earlier:

```
ga4_lp_agg_month_stats = ga4_lp_agg_month_basic.merge(ga4_lp_agg_
month_mean,
        on = ['landingPage', 'month_year'], how = 'left')
```

```
ga4_lp_agg_month_stats = ga4_lp_agg_month_stats.merge(ga4_lp_agg_month_std,
        on = ['landingPage', 'month_year'], how = 'left')

ga4_lp_agg_month_stats.head()
```

This results in the following:

screenPageViewsPerSession	bounceRate	averageSessionDuration	userConversionRate	ecommercePurchases	activeUsers_avg	activeUsers_std	bounceRate_std
0.000000	0.977922	0.436364	0.0	0.0	10.000000	7.745967	0.050046
0.000000	0.989683	0.183333	0.0	0.0	6.500000	3.059750	0.039389
0.000000	0.977181	0.432258	0.0	0.0	6.322581	3.155810	0.051246
0.000000	1.000000	0.032143	0.0	0.0	7.857143	3.922530	0.000000
2.323391	0.218771	3.418182	0.0	0.0	159.363636	111.968096	0.046991

We can now see the additional columns created earlier.

The next block of code calculates the month-on-month on the various performance data.

First, we sort values to get the rows in month order for each landing page as the month-on-month calculation will be dependent on the row positioning.:

```
ga4_lp_agg_month_moms = ga4_lp_agg_month_stats.sort_values(['landingPage',
'month_year'])
```

We're just calculating the monthly stats for activeUsers and bounceRate; however, you can use the same methods on all of the other metric columns.

First, we start by calculating the absolute change from the current row to the previous row (the previous month) using the shift() function.

Note that "1" was entered as a parameter to the shift() function, which means 1 row. If you wanted to calculate the year-on-year difference, then you would enter "12" (i.e., .shift(12)), which would look at the value 12 rows before.

```
ga4_lp_agg_month_moms['activeUsers_delta'] = ga4_lp_agg_month_
moms['activeUsers'] - ga4_lp_agg_month_moms['activeUsers'].shift(1)
```

The month-on-month is then calculated by dividing the absolute change by the current month value and multiplied by 100:

```
ga4_lp_agg_month_moms['activeUsers_mom'] = ((ga4_lp_agg_month_
moms['activeUsers_delta'] / ga4_lp_agg_month_moms['activeUsers'].shift(1) *
100)).round(1)
```

This procedure is repeated for the bounce rate:

```
ga4_lp_agg_month_moms['bounceRate_delta'] = ga4_lp_agg_month_
moms['bounceRate'] - ga4_lp_agg_month_moms['bounceRate'].shift(1)
ga4_lp_agg_month_moms['bounceRate_mom'] = ((ga4_lp_agg_month_
moms['bounceRate_delta'] / ga4_lp_agg_month_moms['bounceRate'].shift(1) *
100)).round(1)
```

```
ga4_lp_agg_month_moms
```

This results in the following:

nversionRate	ecommercePurchases	activeUsers_avg	activeUsers_std	bounceRate_std	activeUsers_delta	activeUsers_mom	bounceRate_delta	bounceRate_mom
0.0	0.0	10.000000	7.745967	0.050046	NaN	NaN	NaN	NaN
0.0	0.0	6.500000	3.059750	0.039389	85.0	77.3	0.011760	1.2
0.0	0.0	6.322581	3.155810	0.051246	1.0	0.5	-0.012501	-1.3
0.0	0.0	7.857143	3.922530	0.000000	24.0	12.2	0.022819	2.3
0.0	0.0	159.363636	111.968096	0.046991	1533.0	696.8	-0.781229	-78.1

The delta and month-on-month columns are added. Note the NaN for the first row which is because no previous row existed for the shift() function to work.

To overwrite NaNs, you could use the np.where() function to replace .isnull() with zero.

An alternative approach would be to use a special function to avoid ordering the rows. However, this could be computationally more expensive to run in the cloud if you're planning to automate this as an all-encompassing SEO data warehouse dashboard reporting system.

Once done, you're ready to upload to your data warehouse of choice.

Loading Data

As mentioned earlier in the chapter, loading involves moving the transformed data into the data warehouse. Once uploaded, it's a good idea to check your data schema and preview what you've uploaded.

The following SQL will produce user trends by month and channel:

```
select yearMonth, channel, sum(users) as users from (
  select yearMonth, 'organic' as channel, users_sum  as users from
  google_analytics.multichannel_ga_monthly
  where channel in ('Organic Traffic')
```

```
  and
DATE_DIFF(CURRENT_DATE(), PARSE_DATE('%Y-%m-%d',
CONCAT(SUBSTR(CAST(yearMonth AS STRING), 1, 4),"-",
       SUBSTR(CAST(yearMonth AS STRING), 5, 2),"-",'01')), MONTH) <= 12
  union all
  select yearMonth, 'non_seo' as channel, all_users - organic as users
  from (
    SELECT yearMonth
      , MAX(IF(channel = 'Organic Traffic', users_sum, 0)) organic
      , MAX(IF(channel = 'All Users', users_sum, 0)) all_users
      from google_analytics.multichannel_ga_monthly
      where channel in ('Organic Traffic', 'All Users')
      group by yearMonth
      )
  )
  where
DATE_DIFF(CURRENT_DATE(), PARSE_DATE('%Y-%m-%d',
CONCAT(SUBSTR(CAST(yearMonth AS STRING), 1, 4),"-",
       SUBSTR(CAST(yearMonth AS STRING), 5, 2),"-",'01')), MONTH) <= 12
  group by yearMonth, channel
order by yearMonth LIMIT 100;
```

The following SQL will produce user traffic stats by month with year-on-year:

```
select
yearMonth,
CASE
  WHEN channel = 'All Users' THEN "non_seo"
  ELSE channel
  END as channel,
users_yoy
from google_analytics.multichannel_ga_monthly
where
channel in ("All Users", "Organic Traffic")
and
```

```
DATE_DIFF(CURRENT_DATE(), PARSE_DATE('%Y-%m-%d',
CONCAT(SUBSTR(CAST(yearMonth AS STRING), 1, 4),"-", SUBSTR(CAST(yearMonth
AS STRING), 5, 2),"-",'01')), MONTH) <= 12
order by yearMonth desc LIMIT 100;
```

The following SQL will produce year-on-year user traffic stats by month with this year vs. last year, for the months year to date:

```
SELECT yearMonth
      , year
      , SUBSTR(CAST(yearMonth AS STRING), 5, 6) as mon_x
      , users_sum
      from google_analytics.multichannel_ga_monthly
      where
      channel in ("Organic Traffic")
      order by mon_x, year desc;
```

This results in the following:

Once run, the result is generated by Google BigQuery under the "Results" tab (Figure 8-1).

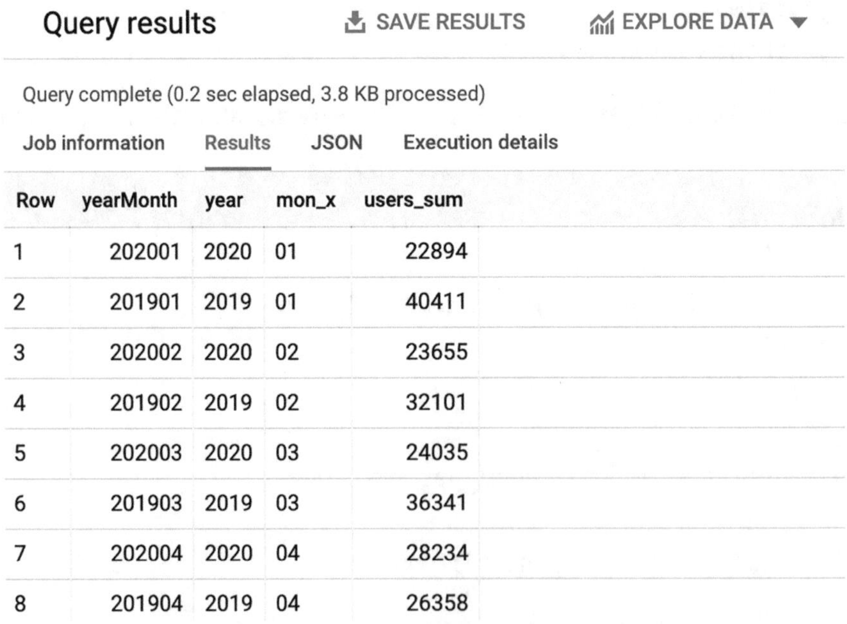

Query results				SAVE RESULTS	EXPLORE DATA ▼

Query complete (0.2 sec elapsed, 3.8 KB processed)

Job information   Results   JSON   Execution details

Row	yearMonth	year	mon_x	users_sum
1	202001	2020	01	22894
2	201901	2019	01	40411
3	202002	2020	02	23655
4	201902	2019	02	32101
5	202003	2020	03	24035
6	201903	2019	03	36341
7	202004	2020	04	28234
8	201904	2019	04	26358

Figure 8-1. *Preview of Google BigQuery results following SQL execution*

Visualization

If you're satisfied with the SQL results, you can use the same queries and visualize these in your front end such as Looker Studio, Tableau, etc., as shown in the following. How does organic search compare to other channels? By volume (top) and YoY (bottom) over time, shown in the Looker Studio graph (Figure 8-2).

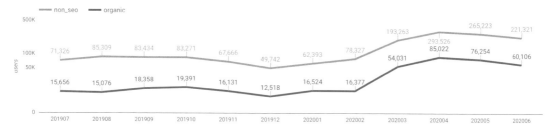

Figure 8-2. *Looker Studio graph showing organic vs. non-SEO channels over the last 12 months*

How does organic search compare to other channels year-on-year? The Looker Studio graph in Figure 8-3 visualizes the SQL statement which calculates the year-on-year traffic numbers for both organic and non-SEO channels. This is useful for seeing how well the SEO is performing for the time of the year (i.e., independent of seasonality). It also gives some measure of how SEO has performed relative to non-SEO channels for the same period.

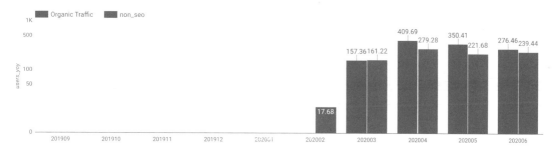

Figure 8-3. *Looker Studio bar chart of year-on-year traffic numbers for both organic and non-SEO channels by month*

How do organic search users compare to last year? The Looker Studio graph in Figure 8-4 shows this year and last year traffic numbers for organic traffic only. This is useful for comparing this year's SEO performance vs. last year's SEO performance in isolation.

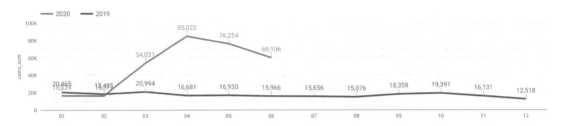

Figure 8-4. *Looker Studio graph showing this year and last year traffic numbers for organic traffic only*

Automation

Naturally, this can all be automated; you just need a team of competent cloud software engineers to automate

- Polling of data via APIs (extract)

- Cleaning, restructuring, and creating new calculated fields (transform)

- Loading (into BigQuery, Amazon Redshift, or others)

The result is simple, although the execution in reality is far more complicated, which relies on cloud engineering skills.

Summary

When putting dashboards together, it's important to begin with the end in mind and think about what the purpose of the dashboard is and who it is for (that is your audience). Once you know the outputs, then work backward.

Dashboards are driven by the data, so you'll need to consider which data sources you'll need. Raw data is seldom a good idea to plug straight into the front end like Looker Studio as it's likely to overwhelm the front end and thus load slowly or crash. Instead, you'll want to summarize the data into meaningful trends.

Extract, transform, and load (ETL) is the process of automating the data collection, summarizing the data, and then loading it into a system. We provided code to help you

- Extract SEO data from common SEO sources including Google Analytics

- Transform to summarize by channel

- See what the data could look like when loaded into Looker Studio

CHAPTER 9

Site Migration Planning

This chapter covers site migration mapping so that you could set the structure of your new site and semiautomate the formation of your migration URLs. The following are some of the techniques we'll be using:

- String manipulation

- Iterating through dataframe rows by converting these into a list

- Using natural language processing (NLP) to compare URL strings

While these techniques will speed up the processing of data for a site migration, they can easily be applied to other use cases.

Verifying Traffic and Ranking Changes

Though the step of verifying the traffic and/or ranking changes following relaunch is not strictly necessary, it's good to go through in case any colleagues are doubtful as to whether the changes in traffic or rankings were attributable to the date you claim. If you're pushed for time however, you can skip this step.

```
import re
import time
import random
import pandas as pd
import numpy as np
import datetime
from textdistance import sorensen_dice
from textdistance import jaccard
pd.set_option('display.max_colwidth', None)
```

© Andreas Voniatis 2023
A. Voniatis, *Data-Driven SEO with Python*, https://doi.org/10.1007/978-1-4842-9175-7_9

```
target_site_search = 'Saga travel'
first_gen = ['Holidays', 'Cruises', 'Travel Updates', 'Accessibility and
Support', 'Brochure Request', 'My Travel', 'Trade']
target_roots = first_gen
source_root_url = 'https://travel.saga.co.uk/'
target_root_url = 'https://www.saga.co.uk/'
```

The data comes from a spreadsheet which is a representation of the site taxonomy or hierarchy, that is, folders and subfolders with the site levels organized in columns:

```
hierarchy_raw = pd.read_csv('data/saga_hierarchy.csv')
hierarchy_raw
```

This results in the following:

	1	2	3	4	5	6	7
0	Existing Customers	NaN	NaN	NaN	NaN	NaN	NaN
1	Homepage	NaN	NaN	NaN	NaN	NaN	NaN
2	Homepage	Insurance	NaN	NaN	NaN	NaN	NaN
3	NaN	Insurance	Contact Us	NaN	NaN	NaN	NaN
4	NaN	Insurance	Refer a Friend T&Cs	NaN	NaN	NaN	NaN
...	...	...	...	...	...	...	...
367	NaN	NaN	Updates	Coronavirus Holidays Guests	NaN	NaN	NaN
368	NaN	NaN	Updates	Cruise Coronavirus Vaccine	NaN	NaN	NaN
369	NaN	NaN	Updates	Holidays Coronavirus Vaccine	NaN	NaN	NaN
370	NaN	NaN	Updates	Safe Travels	NaN	NaN	NaN
371	NaN	Travel	Brochures	NaN	NaN	NaN	NaN

372 rows × 7 columns

In the preceding table, we can see how the spreadsheet looks with numbers across the top denoting the site levels and the page (we'll call them nodes) with names per row with their immediate parent.

Let's get the site levels for each of the parent nodes:

```
site_levels = pd.DataFrame(hierarchy_raw.unstack())
site_levels = site_levels.dropna().drop_duplicates()
```

```
site_levels = site_levels.rename(columns = {0 : 'node'})
site_levels = site_levels.reset_index()
site_levels = site_levels[['level_0', 'node']]
site_levels = site_levels.rename(columns = {'level_0': 'level'})
site_levels
```

This results in the following:

	level	node
0	1	Existing Customers
1	1	Homepage
2	2	Insurance
3	2	Boat Insurance
4	2	Caravan Insurance
...	...	...
325	7	Noordam
326	7	Suites
327	7	Jools Holland
328	7	Artists
329	7	Build Progress

330 rows × 2 columns

With the site nodes defined, which will come in handy later, we're going to find the pairs of parent and child nodes.

Identifying the Parent and Child Nodes

Child nodes are the immediate pages that are a single click away from the parent node. To do this, we'll need a couple of functions. The apply_pcn function will treat the dataframe as a collection of rows and apply the second function. This approach is faster than iterating through a dataframe row by row using .iterrows(), which the latter is known for being very slow.

The parent_child_nodes will take the row, convert it to a list, and then use a list comprehension to ignore blank cells (NaNs, short for "not a number") and append the contents to the list "pairs."

379

Once done, the "pairs" list will be put into a new dataframe "parent_child_map."
Let's iterate by row to pick pairs:

```
pairs = []

def parent_child_nodes(row):
    data = row.values.tolist()
    data = [e for e in data if str(e) not in ('nan')]
    print(data)
    pairs.append(data)

def apply_pcn(df):
  return df.apply(
      lambda row:
        parent_child_nodes(
           row),
      axis=1
  )

apply_pcn(hierarchy_raw)

parent_child_map = pd.DataFrame(pairs,columns=['parent', 'child'])
parent_child_map
```

We now have a table showing the parent and child nodes:

	parent	child
0	Existing Customers	None
1	Homepage	None
2	Homepage	Insurance
3	Insurance	Contact Us
4	Insurance	Refer a Friend T&Cs
...	...	...
367	Updates	Coronavirus Holidays Guests
368	Updates	Cruise Coronavirus Vaccine
369	Updates	Holidays Coronavirus Vaccine
370	Updates	Safe Travels
371	Travel	Brochures

372 rows × 2 columns

Of course, if we want the full URL path, we need to process the data further using a copy of hierarchy_raw. Start with a downward fill of the first column for the home page and then populate the cell should the adjacent right cell not be blank (checked using the function has_data_right):

```
hierarchy_fp = hierarchy_raw
```

```
# Forward Fill HOMEPAGE
hierarchy_fp['1'] = hierarchy_fp['1'].ffill()
```

Here's the function to check for cells on the right to see if populated with data or NANs:

```
def has_data_right(idx):
    return hierarchy_fp[hierarchy_fp.columns[idx:]].notnull().
    apply(any, axis=1)
```

```
for c in hierarchy_fp.columns[1:]:
    hierarchy_fp.loc[has_data_right(int(c)), c] = hierarchy_fp.loc[has_
    data_right(int(c)), c].ffill()
```

```
hierarchy_fp
```

The following shows the resulting hierarchy_fp dataframe with all the folder names needed to construct a full path to the URL:

	1	2	3	4	5	6	7
0	Existing Customers	NaN	NaN	NaN	NaN	NaN	NaN
1	Homepage	NaN	NaN	NaN	NaN	NaN	NaN
2	Homepage	Insurance	NaN	NaN	NaN	NaN	NaN
3	Homepage	Insurance	Contact Us	NaN	NaN	NaN	NaN
4	Homepage	Insurance	Refer a Friend T&Cs	NaN	NaN	NaN	NaN
...	...	...	...	...	...	...	...
367	Homepage	Travel	Updates	Coronavirus Holidays Guests	NaN	NaN	NaN
368	Homepage	Travel	Updates	Cruise Coronavirus Vaccine	NaN	NaN	NaN
369	Homepage	Travel	Updates	Holidays Coronavirus Vaccine	NaN	NaN	NaN
370	Homepage	Travel	Updates	Safe Travels	NaN	NaN	NaN
371	Homepage	Travel	Brochures	NaN	NaN	NaN	NaN

372 rows × 7 columns

With this in mind, we can now iterate row by row in the dataframe to remove blanks (NaNs) and join them with a forward slash (/):

```
min_fp_nonnan = hierarchy_fp
full_paths = []

def find_full_paths(row):
    data = row.values.tolist()
    data = [e for e in data if str(e) not in ('nan')]
    data = '/'.join(data)
    print(data)
    full_paths.append(data)

def apply_ffp(df):
  return df.apply(
      lambda row:
        find_full_paths(
          row),
      axis=1
  )
```

```
apply_ffp(min_fp_nonnan)
full_paths
```

This results in the following:

```
['Existing Customers',
 'Homepage',
 'Homepage/Insurance',
 'Homepage/Insurance/Contact Us',
 'Homepage/Insurance/Refer a Friend T&Cs',
 'Homepage/Insurance/Saga Credit Agreement',
 'Homepage/Boat Insurance',
 'Homepage/Boat Insurance/How to Make a Claim',
 'Homepage/Caravan Insurance',
 'Homepage/Caravan Insurance/Additional Cover',
 'Homepage/Caravan Insurance/Cover at a Glance',
 'Homepage/Caravan Insurance/Discounts and Excesses',
 'Homepage/Caravan Insurance/Frequently Asked Questions',
 'Homepage/Caravan Insurance/How to Make a Claim',
 'Homepage/Caravan Insurance/Policy Booklets',
 'Homepage/Car Insurance',
 'Homepage/Car Insurance/Over 70s',
 'Homepage/Car Insurance/Buildings Insurance',
```

Now that we have the full folder names joined, a bit of string formatting is required to get them to resemble URL paths. This is what we'll do here:

```
#full_paths
full_path_df =  pd.DataFrame(full_paths,columns=['full_path'])
full_path_df['full_path'] =  full_path_df.full_path.str.
replace('Homepage/', '')
full_path_df['full_path'] =  full_path_df.full_path.str.
replace('Homepage', '')
full_path_df['full_path'] =  full_path_df.full_path.str.replace(' ', '-')
full_path_df['full_path'] =  full_path_df.full_path.str.replace('&', 'and')
full_path_df['full_path'] =  full_path_df.full_path.str.lower()
full_path_df['full_path'] =  target_root_url + full_path_df.full_path
full_path_df
```

This results in the following:

	full_path
0	https://www.saga.co.uk/existing-customers
1	https://www.saga.co.uk/
2	https://www.saga.co.uk/insurance
3	https://www.saga.co.uk/insurance/contact-us
4	https://www.saga.co.uk/insurance/refer-a-friend-tandcs
...	...
367	https://www.saga.co.uk/travel/updates/coronavirus-holidays-guests
368	https://www.saga.co.uk/travel/updates/cruise-coronavirus-vaccine
369	https://www.saga.co.uk/travel/updates/holidays-coronavirus-vaccine
370	https://www.saga.co.uk/travel/updates/safe-travels
371	https://www.saga.co.uk/travel/brochures

372 rows × 1 columns

The full URL path has now been constructed and pushed into a dataframe, so we can now add this to the parent_child_map dataframe created earlier:

```
# Join Parent nodes to Full paths
full_node_map = pd.concat([parent_child_map, full_path_df], axis=1)
full_node_map
```

Now we have a table with the nodes and the full path:

	parent	child	full_path
0	Existing Customers	None	https://www.saga.co.uk/existing-customers
1	Homepage	None	https://www.saga.co.uk/
2	Homepage	Insurance	https://www.saga.co.uk/insurance
3	Insurance	Contact Us	https://www.saga.co.uk/insurance/contact-us
4	Insurance	Refer a Friend T&Cs	https://www.saga.co.uk/insurance/refer-a-friend-tandcs
...	...	...	...
367	Updates	Coronavirus Holidays Guests	https://www.saga.co.uk/travel/updates/coronavirus-holidays-guests
368	Updates	Cruise Coronavirus Vaccine	https://www.saga.co.uk/travel/updates/cruise-coronavirus-vaccine
369	Updates	Holidays Coronavirus Vaccine	https://www.saga.co.uk/travel/updates/holidays-coronavirus-vaccine
370	Updates	Safe Travels	https://www.saga.co.uk/travel/updates/safe-travels
371	Travel	Brochures	https://www.saga.co.uk/travel/brochures

372 rows × 3 columns

Separating Migration Documents

Often, it's quite common in large organizations for different business units wanting separate migration documents for their particular website. In the following, we will iterate by row to find all of the child nodes for the "Travel" division and append these to a list "target_roots":

```
# filter for target BU

target_node_map = full_node_map

def append_target_pairs(row):
    data = row.values.tolist()
    [target_roots.append(data[1]) for e in data if str(e) in target_roots]

def apply_atp(df):
  return df.apply(
      lambda row:
        append_target_pairs(
          row),
      axis=1
  )
```

```
apply_atp(target_node_map)
target_roots = list(set(target_roots))
target_roots
```

This results in the following:

```
['Mauritius',
 'Guide to the Dolomites',
 'Meet the Team',
 'Big Ship Cruising',
 'Christmas Breaks',
 'Suites',
 'Coronavirus Cruise Passengers',
 'FAQs',
 'Brexit Travel Advice',
 'Holiday Creators',
 'Hassle Free Travel Service',
 'All Inclusive',
 'At the Airport',
 'My Travel',
 'Britannia Club',
 'Bespoke Tours',
 'Spirit of Adventure Blog',
 'Hosts',
```

Having extracted the Travel nodes in target_roots, we can now filter for Travel URLs only, starting with child nodes:

```
#Target Children
stop_strings = ['insurance', 'breakdown-cover']
target_parent_nodes = target_node_map[~target_node_map.full_path.str.
contains('|'.join(stop_strings))]
target_parent_nodes
```

This results in the following:

	parent	child	full_path
111	Holidays	Destinations	https://www.saga.co.uk/holidays/destinations
112	Destinations	Africa	https://www.saga.co.uk/holidays/destinations/africa
113	Africa	Egypt	https://www.saga.co.uk/holidays/destinations/africa/egypt
114	Egypt	Ancient Egypt Revealed	https://www.saga.co.uk/holidays/destinations/africa/egypt/ancient-egypt-revealed
115	Egypt	Ancient Wonders of Egypt	https://www.saga.co.uk/holidays/destinations/africa/egypt/ancient-wonders-of-egypt
...	...	...	...
359	Accessibility and Support	Assistance Dogs	https://www.saga.co.uk/travel/accessibility-and-support/assistance-dogs
360	Accessibility and Support	Mobility Aids	https://www.saga.co.uk/travel/accessibility-and-support/mobility-aids
361	Accessibility and Support	Saga Cruises	https://www.saga.co.uk/travel/accessibility-and-support/saga-cruises
362	Accessibility and Support	Transport and Transfers	https://www.saga.co.uk/travel/accessibility-and-support/transport-and-transfers
363	Accessibility and Support	Travelling by Air	https://www.saga.co.uk/travel/accessibility-and-support/travelling-by-air

247 rows × 3 columns

And now the parent nodes:

```
target_parent_nodes = target_node_map[target_node_map.child.
isin(first_gen)]
target_parent_nodes
```

This results in the following:

	parent	child	full_path
110	Homepage	Holidays	https://www.saga.co.uk/holidays
111	Holidays	Destinations	https://www.saga.co.uk/holidays/destinations
112	Destinations	Africa	https://www.saga.co.uk/holidays/destinations/africa
113	Africa	Egypt	https://www.saga.co.uk/holidays/destinations/africa/egypt
114	Egypt	Ancient Egypt Revealed	https://www.saga.co.uk/holidays/destinations/africa/egypt/ancient-egypt-revealed
...	...	...	...
366	Updates	Coronavirus Cruise Passengers	https://www.saga.co.uk/travel/updates/coronavirus-cruise-passengers
367	Updates	Coronavirus Holidays Guests	https://www.saga.co.uk/travel/updates/coronavirus-holidays-guests
368	Updates	Cruise Coronavirus Vaccine	https://www.saga.co.uk/travel/updates/cruise-coronavirus-vaccine
369	Updates	Holidays Coronavirus Vaccine	https://www.saga.co.uk/travel/updates/holidays-coronavirus-vaccine
370	Updates	Safe Travels	https://www.saga.co.uk/travel/updates/safe-travels

183 rows × 3 columns

The next job is to concatenate both of these into a single table:

```
target_node_map = pd.concat([target_parent_nodes, target_kid_nodes])
target_node_map
```

This results in the following:

	parent	child	full_path
110	Homepage	Holidays	https://www.saga.co.uk/holidays
111	Holidays	Destinations	https://www.saga.co.uk/holidays/destinations
112	Destinations	Africa	https://www.saga.co.uk/holidays/destinations/africa
113	Africa	Egypt	https://www.saga.co.uk/holidays/destinations/africa/egypt
114	Egypt	Ancient Egypt Revealed	https://www.saga.co.uk/holidays/destinations/africa/egypt/ancient-egypt-revealed
...	...	...	...
359	Accessibility and Support	Assistance Dogs	https://www.saga.co.uk/travel/accessibility-and-support/assistance-dogs
360	Accessibility and Support	Mobility Aids	https://www.saga.co.uk/travel/accessibility-and-support/mobility-aids
361	Accessibility and Support	Saga Cruises	https://www.saga.co.uk/travel/accessibility-and-support/saga-cruises
362	Accessibility and Support	Transport and Transfers	https://www.saga.co.uk/travel/accessibility-and-support/transport-and-transfers
363	Accessibility and Support	Travelling by Air	https://www.saga.co.uk/travel/accessibility-and-support/travelling-by-air

430 rows × 3 columns

With the Travel site URLs successfully filtered, we will now join the site levels:

```
nodes_levelled = pd.merge(target_node_map, site_levels, how='left', left_on='child', right_on='node')
del nodes_levelled['node']
nodes_levelled['child'] = np.where(nodes_levelled.child.isnull().values, '', nodes_levelled.child)
nodes_levelled['level'] = np.where(nodes_levelled.level.isnull().values, 0, nodes_levelled.level)
nodes_levelled
```

This results in the following:

	parent	child	full_path	level
0	Homepage	Holidays	https://www.saga.co.uk/holidays	2
1	Holidays	Destinations	https://www.saga.co.uk/holidays/destinations	3
2	Destinations	Africa	https://www.saga.co.uk/holidays/destinations/africa	4
3	Africa	Egypt	https://www.saga.co.uk/holidays/destinations/africa/egypt	5
4	Egypt	Ancient Egypt Revealed	https://www.saga.co.uk/holidays/destinations/africa/egypt/ancient-egypt-revealed	6
...	...	...	...	...
425	Accessibility and Support	Assistance Dogs	https://www.saga.co.uk/travel/accessibility-and-support/assistance-dogs	4
426	Accessibility and Support	Mobility Aids	https://www.saga.co.uk/travel/accessibility-and-support/mobility-aids	4
427	Accessibility and Support	Saga Cruises	https://www.saga.co.uk/travel/accessibility-and-support/saga-cruises	4
428	Accessibility and Support	Transport and Transfers	https://www.saga.co.uk/travel/accessibility-and-support/transport-and-transfers	4
429	Accessibility and Support	Travelling by Air	https://www.saga.co.uk/travel/accessibility-and-support/travelling-by-air	4

430 rows × 4 columns

The site levels were joined using Pandas Merge which is equivalent to Microsoft Excel's vlookup function. Because the column names were different in both tables, this had to be specified under left_on and right_on as shown earlier.

Finding the Closest Matching Category URL

Now that we have the new category level URL structures for the travel division, we're ready to find the closest matching live site URL.

You could try using natural language processing (NLP) techniques here, but these will be quite fruitless. That's because the live URL strings may not reflect the proposed URL structure, especially if you're looking to "clean house" and put in more sensible new ideas for the site relaunch.

The approach we take here is to let Google do the work by taking the highest ranking URL for a search (as displayed in Google's search results) on a combination of the parent child site names. After all, Google has far better tools at its disposal, so it only makes sense to use them.

```
nodes_levelled['search_query'] = target_site_search + ' ' + nodes_levelled.
parent + ' ' + nodes_levelled.child
nodes_levelled['search_query'] = target_site_search + ' "' + nodes_
levelled.parent + ' ' + nodes_levelled.child + '"'
nodes_levelled
```

This results in the following:

	parent	child	full_path	level	search_query
0	Homepage	Holidays	https://www.saga.co.uk/holidays	2	Saga travel "Homepage Holidays"
1	Holidays	Destinations	https://www.saga.co.uk/holidays/destinations	3	Saga travel "Holidays Destinations"
2	Destinations	Africa	https://www.saga.co.uk/holidays/destinations/africa	4	Saga travel "Destinations Africa"
3	Africa	Egypt	https://www.saga.co.uk/holidays/destinations/africa/egypt	5	Saga travel "Africa Egypt"
4	Egypt	Ancient Egypt Revealed	https://www.saga.co.uk/holidays/destinations/africa/egypt/ancient-egypt-revealed	6	Saga travel "Egypt Ancient Egypt Revealed"
...	...	...	...	...	...
425	Accessibility and Support	Assistance Dogs	https://www.saga.co.uk/travel/accessibility-and-support/assistance-dogs	4	Saga travel "Accessibility and Support Assistance Dogs"
426	Accessibility and Support	Mobility Aids	https://www.saga.co.uk/travel/accessibility-and-support/mobility-aids	4	Saga travel "Accessibility and Support Mobility Aids"
427	Accessibility and Support	Saga Cruises	https://www.saga.co.uk/travel/accessibility-and-support/saga-cruises	4	Saga travel "Accessibility and Support Saga Cruises"
428	Accessibility and Support	Transport and Transfers	https://www.saga.co.uk/travel/accessibility-and-support/transport-and-transfers	4	Saga travel "Accessibility and Support Transport and Transfers"
429	Accessibility and Support	Travelling by Air	https://www.saga.co.uk/travel/accessibility-and-support/travelling-by-air	4	Saga travel "Accessibility and Support Travelling by Air"

430 rows × 5 columns

Having taken a combination of the site name and parent and child nodes, these have formed the search strings we will use to get SERPs data for:

```
serptool_queries = nodes_levelled['search_query'].to_list()
serptool_queries
```

This results in the following:

```
['Saga travel homepage holidays',
 'Saga travel homepage cruises',
 'Saga travel homepage travel updates',
 'Saga travel homepage accessibility and support',
 'Saga travel homepage brochure request',
 'Saga travel homepage trade',
 'Saga travel holidays destinations',
 'Saga travel destinations africa',
 'Saga travel africa egypt',
 'Saga travel egypt ancient egypt revealed',
 'Saga travel egypt ancient wonders of egypt',
 'Saga travel egypt el quseir',
 'Saga travel el quseir radisson blu resort',
 'Saga travel africa ethiopia', ...
```

Using the preceding data, these could be checked using your favorite SEO rank checking tool API and then loaded into the notebook:

```
saga_serps = pd.read_csv('client_serps.csv')
saga_serps
```

This results in the following:

	keyword	rank_absolute	url
0	Saga travel england art in the north east	1	https://travel.saga.co.uk/holidays/destinations/europe/uk/england/art-in-the-north-east.aspx
1	Saga travel england art in the north east	2	https://travel.saga.co.uk/holidays/destinations/europe/uk/england/art-in-bournemouth.aspx
2	Saga travel england art in the north east	3	https://travel.saga.co.uk/holidays/destinations/europe/uk/england/liverpools-art-collections.aspx
4	Saga travel england art in the north east	5	https://travel.saga.co.uk/holidays/destinations/europe/uk/england.aspx
5	Saga travel england art in the north east	6	https://travel.saga.co.uk/holidays/destinations/europe/uk/england/gardens-of-the-north-east.aspx
...	...	...	...
20423	Saga travel uk northern ireland	16	https://travel.saga.co.uk/cruises/ocean/where-we-go/british-isles-cruises/secrets-of-the-emerald-isle.aspx
20424	Saga travel uk northern ireland	17	https://travel.saga.co.uk/forms/request-a-brochure.aspx
20425	Saga travel uk northern ireland	18	https://travel.saga.co.uk/travel-insurance.aspx
20426	Saga travel uk northern ireland	19	https://travel.saga.co.uk/faqs.aspx
20449	Saga travel uk northern ireland	42	https://travel.saga.co.uk/cruises/ocean/where-we-go/british-isles-cruises.aspx

2434 rows × 3 columns

Earlier, we have the SERPs loaded into the notebook, showing the keyword, rank position, and URL.

It is now time to extract the top ranking URL, by grouping the SERPs dataframe by keyword and then selecting the top ranked URL (if it hasn't already been selected). The reason is a single URL cannot be simultaneously redirected to two different URLs; hence, the "while" clause used in the Python code is checking whether the URL hasn't already been used for a previous keyword:

```
serps_grp = saga_serps.groupby('keyword')

current_allocated = []

def filter_top_serp(df):
    del df['keyword']
    i = 0
    while not df.iloc[i]['url'] in current_allocated:
        if not df.iloc[i]['url'] in current_allocated:
            current_allocated.append(df.iloc[i]['url'])
            return df.iloc[i]
```

```
    else:
        i += 1
```

```
current_map = serps_grp.apply(filter_top_serp)
```

Concatenate with the initial dataframe:

```
current_map_df = pd.concat([current_map],axis=0).reset_index()
del current_map_df['rank_absolute']
```

```
current_map_df = current_map_df.rename(columns = {'keyword': 'search_
query', 'url': 'current_url'})
```

```
current_map_df['current_alloc'] = pd.DataFrame({'current_alloc':current_
allocated})
```

```
current_map_df['current_url'] = np.where(current_map_df.current_url.
isnull(), current_map_df.current_alloc, current_map_df.current_url)
```

```
del current_map_df['current_alloc']
current_map_df['search_query'] = current_map_df.search_query.str.lower()
current_map_df
```

The result is a dataframe with the current URLs which we can now join to the main table:

	search_query	current_url
111	saga travel holidays book with confidence	https://travel.saga.co.uk/holidays/book-with-confidence.aspx
112	saga travel holidays deals	https://travel.saga.co.uk/cruises.aspx
113	saga travel holidays destinations	https://travel.saga.co.uk/holidays/destinations.aspx
114	saga travel holidays flying from your local airport	https://travel.saga.co.uk/travel-updates.aspx
115	saga travel holidays hassle free travel service	https://travel.saga.co.uk/holidays.aspx
116	saga travel holidays holiday types	https://travel.saga.co.uk/holidays/holiday-types.aspx
117	saga travel holidays meet the team	https://travel.saga.co.uk/holidays/holiday-types/hotel-holidays/signature-hotels.aspx
118	saga travel holidays our awards	https://travel.saga.co.uk/inspire-me/bestsellers.aspx
119	saga travel holidays reasons to choose saga	https://travel.saga.co.uk/inspire-me/bucket-list-tour-experiences.aspx
120	saga travel holidays reassurance promise	https://travel.saga.co.uk/holidays/reassurance-promise.aspx
121	saga travel holidays vip travel service	https://travel.saga.co.uk/holidays/vip-travel-service.aspx
122	saga travel holidays what's included	https://travel.saga.co.uk/inspire-me/travel-guides.aspx
123	saga travel holidays when to go	https://travel.saga.co.uk/inspire-me/wonders-of-the-world.aspx

Mapping Current URLs to the New Category URLs

The following code joins the current live category level URLs to the proposed new site URLs:

```
nodes_levelled['search_query'] = nodes_levelled.search_query.str.lower()
ia_current_mapping = pd.merge(nodes_levelled, current_map_df, on = 'search_
query', how = 'left')
ia_current_mapping = ia_current_mapping[['parent', 'child', 'level',
'current_url', 'full_path']]
ia_current_mapping
```

Here, we can see that neither this method nor Google is perfect. Nonetheless, it's a good start and saves a lot of manual work.

	parent	child	level	current_url	full_path
0	Existing Customers		0	https://www.saga.co.uk/insurance/travel-insurance/existing-customers	https://www.saga.co.uk/existing-customers
1	Homepage		0	https://travel.saga.co.uk/meet-the-team/travel-consultants.aspx	https://www.saga.co.uk/
2	Homepage	Holidays	2	https://www.fredholidays.co.uk/	https://www.saga.co.uk/holidays
3	Holidays	Destinations	3	https://travel.saga.co.uk/holidays/destinations.aspx	https://www.saga.co.uk/holidays/destinations
4	Destinations	Africa	4	https://travel.saga.co.uk/holidays/destinations/africa.aspx	https://www.saga.co.uk/holidays/destinations/africa
...	...	...	...	...	...
492	Accessibility and Support	Assistance Dogs	4	https://travel.saga.co.uk/accessibility-and-support/assistance-dogs.aspx	https://www.saga.co.uk/travel/accessibility-and-support/assistance-dogs
493	Accessibility and Support	Mobility Aids	4	https://travel.saga.co.uk/accessibility-and-support/mobility-aids.aspx	https://www.saga.co.uk/travel/accessibility-and-support/mobility-aids
494	Accessibility and Support	Saga Cruises	4	https://travel.saga.co.uk/accessibility-and-support/saga-cruises.aspx	https://www.saga.co.uk/travel/accessibility-and-support/saga-cruises
495	Accessibility and Support	Transport and Transfers	4	https://travel.saga.co.uk/accessibility-and-support/transport-and-transfers.aspx	https://www.saga.co.uk/travel/accessibility-and-support/transport-and-transfers
496	Accessibility and Support	Travelling by Air	4	https://travel.saga.co.uk/accessibility-and-support/travelling-by-air.aspx	https://www.saga.co.uk/travel/accessibility-and-support/travelling-by-air

497 rows × 5 columns

Let's tidy the table up by renaming a few columns and replacing NaNs with blanks:

```
# rearrange columns
ia_current_mapping = ia_current_mapping[['parent', 'child',
'level','search_query', 'current_url', 'full_path']]
ia_current_mapping = ia_current_mapping.rename(columns = {'full_path':
'migration_url'})
ia_current_mapping['current_url'] = np.where(ia_current_mapping.current_
url.isnull(), '', ia_current_mapping.current_url)
ia_current_mapping
```

393

In the following table, you can see that the first five lines have a simi value of zero, because the current URLs are blank, so of course there is zero string similarity between the proposed migration URL and the current URL.

	parent	child	level	current_url	migration_url
0	Existing Customers		0	https://www.saga.co.uk/insurance/travel-insurance/existing-customers	https://www.saga.co.uk/existing-customers
1	Homepage		0	https://travel.saga.co.uk/meet-the-team/travel-consultants.aspx	https://www.saga.co.uk/
2	Homepage	Holidays	2	https://www.fredholidays.co.uk/	https://www.saga.co.uk/holidays
3	Holidays	Destinations	3	https://travel.saga.co.uk/holidays/destinations.aspx	https://www.saga.co.uk/holidays/destinations
4	Destinations	Africa	4	https://travel.saga.co.uk/holidays/destinations/africa.aspx	https://www.saga.co.uk/holidays/destinations/africa
...	...	...	...	...	...
492	Accessibility and Support	Assistance Dogs	4	https://travel.saga.co.uk/accessibility-and-support/assistance-dogs.aspx	https://www.saga.co.uk/travel/accessibility-and-support/assistance-dogs
493	Accessibility and Support	Mobility Aids	4	https://travel.saga.co.uk/accessibility-and-support/mobility-aids.aspx	https://www.saga.co.uk/travel/accessibility-and-support/mobility-aids
494	Accessibility and Support	Saga Cruises	4	https://travel.saga.co.uk/accessibility-and-support/saga-cruises.aspx	https://www.saga.co.uk/travel/accessibility-and-support/saga-cruises
495	Accessibility and Support	Transport and Transfers	4	https://travel.saga.co.uk/accessibility-and-support/transport-and-transfers.aspx	https://www.saga.co.uk/travel/accessibility-and-support/transport-and-transfers
496	Accessibility and Support	Travelling by Air	4	https://travel.saga.co.uk/accessibility-and-support/travelling-by-air.aspx	https://www.saga.co.uk/travel/accessibility-and-support/travelling-by-air

497 rows × 5 columns

With the table tidied, we will use NLP methods to compare the string similarity of the current URL:

```
ia_current_simi = ia_current_mapping
ia_current_simi = ia_current_simi.drop_duplicates()
ia_current_simi['simi'] = ia_current_simi.loc[
    :, ['current_url', 'migration_url']].apply(
    lambda x: sorensen_dice(*x), axis=1)

ia_current_simi
```

The string similarity is helpful because when we review the migration URLs in a spreadsheet app like Microsoft Excel, we can filter for URLs that are not very similar, for instance, less than 0.9, which shows us current URLs that might not be a good match for the migration URLs. Rows with missing current URLs will need to be manually fixed, and the ones deduced from the SERPs will require a review.

```
ia_current_mapping.to_csv('exports/' + target_site_search + '_ia_current_
mapping.csv')
```

	parent	child	level	current_url	migration_url	simi
0	Existing Customers		0		https://www.saga.co.uk/existing-customers	0.000000
1	Homepage		0		https://www.saga.co.uk/	0.000000
2	Homepage	Holidays	2	https://travel.saga.co.uk/cruises/ocean/find-your-cruise.aspx	https://www.saga.co.uk/holidays	0.586957
3	Holidays	Destinations	3	https://travel.saga.co.uk/holidays/destinations.aspx	https://www.saga.co.uk/holidays/destinations	0.854167
4	Destinations	Africa	4	https://travel.saga.co.uk/holidays/destinations/africa.aspx	https://www.saga.co.uk/holidays/destinations/africa	0.872727
...	...	...	...	...	...	...
259	Updates	Coronavirus Holidays Guests	4		https://www.saga.co.uk/travel/updates/coronavirus-holidays-guests	0.000000
260	Updates	Cruise Coronavirus Vaccine	4		https://www.saga.co.uk/travel/updates/cruise-coronavirus-vaccine	0.000000
261	Updates	Holidays Coronavirus Vaccine	4		https://www.saga.co.uk/travel/updates/holidays-coronavirus-vaccine	0.000000
262	Updates	Safe Travels	4		https://www.saga.co.uk/travel/updates/safe-travels	0.000000
263	Travel	Brochures	3		https://www.saga.co.uk/travel/brochures	0.000000

264 rows × 6 columns

Mapping the Remaining URLs to the Migration URL

Now that the category URLs and subcategory URLs have the URL structures set, we're ready to set the migration URLs for the rest of the site. We'll assume that you've edited the ia_current_mapping CSV export generated earlier in Excel, corrected any errors not processed such as the missing current URLs (now not missing), and are thus ready to import:

```
import re
import time
import random
import pandas as pd
import numpy as np
import datetime
from textdistance import sorensen_dice
pd.set_option('display.max_colwidth', None)
import os.path

target_site_search = 'saga'
target_bu = 'travel'
target_roots = ['Holidays', 'Cruises', 'Travel Updates', 'Accessibility and
Support', 'Brochure Request', 'My Travel', 'Trade']
source_root_url = 'https://travel.saga.co.uk/'
```

```
migration_root_url = 'https://www.saga.co.uk/'
source_hostname = 'travel.saga.co.uk'
file_path = 'cases/'+ target_site_search + '/'

latest_mapping_raw = pd.read_csv('data/Saga travel_ia_edited_mapping_
dd.csv')
latest_mapping_raw
```

The imported mapping is an edited Excel file to reflect the business and operational requirements that wouldn't be adjusted for in the previous section.

	parent	child	level	current_url	migration_url
0	Homepage	Holidays	0	https://travel.saga.co.uk/	https://www.saga.co.uk/holidays
1	Homepage	Holidays	2	https://travel.saga.co.uk/holidays	https://www.saga.co.uk/holidays
2	Holidays	Destinations	3	https://travel.saga.co.uk/holidays/destinations.aspx	https://www.saga.co.uk/holidays/destinations
3	Destinations	Africa	4	https://travel.saga.co.uk/holidays/destinations/africa.aspx	https://www.saga.co.uk/holidays/destinations/africa
4	Africa	Egypt	5	https://travel.saga.co.uk/holidays/destinations/africa/egypt.aspx	https://www.saga.co.uk/holidays/destinations/africa/egypt
...	...	...	...	...	...
464	Accessibility and Support	Assistance Dogs	4	https://travel.saga.co.uk/accessibility-and-support/assistance-dogs.aspx	https://www.saga.co.uk/travel/accessibility-and-support/assistance-dogs
465	Accessibility and Support	Mobility Aids	4	https://travel.saga.co.uk/accessibility-and-support/mobility-aids.aspx	https://www.saga.co.uk/travel/accessibility-and-support/mobility-aids
466	Accessibility and Support	Saga Cruises	4	https://travel.saga.co.uk/accessibility-and-support/saga-cruises.aspx	https://www.saga.co.uk/travel/accessibility-and-support/saga-cruises
467	Accessibility and Support	Transport and Transfers	4	https://travel.saga.co.uk/accessibility-and-support/transport-and-transfers.aspx	https://www.saga.co.uk/travel/accessibility-and-support/transport-and-transfers
468	Accessibility and Support	Travelling by Air	4	https://travel.saga.co.uk/accessibility-and-support/travelling-by-air.aspx	https://www.saga.co.uk/travel/accessibility-and-support/travelling-by-air

469 rows × 5 columns

With the URL structures set for the category and subcategory URLs, we're now going to break down the current URLs and the migration URLs, so that we can create a mapping formula.

When we import the rest of the site URLs, the script will use their folder structure to convert them to the new migration URL structure:

```
latest_mapping_full_branch = latest_mapping_raw[['parent', 'child',
'level', 'current_url','migration_url']]

latest_mapping_full_branch['current_url'] = np.where(latest_mapping_full_
branch.current_url.isnull(), '', latest_mapping_full_branch.current_url)
```

To create the new URL structures, create a template variable called "new_branch"; we simply take the migration URLs and grab the folders between the root domain and the web page URL string.

For example, the new_branch value of https://travel.saga.co.uk/holidays/
destinations.aspx becomes /holidays/.

To extract the folders in between, we remove the root domain, split the string by
forward slashes ('/'), then extract everything apart from the last element.

Set the new URL structure:

```
latest_mapping_full_branch['new_branch'] = latest_mapping_full_
branch['migration_url'].str.replace(migration_root_url, '', regex = False)
latest_mapping_full_branch['new_branch'] = latest_mapping_full_branch['new_
branch'].str.split('/').str[:-1]

latest_mapping_full_branch['new_branch'] = ['/'.join(map(str, l)) for l in
latest_mapping_full_branch['new_branch']]
```

Similar principles are applied to the following old_branch, which is the URL
structure for the current URLs:

```
latest_mapping_full_branch['old_branch'] = latest_mapping_full_
branch['current_url'].str.replace(migration_root_url, '', regex = False)
```

To make the new URLs more evergreen and thus without dates, we stick all of the
undesirables into a list and tell Python to remove everything from the list:

```
remove_strs = ['2018/', '2019/', '2020/','2021/', 'jan/', 'feb/', 'mar/',
'apr/', 'may/','jun/', 'jul/', 'aug/','sep/','oct/', 'nov/', 'dec/']
latest_mapping_full_branch['old_branch'] = latest_mapping_full_branch['old_
branch'].str.replace('|'.join(remove_strs), '')
```

Remove the node:

```
latest_mapping_full_branch['old_branch'] = latest_mapping_full_branch['old_
branch'].str.split('/').str[:-1]
latest_mapping_full_branch['old_branch'] = latest_mapping_full_branch['old_
branch'].apply(lambda x: '/'.join(map(str, x)))
```

The node is the URL string that is specific to the page itself, which we're extracting as follows:

```
# set the NODE
latest_mapping_full_branch['node'] = latest_mapping_full_branch['current_
url'].str.split('/').str[-1]
latest_mapping_full_branch['node'] = latest_mapping_full_branch['node'].
str.replace('.aspx', '', regex = False)
latest_mapping_full_branch['node'] = latest_mapping_full_branch['node'].
str.replace(' ', '-', regex = False)
```

If any old_branch values are empty, then we just substitute the node. np.where is the Pandas equivalent of Excel's if statement:

```
latest_mapping_full_branch['old_branch'] = np.where(latest_mapping_full_
branch['old_branch'] == '', latest_mapping_full_branch.node, latest_
mapping_full_branch['old_branch'])
latest_mapping_full_branch
```

This results in the following:

	parent	child	level	current_url	migration_url	new_branch	old_branch	node
0	Homepage	Holidays	0	https://travel.saga.co.uk/	https://www.saga.co.uk/hol...			
1	Homepage	Holidays	2	https://travel.saga.co.uk/...	https://www.saga.co.uk/hol...		holidays	holidays
2	Holidays	Destinations	3	https://travel.saga.co.uk/...	https://www.saga.co.uk/hol...	holidays	holidays	destinations
3	Destinations	Africa	4	https://travel.saga.co.uk/...	https://www.saga.co.uk/hol...	holidays/destinations	holidays/destinations	africa
4	Africa	Egypt	5	https://travel.saga.co.uk/...	https://www.saga.co.uk/hol...	holidays/destinations/africa	holidays/destinations/africa	egypt
...	...	...	...	...	...	...	...	...
464	Accessibility and Support	Assistance Dogs	4	https://travel.saga.co.uk/...	https://www.saga.co.uk/tra...	travel/accessibility-and-s...	accessibility-and-support	assistance-dogs
465	Accessibility and Support	Mobility Aids	4	https://travel.saga.co.uk/...	https://www.saga.co.uk/tra...	travel/accessibility-and-s...	accessibility-and-support	mobility-aids
466	Accessibility and Support	Saga Cruises	4	https://travel.saga.co.uk/...	https://www.saga.co.uk/tra...	travel/accessibility-and-s...	accessibility-and-support	saga-cruises
467	Accessibility and Support	Transport and Transfers	4	https://travel.saga.co.uk/...	https://www.saga.co.uk/tra...	travel/accessibility-and-s...	accessibility-and-support	transport-and-transfers
468	Accessibility and Support	Travelling by Air	4	https://travel.saga.co.uk/...	https://www.saga.co.uk/tra...	travel/accessibility-and-s...	accessibility-and-support	travelling-by-air

469 rows × 8 columns

With URL structures broken down and reconstituted, we're ready to put a mapping together. A bit of cleanup happens as follows, where we drop duplicate rows and remove blank rows with empty current_url values. We don't expect any anomalies at this stage, but just in case.

```
branch_map = latest_mapping_full_branchbranch_map = branch_map.sort_
values('old_branch')
branch_map = branch_map.drop_duplicates(subset = 'old_branch')
branch_map = branch_map[~branch_map['current_url'].isnull()]
branch_map = branch_map[['old_branch', 'new_branch']]
branch_map['new_branch'] = np.where(branch_map['new_branch'] == '',
'holidays', branch_map['new_branch'])
branch_map
```

This results in the following:

	old_branch	new_branch
0		holidays
466	accessibility-and-support	travel/accessibility-and-s...
173	cruises	cruises
397	cruises/ocean	cruises/ocean
368	cruises/ocean/cruise-excur...	cruises/ocean/cruise-excur...
...	...	...
356	meet-the-team	holidays/meet-the-team
360	offers	holidays/deals
362	offers/flight-offers	holidays/deals/flight-offers
461	trade	travel/trade
460	travel-updates	travel/trade

81 rows × 2 columns

Importing the URLs

With the mapping in place, we're ready to import the URLs and fit them to the new migration structure:

```
DETERMINE the unallocated URLs
target_crawl_raw = pd.read_csv('data/crawl_urls.csv')
target_crawl_raw
```

We're removing URLs and subfolders that won't move as part of the site relaunch:

```
target_crawl_urls = target_crawl_raw
stop_folders = ['/membership/', '/magazine/', '/saga-charities/', '/
legal/', '/money/', '/my/', '/care/', '/magazine-subscriptions/',
                '/membership', '/magazine', '/saga-charities', '/legal',
                '/money', '/my', '/care', '/magazine-subscriptions',
                'boardbasis'
                '/antiquity', '/pharaoh', '/orca', '/italy-splendour', '/
                walking', '/archaeology',
                '/gardens', '/music', '/MyS', '/404', '/contentli']
target_crawl_urls = target_crawl_urls[~target_crawl_urls['URL'].str.
contains('|'.join(stop_folders))]
target_crawl_urls = target_crawl_urls[target_crawl_urls['Host'].str.
contains('www.saga.co.uk')]
target_crawl_urls
```

This results in the following:

	deeprank	level	http_status_code	indexable	page_title	url	description	found_at_url	primary_
0	10.00	1.0	200	True	Saga Holidays - Over 50s Holidays & Cruises - ...	https://travel.saga.co.uk/	Find your perfect holiday from over 50s cruise...	https://travel.saga.co.uk/faqs.aspx	
1	9.13	2.0	200	True	European River Cruise Holidays For Over 50s - ...	https://travel.saga.co.uk/cruises/river.aspx	Relax in the splendour of Europe's most beauti...	https://travel.saga.co.uk/	
2	9.12	2.0	200	True	Ocean Cruises - Boutique Cruise Holidays For O...	https://travel.saga.co.uk/cruises/ocean.aspx	Saga boutique cruising uses smaller ships, get...	https://travel.saga.co.uk/	
3	8.85	2.0	200	True	Holidays In The UK - Over 50s Holidays & Cruis...	https://travel.saga.co.uk/holidays/destination...	Discover the beautiful sights found close to h...	https://travel.saga.co.uk/	
4	8.77	2.0	200	True	Holiday Destinations & Packages For Over 50s W...	https://travel.saga.co.uk/holidays/destination...	Saga holidays can take you to some of the best...	https://travel.saga.co.uk/	
...	...	...	...	...	...	...	...	...	...

The crawl data is in and now subsetted for the URLs we want to migrate. Next, we're sticking these into a list to ensure they are unique:

```
current_url_lst = latest_mapping_full_branch.current_url.to_list()
mapped_url_lst = list(set(current_url_lst))
print(len(mapped_url_lst))
```

This results in the following:

```
469
258
```

Then we create a new dataframe "target_crawl_unmigrated" which excludes URLs already mapped (i.e., the category and subcategory URLs):

```
target_crawl_unmigrated = target_crawl_urls[~target_crawl_urls['URL'].
isin(mapped_url_lst)]
target_crawl_unmigrated
```

At this stage, it's sensible to check if we have any redirects (300 responses) and other non-"200" server status URLs:

```
target_crawl_unmigrated.groupby('HTTP Status Code').agg({'HTTP Status
Code': 'count'})
```

This results in the following:

	http_status_code
http_status_code	
200	1140

We can see that all of the filtered URLs we've yet to migrate all serve live pages (returning a 200 response). If we did have 301s, we could use the following code to inspect those 301 URLs:

```
target_crawl_unmigrated[target_crawl_unmigrated['HTTP Status Code'] ==
'301'][[url]]
```

To handle redirecting URLs, we want to ensure they are included in the mapping so that we can avoid redirect chains when migrating the site URLs. A redirect chain is when there are multiple redirects between the initial URL requested and the final destination URL. We'll achieve this by ensuring these are listed as current URLs.

Mutate the old_branch:

```
target_crawl_mutate = target_crawl_unmigrated
target_crawl_mutate = target_crawl_mutate.rename(columns = {'url':
'current_url'})
```

Create a list of our conditions:

```
redirect_conds = [
    target_crawl_mutate['http_status_code'].isin(['200', '204', '404',
    '410', '500']),
    target_crawl_mutate['http_status_code'].isin(['301', '302',
    '307', '308'])
]
```

Create a list of the values we want to assign for each condition:

```
desturl_values = [target_crawl_mutate['current_url'],
                  target_crawl_mutate['redirected_to_url'],
                  ]
```

Create a new column and use np.select to assign values to it using our lists as arguments:

```
target_crawl_mutate['dest_url'] = np.select(redirect_conds, desturl_values)
target_crawl_mutate = target_crawl_mutate[['dest_url']]
target_crawl_mutate = target_crawl_mutate.rename(columns = {'dest_url':
'current_url'})
```

Redirects notwithstanding, at this point these are dealt with. The following code will now break down the URL into structures ready for mapping using the table created in earlier steps:

```
target_crawl_mutate['old_branch'] = target_crawl_mutate['current_url'].str.
replace(source_root_url, '', regex = False)
```

```
remove_strs = ['2018/', '2019/', '2020/','2021/', 'jan/', 'feb/', 'mar/',
'apr/', 'may/','jun/',
                'jul/', 'aug/','sep/','oct/', 'nov/', 'dec/']
target_crawl_mutate['old_branch'] = target_crawl_mutate['old_branch'].str.
replace('|'.join(remove_strs), '')
target_crawl_mutate['old_branch'] = target_crawl_mutate['old_branch'].str.
split('/').str[:-1]

target_crawl_mutate['node'] = target_crawl_mutate['current_url'].str.
split('/').str[-1] # node only
target_crawl_mutate['node'] = target_crawl_mutate['node'].str.replace('.
aspx', '', regex = False)
target_crawl_mutate['node'] = target_crawl_mutate['node'].str.replace(' ',
'-', regex = False)

target_crawl_mutate['old_branch'] = target_crawl_mutate.old_branch.
apply(lambda x: '/'.join(map(str, x)))

target_crawl_mutate
```

The following crawl data "target_crawl_mutate" now has old branches, which means after a bit of cleanup, removing unnecessary column names, we can merge these with the branch map created earlier to help formulate the migration URLs.

ted_to_url_digest	url_alias_digest	url_digest	old_branch	node
NaN	/s0bE3zPuFU2qCUZQvgFPuNCYwEjzVS/oSfH...	/s0bE3zPuFU2qCUZQvgFPuNCYwEjzVS/oSfH...	holidays/holiday-types	singles-holidays
NaN	sAlZ3mR6UGcyrPyl/y8AoYapRRnWqMiX/1Ny...	sAlZ3mR6UGcyrPyl/y8AoYapRRnWqMiX/1Ny...	holidays/destinations	europe
NaN	N9xRIfewr+ZNhjTeeENQZcw6oiJcHrNMYfuG...	N9xRIfewr+ZNhjTeeENQZcw6oiJcHrNMYfuG...	holidays/destinations/europe	spain
NaN	tRUc/HuzDLljgmvwHrbq+2GVpLTWnJ5Hg4nX...	tRUc/HuzDLljgmvwHrbq+2GVpLTWnJ5Hg4nX...	forms	request-a-brochure
NaN	XzhiT3V3Uk6hIWZ82ZCD3wdPdWH/99xkhV1A...	XzhiT3V3Uk6hIWZ82ZCD3wdPdWH/99xkhV1A...		cruises
...	...	...	...	...

Let's now look up a new branch as there may be old branches not quite covered in the remaining URLs, that is, unmatched exceptions:

```
unallocated_branch = target_crawl_mutate[['current_url', 'http_status_
code', 'old_branch', 'node']]
unallocated_branch = unallocated_branch.merge(branch_map, on = 'old_
branch', how = 'left')
unallocated_branch
```

This results in the following:

	current_url	http_status_code	old_branch	node	new_branch
0	https://travel.saga.co.uk/holidays/h...	200	holidays/holiday-types	singles-holidays	holidays
1	https://travel.saga.co.uk/holidays/d...	200	holidays/destinations	europe	holidays/destinations
2	https://travel.saga.co.uk/holidays/d...	200	holidays/destinations/europe	spain	holidays/destinations/europe
3	https://travel.saga.co.uk/forms/requ...	200	forms	request-a-brochure	NaN
4	https://travel.saga.co.uk/cruises.aspx	200		cruises	holidays
...	...	...	...	...	...
1135	https://travel.saga.co.uk/holidays/h...	200	holidays/holiday-types/escorted-tour...	rv-mekong-pandaw	holidays/types/escorted-tours/small-...
1136	https://travel.saga.co.uk/holidays/d...	200	holidays/destinations/north america/...	rocky-mountains-and-alaskan-adventure	NaN
1137	https://travel.saga.co.uk/holidays/d...	200	holidays/destinations/europe/uk/england	gilbert-and-sullivan-festival-midwee...	holidays/destinations/europe/uk/england
1138	https://travel.saga.co.uk/holidays/d...	200	holidays/destinations/europe/uk/england	christmas-on-lake-windermere	holidays/destinations/europe/uk/england
1139	https://travel.saga.co.uk/holidays/d...	200	holidays/destinations/north-america/...	treasures-of-the-yucatan?pid=ppsg&sc...	NaN

1140 rows × 5 columns

The unmigrated URLs now have the suggested URL structure which can be used to create a new column forming the suggested migration URL. We will start with a bit of cleanup to handle blank new branch values:

```
allocated_fillnb = unallocated_branch

allocated_fillnb['new_branch'] = np.where(allocated_fillnb.new_branch.
isnull(),
                                          '',
                                          allocated_fillnb.new_branch)
allocated_fillnb['new_branch'] = np.where(allocated_fillnb.new_
branch == '',
                                          allocated_fillnb.old_branch,
                                          allocated_fillnb.new_branch)
allocated_fillnb = allocated_fillnb[allocated_fillnb.new_branch != '']
allocated_fillnb.sort_values('new_branch')
```

This results in the following:

	current_url	http_status_code	old_branch	node	new_branch
968	https://travel.saga.co.uk/brochure-r...	200	brochure-request/2021-trade-uk	TRADE_UKbrochure.html	brochure-request/2021-trade-uk
915	https://travel.saga.co.uk/brochure-r...	200	brochure-request/artwork/2020-sofa	spirit-of-adventure-onboard-artwork-...	brochure-request/artwork/2020-sofa
970	https://travel.saga.co.uk/brochure-r...	200	brochure-request/nhb-em6916	europe-and-the-mediterranean-collect...	brochure-request/nhb-em6916
896	https://travel.saga.co.uk/brochure-r...	200	brochure-request/nhb-em6954	europe-and-the-mediterranean-collect...	brochure-request/nhb-em6954
1117	https://travel.saga.co.uk/brochure-r...	200	brochure-request/nhb-em6968	winter-sun-collection.html	brochure-request/nhb-em6968
...	...	...	...	...	...
216	https://travel.saga.co.uk/travel-upd...	200	travel-updates	traffic-light-system	travel/trade
1104	https://travel.saga.co.uk/trade/mana...	200	trade	managebooking	travel/trade
1004	https://travel.saga.co.uk/trade/prod...	200	trade	product-range	travel/trade
938	https://travel.saga.co.uk/trade/faqs...	200	trade	faqs	travel/trade
898	https://travel.saga.co.uk/trade/ince...	200	trade	incentives	travel/trade

1140 rows × 5 columns

More cleanup ensues to handle URL nodes that contain parameter characters such as "?" and "=". Then we attempt to create columns showing their Parent and Child URL node folders based on the text position within the overall URL string:

```
allocated_draft = allocated_fillnb

allocated_draft['node'] = np.where(allocated_draft['node'].str.
contains('(\?|=)'),

                                    '',
                                    allocated_draft['node'])

allocated_draft['Parent'] = allocated_draft['new_branch']
allocated_draft['Parent'] = allocated_draft['Parent'].str.split('/').str[0]

allocated_draft['Child'] = allocated_draft['new_branch'].str.
split('/').str[1]
allocated_draft['Child'] = np.where(allocated_draft['Child'].isnull(),
                                    allocated_draft['node'],
                                    allocated_draft['Child']
                                    )

allocated_draft
```

This results in the following:

	current_url	http_status_code	old_branch	node	new_branch	Parent	Child
0	https://travel.saga.co.uk/holid...	200	holidays/holiday-types	singles-holidays	holidays	holidays	singles-holidays
1	https://travel.saga.co.uk/holid...	200	holidays/destinations	europe	holidays/destinations	holidays	destinations
2	https://travel.saga.co.uk/holid...	200	holidays/destinations/europe	spain	holidays/destinations/europe	holidays	destinations
3	https://travel.saga.co.uk/forms...	200	forms	request-a-brochure	forms	forms	request-a-brochure
4	https://travel.saga.co.uk/cruis...	200		cruises	holidays	holidays	cruises
...	...	...	...	...	...	...	...
1135	https://travel.saga.co.uk/holid...	200	holidays/holiday-types/escorted...	rv-mekong-pandaw	holidays/types/escorted-tours/s...	holidays	types
1136	https://travel.saga.co.uk/holid...	200	holidays/destinations/north ame...	rocky-mountains-and-alaskan-adv...	holidays/destinations/north ame...	holidays	destinations
1137	https://travel.saga.co.uk/holid...	200	holidays/destinations/europe/uk...	gilbert-and-sullivan-festival-m...	holidays/destinations/europe/uk...	holidays	destinations
1138	https://travel.saga.co.uk/holid...	200	holidays/destinations/europe/uk...	christmas-on-lake-windermere	holidays/destinations/europe/uk...	holidays	destinations
1139	https://travel.saga.co.uk/holid...	200	holidays/destinations/north-ame...		holidays/destinations/north-ame...	holidays	destinations

1140 rows × 7 columns

The preceding table now has the Parent and Child folders. At this stage, we're looking to ensure the new URL structure (new_branch) fall into one of the major sections of the new travel site before putting together the migration URLs.

Convert the root parent folder names to lowercase:

```
target_roots_urled = [elem.lower() for elem in target_roots]
target_roots_urled = [elem.replace(' ', '-') for elem in target_
roots_urled]
print(target_roots_urled)
```

Sort out the branches:

```
sorted_branches = []

def change_urls(row):
    data = row.values.tolist()
    #print(data)
    if not data[-2] in target_roots_urled:
        data[-3] = 'holidays/' + str(data[-3])
    sorted_branches.append(data)
```

```
def apply_cip(df):
  return df.apply(lambda row: change_urls(row), axis=1)

apply_cip(allocated_draft)

allocated_drafted =  pd.DataFrame(sorted_branches, columns=allocated_draft.
columns.tolist())
pd.set_option('display.max_colwidth', 35)
allocated_drafted
```

This results in the following:

```
['holidays', 'cruises', 'travel-updates', 'accessibility-and-support', 'brochure-request', 'my-travel', 'trade']
```

	current_url	http_status_code	old_branch	node	new_branch	Parent	Child
0	https://travel.saga.co.uk/holid...	200	holidays/holiday-types	singles-holidays	holidays	holidays	singles-holidays
1	https://travel.saga.co.uk/holid...	200	holidays/destinations	europe	holidays/destinations	holidays	destinations
2	https://travel.saga.co.uk/holid...	200	holidays/destinations/europe	spain	holidays/destinations/europe	holidays	destinations
3	https://travel.saga.co.uk/forms...	200	forms	request-a-brochure	holidays/forms	forms	request-a-brochure
4	https://travel.saga.co.uk/cruis...	200		cruises	holidays	holidays	cruises
...	...	...	...	...	...	...	...
1135	https://travel.saga.co.uk/holid...	200	holidays/holiday-types/escorted...	rv-mekong-pandaw	holidays/types/escorted-tours/s...	holidays	types
1136	https://travel.saga.co.uk/holid...	200	holidays/destinations/north ame...	rocky-mountains-and-alaskan-adv...	holidays/destinations/north ame...	holidays	destinations
1137	https://travel.saga.co.uk/holid...	200	holidays/destinations/europe/uk...	gilbert-and-sullivan-festival-m...	holidays/destinations/europe/uk...	holidays	destinations
1138	https://travel.saga.co.uk/holid...	200	holidays/destinations/europe/uk...	christmas-on-lake-windermere	holidays/destinations/europe/uk...	holidays	destinations
1139	https://travel.saga.co.uk/holid...	200	holidays/destinations/north-ame...		holidays/destinations/north-ame...	holidays	destinations

1140 rows × 7 columns

Any folders that didn't have a parent node in the list printed earlier are allocated to holidays. This should be right 90% of the time. Time to form the draft migration URL:

```
allocated_drafted['Migration URL'] = migration_root_url + allocated_
drafted.new_branch + '/' + allocated_drafted.node

allocated_drafted['Migration URL'] = np.where(allocated_drafted['Migration
URL'].str.endswith('/'),
                                              allocated_drafted['Migration
                                              URL'].str[:-1],
                                              allocated_drafted
                                              ['Migration URL'])
```

```
allocated_drafted['Parent'] = allocated_drafted['Parent'].str.
replace('-', ' ')
allocated_drafted['Parent'] = allocated_drafted['Parent'].str.title()

allocated_drafted['Child'] = allocated_drafted['Child'].str.
replace('-', ' ')
allocated_drafted['Child'] = allocated_drafted['Child'].str.title()
```

Set to lowercase:

```
allocated_drafted['Migration URL'] = allocated_drafted['Migration URL'].
str.lower()
pd.set_option('display.max_colwidth', 25)

allocated_drafted
```

This results in the following:

	current_url	http_status_code	old_branch	node	new_branch	Parent	Child	Migration URL
0	https://travel.saga.c...	200	holidays/holiday-types	singles-holidays	holidays	Holidays	Singles Holidays	https://www.saga.co.u...
1	https://travel.saga.c...	200	holidays/destinations	europe	holidays/destinations	Holidays	Destinations	https://www.saga.co.u...
2	https://travel.saga.c...	200	holidays/destinations...	spain	holidays/destinations...	Holidays	Destinations	https://www.saga.co.u...
3	https://travel.saga.c...	200	forms	request-a-brochure	holidays/forms	Forms	Request A Brochure	https://www.saga.co.u...
4	https://travel.saga.c...	200		cruises	holidays	Holidays	Cruises	https://www.saga.co.u...
...	...	...	...	...	...	...	...	...
1135	https://travel.saga.c...	200	holidays/holiday-type...	rv-mekong-pandaw	holidays/types/escort...	Holidays	Types	https://www.saga.co.u...
1136	https://travel.saga.c...	200	holidays/destinations...	rocky-mountains-and-a...	holidays/destinations...	Holidays	Destinations	https://www.saga.co.u...
1137	https://travel.saga.c...	200	holidays/destinations...	gilbert-and-sullivan-...	holidays/destinations...	Holidays	Destinations	https://www.saga.co.u...
1138	https://travel.saga.c...	200	holidays/destinations...	christmas-on-lake-win...	holidays/destinations...	Holidays	Destinations	https://www.saga.co.u...
1139	https://travel.saga.c...	200	holidays/destinations...		holidays/destinations...	Holidays	Destinations	https://www.saga.co.u...

1140 rows × 8 columns

By concatenating the domain, new branch, and node, the migration URLs are now fully formed:.

```
allocated_drafted['migration_url'] = migration_root_url + allocated_
drafted.new_branch + '/' + allocated_drafted.node

allocated_drafted['migration_url'] = np.where(allocated_drafted['migration_
url'].str.endswith('/'),
```

```
                                          allocated_drafted['migration_
                                          url'].str[:-1],
                                          allocated_drafted
                                          ['migration_url'])

allocated_drafted['Parent'] = allocated_drafted['Parent'].str.
replace('-', ' ')
allocated_drafted['Parent'] = allocated_drafted['Parent'].str.title()

allocated_drafted['Child'] = allocated_drafted['Child'].str.
replace('-', ' ')
allocated_drafted['Child'] = allocated_drafted['Child'].str.title()
```

Set to lowercase:

```
allocated_drafted['migration_url'] = allocated_drafted['migration_url'].
str.lower()
allocated_drafted.columns = allocated_drafted.columns.str.lower()
pd.set_option('display.max_colwidth', 25)

allocated_drafted
```

This results in the following:

	current_url	http_status_code	old_branch	node	new_branch	parent	child	migration_url
0	https://travel.saga.c...	200	holidays/holiday-types	singles-holidays	holidays	Holidays	Singles Holidays	https://www.saga.co.u...
1	https://travel.saga.c...	200	holidays/destinations	europe	holidays/destinations	Holidays	Destinations	https://www.saga.co.u...
2	https://travel.saga.c...	200	holidays/destinations...	spain	holidays/destinations...	Holidays	Destinations	https://www.saga.co.u...
3	https://travel.saga.c...	200	forms	request-a-brochure	holidays/forms	Forms	Request A Brochure	https://www.saga.co.u...
4	https://travel.saga.c...	200		cruises	holidays	Holidays	Cruises	https://www.saga.co.u...
...	...	...	...	...	...	...	...	...
1135	https://travel.saga.c...	200	holidays/holiday-type...	rv-mekong-pandaw	holidays/types/escort...	Holidays	Types	https://www.saga.co.u...
1136	https://travel.saga.c...	200	holidays/destinations...	rocky-mountains-and-a...	holidays/destinations...	Holidays	Destinations	https://www.saga.co.u...
1137	https://travel.saga.c...	200	holidays/destinations...	gilbert-and-sullivan-...	holidays/destinations...	Holidays	Destinations	https://www.saga.co.u...
1138	https://travel.saga.c...	200	holidays/destinations...	christmas-on-lake-win...	holidays/destinations...	Holidays	Destinations	https://www.saga.co.u...
1139	https://travel.saga.c...	200	holidays/destinations...		holidays/destinations...	Holidays	Destinations	https://www.saga.co.u...

1140 rows × 8 columns

To prepare the combining of the remaining URLs to the original latest mapping, we need to add site levels and some basic checks such as removing duplicate current URLs (after all, the same URL can't be redirected to two or more different URLs).

The site level is calculated by counting the number of slashes in the migration URL and subtracting one from it. This means the home page is one, and all other pages are referenced from there.

```
allocated_distinct = allocated_drafted
allocated_distinct = allocated_distinct.drop_duplicates(subset =
'current_url')
allocated_distinct['migration_url'] = allocated_distinct['migration_url'].
str.replace('/holidays/cruises/', '/cruises/')
allocated_distinct['migration_url'] = allocated_distinct['migration_url'].
str.replace(' ', '-')
allocated_level = allocated_distinct
allocated_level['level'] = allocated_level.migration_url.str.count('/') - 1
allocated_level = allocated_level[['parent', 'child', 'level', 'current_
url', 'migration_url']]
pd.set_option('display.max_colwidth', 65)
allocated_level
```

This results in the following:

	parent	child	level	current_url	migration_url
0	Holidays	Singles Holidays	3	https://travel.saga.co.uk/holidays/holiday-types/singles-holi...	https://www.saga.co.uk/holidays/singles-holidays
1	Holidays	Destinations	4	https://travel.saga.co.uk/holidays/destinations/europe.aspx	https://www.saga.co.uk/holidays/destinations/europe
2	Holidays	Destinations	5	https://travel.saga.co.uk/holidays/destinations/europe/spain....	https://www.saga.co.uk/holidays/destinations/europe/spain
3	Forms	Request A Brochure	4	https://travel.saga.co.uk/forms/request-a-brochure.aspx	https://www.saga.co.uk/holidays/forms/request-a-brochure
4	Holidays	Cruises	3	https://travel.saga.co.uk/cruises.aspx	https://www.saga.co.uk/holidays/cruises
...	...	...	...	...	...
1135	Holidays	Types	6	https://travel.saga.co.uk/holidays/holiday-types/escorted-tou...	https://www.saga.co.uk/holidays/types/escorted-tours/small-sh...
1136	Holidays	Destinations	6	https://travel.saga.co.uk/holidays/destinations/north america...	https://www.saga.co.uk/holidays/destinations/north-america/ca...
1137	Holidays	Destinations	7	https://travel.saga.co.uk/holidays/destinations/europe/uk/eng...	https://www.saga.co.uk/holidays/destinations/europe/uk/englan...
1138	Holidays	Destinations	7	https://travel.saga.co.uk/holidays/destinations/europe/uk/eng...	https://www.saga.co.uk/holidays/destinations/europe/uk/englan...
1139	Holidays	Destinations	5	https://travel.saga.co.uk/holidays/destinations/north-america...	https://www.saga.co.uk/holidays/destinations/north-america/me...

1140 rows × 5 columns

With the columns now matching the original imported travel mapping, we're ready to combine:

```
total_mapping = pd.concat([latest_mapping_raw, allocated_level])
total_mapping
```

This results in the following:

	parent	child	level	current_url	migration_url
0	Homepage	Holidays	0	https://travel.saga.co.uk/	https://www.saga.co.uk/holidays
1	Homepage	Holidays	2	https://travel.saga.co.uk/holidays	https://www.saga.co.uk/holidays
2	Holidays	Destinations	3	https://travel.saga.co.uk/holidays/destinations.aspx	https://www.saga.co.uk/holidays/destinations
3	Destinations	Africa	4	https://travel.saga.co.uk/holidays/destinations/africa.aspx	https://www.saga.co.uk/holidays/destinations/africa
4	Africa	Egypt	5	https://travel.saga.co.uk/holidays/destinations/africa/egypt....	https://www.saga.co.uk/holidays/destinations/africa/egypt
...	...	...	...	...	...
1135	Holidays	Types	6	https://travel.saga.co.uk/holidays/holiday-types/escorted-tou...	https://www.saga.co.uk/holidays/types/escorted-tours/small-sh...
1136	Holidays	Destinations	6	https://travel.saga.co.uk/holidays/destinations/north america...	https://www.saga.co.uk/holidays/destinations/north-america/ca...
1137	Holidays	Destinations	7	https://travel.saga.co.uk/holidays/destinations/europe/uk/eng...	https://www.saga.co.uk/holidays/destinations/europe/uk/englan...
1138	Holidays	Destinations	7	https://travel.saga.co.uk/holidays/destinations/europe/uk/eng...	https://www.saga.co.uk/holidays/destinations/europe/uk/englan...
1139	Holidays	Destinations	5	https://travel.saga.co.uk/holidays/destinations/north-america...	https://www.saga.co.uk/holidays/destinations/north-america/me...

1609 rows × 5 columns

The rows are now combined. Now we will drop duplicate rows and calculate the string similarity between the current URL and the migration URL, which will help us in the manual review of the export file:

```
total_mapping_simi = total_mapping.drop_duplicates(subset = 'current_url')

total_mapping_simi['simi'] = total_mapping_simi.loc[
    :, ['current_url', 'migration_url']].apply(
    lambda x: sorensen_dice(*x), axis=1)

total_mapping_simi.to_csv('exports/' + target_site_search + '_' +
target_bu  + '_total_mapping_simi.csv')

total_mapping_simi
```

This results in the following:

	parent	child	level	current_url	migration_url	simi
0	Homepage	Holidays	0	https://travel.saga.co.uk/	https://www.saga.co.uk/holidays	0.771930
1	Homepage	Holidays	2	https://travel.saga.co.uk/holidays	https://www.saga.co.uk/holidays	0.861538
2	Holidays	Destinations	3	https://travel.saga.co.uk/holidays/destinations.aspx	https://www.saga.co.uk/holidays/destinations	0.854167
3	Destinations	Africa	4	https://travel.saga.co.uk/holidays/destinations/africa.aspx	https://www.saga.co.uk/holidays/destinations/africa	0.872727
4	Africa	Egypt	5	https://travel.saga.co.uk/holidays/destinations/africa/e...	https://www.saga.co.uk/holidays/destinations/africa/egypt	0.885246
...	...	...	...	...	...	...
1135	Holidays	Types	6	https://travel.saga.co.uk/holidays/holiday-types/escorte...	https://www.saga.co.uk/holidays/types/escorted-tours/sma...	0.838710
1136	Holidays	Destinations	6	https://travel.saga.co.uk/holidays/destinations/north am...	https://www.saga.co.uk/holidays/destinations/north-ameri...	0.887850
1137	Holidays	Destinations	7	https://travel.saga.co.uk/holidays/destinations/europe/u...	https://www.saga.co.uk/holidays/destinations/europe/uk/e...	0.943089
1138	Holidays	Destinations	7	https://travel.saga.co.uk/holidays/destinations/europe/u...	https://www.saga.co.uk/holidays/destinations/europe/uk/e...	0.926316
1139	Holidays	Destinations	5	https://travel.saga.co.uk/holidays/destinations/north-am...	https://www.saga.co.uk/holidays/destinations/north-ameri...	0.245283

1398 rows × 6 columns

We now have the migration mapping ready to review in Excel. Note the row number has reduced as duplicate current URLs have been eliminated. We've also put a new column "simi" to help flag any URLs that have "migration URLs less than 75% similar to their current URL counterpart." Although not foolproof, this will help provide a quick way of finding and sorting any anomalies.

Migration planning can inspire challenge and dread for a lot of SEOs. AI and data science have yet to advance far enough to fully automate, let alone semiautomate, most of the site migration planning process.

Much of the advance will depend on the NLP models at the AI level available to the SEO industry to reliably understand, reduce, and map existing content URLs to new URLs.

The next section will now address troubleshooting traffic losses following a site migration.

Migration Forensics

At this point, we're here to work out what changed and which content was affected following a migration. We'll be taking the following steps:

1. Traffic trends

2. Determine the change point

3. Determine the winning and losing content

4. Gather a list of URLs before and after for crawling

5. Group and segment URLs

6. Diagnose

7. Road map of recommendations

As usual, we start importing our libraries. You'll notice that some of the packages include some string distance functions from textdistance and timedelta to help us work with time series data:

```
import pandas as pd
import numpy as np
from textdistance import sorensen_dice
```

```
from plotnine import *
import datetime
from datetime import timedelta
from textdistance import sorensen_dice

root_url = 'https://www.saasforecom.com'
root_domain = 'saasforecom.com'
hostname = 'saasforecom'
```

Traffic Trends

With the libraries imported and the variables set, we'll import the data from Google Analytics (GA). We use GA because it gives us a breakdown by date that is not easily found in Google Search Console (GSC) without substantial sampling.

```
ga_orgdatelp_raw = pd.read_csv('data/Analytics www.salesorder.com All
Traffic 20200901-20201231.csv', skiprows = 5)
```

Here, we're getting rid of the rows which are not part of the main table that is default in GA tabular exports:

```
ga_orgdatelp_raw = ga_orgdatelp_raw[~ga_orgdatelp_raw['Landing Page'].
isnull()]
ga_orgdatelp_raw = ga_orgdatelp_raw[~ga_orgdatelp_raw['Sessions'].isnull()]
```

To make the columns cleaner, we'll perform a number of string operations:

```
ga_orgdatelp_raw.columns = ga_orgdatelp_raw.columns.str.lower()
ga_orgdatelp_raw.columns = ga_orgdatelp_raw.columns.str.replace('/', '',
regex = False)
ga_orgdatelp_raw.columns = ga_orgdatelp_raw.columns.str.replace('.', '',
regex = False)
ga_orgdatelp_raw.columns = ga_orgdatelp_raw.columns.str.replace('% ', '',
regex = False)
ga_orgdatelp_raw.columns = ga_orgdatelp_raw.columns.str.replace('  ', ' ',
regex = False)
ga_orgdatelp_raw.columns = ga_orgdatelp_raw.columns.str.replace(' ', '_',
regex = False)
```

```
ga_orgdatelp_raw = ga_orgdatelp_raw[ga_orgdatelp_raw['landing_page'] != '/
pages/login.aspx']
ga_orgdatelp_raw
```

This results in the following:

	landing_page	date	sessions	new_sessions	new_users	bounce_rate	pages_session	avg_session_duration	tra…
0	/	2020-12-08	180	35.56	64	0.56	3.24	124.0	
1	/	2020-12-18	175	24.00	42	0.00	4.60	190.0	
2	/	2020-12-07	169	23.67	40	0.00	4.15	283.0	
3	/	2020-11-30	163	23.31	38	0.00	4.85	271.0	
4	/	2020-12-15	158	27.22	43	0.63	3.86	152.0	
...	...	...	...	...	...	...	...	...	
604	/submit-support-ticket/	2020-12-15	1	0.00	0	0.00	50.00	3122.0	
605	/submit-support-ticket/	2020-12-19	1	0.00	0	0.00	2.00	0.0	
606	/trial-settings-2-set-admin-company-details/	2020-12-27	1	0.00	0	0.00	80.00	4516.0	
607	/zendesk-request-form-test/	2020-12-11	1	0.00	0	0.00	4.00	237.0	
608	/zendesk-request-form-test/?preview_id=15696&preview_nonce=83bf02d08a&preview=true	2020-12-10	1	0.00	0	0.00	12.00	333.0	

609 rows × 11 columns

With the data imported and the column names cleaned and nicely formatted, we'll get to work on cleaning the actual data inside the columns themselves.

This again will make use of string operations to remove special characters and cast the data type as a number as opposed to a string.

Clean the GA data:

```
ga_clean = ga_orgdatelp_raw
```

Format the dates:

```
ga_clean['date'] = pd.to_datetime(ga_clean.date, format='%Y%m%d')

ga_clean['bounce_rate'] = ga_clean.bounce_rate.str.replace('%', '')
ga_clean['bounce_rate'] = ga_clean.bounce_rate.astype(float)
ga_clean['new_sessions'] = ga_clean.new_sessions.str.replace('%', '')
ga_clean['new_sessions'] = ga_clean.new_sessions.astype(float)
ga_clean['ecommerce_conversion_rate'] = ga_clean.ecommerce_conversion_rate.
str.replace('%', '')
```

```
ga_clean['ecommerce_conversion_rate'] = ga_clean.ecommerce_conversion_rate.
astype(float)
ga_clean['revenue'] = ga_clean.revenue.str.replace('$', '')
ga_clean['revenue'] = ga_clean.revenue.astype(float)
ga_clean['avg_session_duration'] = ga_clean.avg_session_duration.str.
replace('<', '')
ga_clean['avg_session_duration'] = pd.to_timedelta(ga_clean.avg_session_
duration).astype(int) / 1e9
ga_clean
```

This results in the following:

landing_page	date	sessions	new_sessions	new_users	bounce_rate	pages_session	avg_session_duration	transactions	revenue	ecommerce_conversio
/	2020-12-08	180	35.56	64	0.56	3.24	1.240000e-07	0.0	0.0	
/	2020-12-18	175	24.00	42	0.00	4.60	1.900000e-07	0.0	0.0	
/	2020-12-07	169	23.67	40	0.00	4.15	2.830000e-07	0.0	0.0	
/	2020-11-30	163	23.31	38	0.00	4.85	2.710000e-07	0.0	0.0	
/	2020-12-15	158	27.22	43	0.63	3.86	1.520000e-07	0.0	0.0	
...	...	...	...	...	...	...	...	...	...	
/submit-support-ticket/	2020-12-15	1	0.00	0	0.00	50.00	3.122000e-06	0.0	0.0	
/submit-support-ticket/	2020-12-19	1	0.00	0	0.00	2.00	0.000000e+00	0.0	0.0	
dmin-company-details/	2020-12-27	1	0.00	0	0.00	80.00	4.516000e-06	0.0	0.0	
desk-request-form-test/	2020-12-11	1	0.00	0	0.00	4.00	2.370000e-07	0.0	0.0	
esk-request-form-test/? of02d08a&preview=true	2020-12-10	1	0.00	0	0.00	12.00	3.330000e-07	0.0	0.0	

```
ga_stats = ga_clean
```

We select the columns we actually want. You may have noticed that some columns were cleaned up and ended up not being used. This may seem like a waste of effort; however, you don't always know what you will need or for what purpose. So cleaning columns is a good standard practice so that the data is ready should you discover that you need it later on.

```
ga_stats = ga_stats[['landing_page', 'date', 'new_sessions', 'avg_session_
duration']]
ga_stats = ga_stats.rename(columns = {'landing_page': 'subpath'})
ga_stats
```

This results in the following:

	subpath	date	new_sessions	avg_session_duration
0	/	2020-12-08	35.56	1.240000e-07
1	/	2020-12-18	24.00	1.900000e-07
2	/	2020-12-07	23.67	2.830000e-07
3	/	2020-11-30	23.31	2.710000e-07
4	/	2020-12-15	27.22	1.520000e-07
...	...	...	...	...
604	/submit-support-ticket/	2020-12-15	0.00	3.122000e-06
605	/submit-support-ticket/	2020-12-19	0.00	0.000000e+00
606	/trial-settings-2-set-admin-company-details/	2020-12-27	0.00	4.516000e-06
607	/zendesk-request-form-test/	2020-12-11	0.00	2.370000e-07
608	/zendesk-request-form-test/?preview_id=15696&preview_nonce=83bf02d08a&preview=true	2020-12-10	0.00	3.330000e-07

609 rows × 4 columns

Import GSC Pages data to grab all of the unique URLs for crawling:

```
all_gsc_raw = pd.read_csv('data/throughout_Pages.csv')
all_gsc_raw.columns = all_gsc_raw.columns.str.lower()
all_gsc_raw.columns = all_gsc_raw.columns.str.replace('/', '',
regex = False)
all_gsc_raw.columns = all_gsc_raw.columns.str.replace('.', '',
regex = False)
all_gsc_raw.columns = all_gsc_raw.columns.str.replace('% ', '',
regex = False)
all_gsc_raw.columns = all_gsc_raw.columns.str.replace('  ', ' ',
regex = False)
all_gsc_raw.columns = all_gsc_raw.columns.str.replace(' ', '_',
regex = False)
all_gsc_raw['ctr'] = all_gsc_raw.ctr.str.replace('%', '', regex = False)
print(all_gsc_raw.head())
all_gsc_urls = all_gsc_raw[['top_pages']]
all_gsc_urls = all_gsc_urls.rename(columns = {'top_pages': 'url'}).drop_
duplicates()
all_gsc_urls
```

This results in the following:

	url
0	https://www.saasforecom.com/
1	https://www.saasforecom.com/documentation/products-services/working-with-items/non-stock-item/
2	https://www.saasforecom.com/documentation/sales/working-with-sales-orders/allocating-stock-to-saasforecoms/pick-pack-ship-process/
3	https://www.saasforecom.com/documentation/sales/working-with-sales-orders/allocating-stock-to-saasforecoms/
4	https://www.saasforecom.com/warehouse-management-system/
...	...
978	https://www.saasforecom.com/documentation/tag/contacts/
979	https://www.saasforecom.com/b2b-ecommerce-solution/#elementor-toc__heading-anchor-7
980	https://www.saasforecom.com/help/sales-accounting/using-sales-invoice/#elementor-toc__heading-anchor-54
981	https://www.saasforecom.com/help/start/setup-document-template/#elementor-toc__heading-anchor-5
982	https://www.saasforecom.com/documentation/tag/tax-codes/

983 rows × 1 columns

The GA URLs will also be extracted and joined to the domain ready for crawling:

```
ga_raw_urls = ga_raw_comb[['landing_page']]
ga_raw_urls = ga_raw_urls.rename(columns = {'landing_page': 'url'}).drop_
duplicates()
ga_raw_urls['url'] = root_url + ga_raw_urls['url']
ga_raw_urls
```

This results in the following:

	url
0	https://www.saasforecom.com/
26	https://www.saasforecom.com/cloud-erp-system-saas-erp-guide-wholesalers/
37	https://www.saasforecom.com/cloud-based-inventory-management/
39	https://www.saasforecom.com/help/
40	https://www.saasforecom.com/cloud-erp-solution/
...	...
588	https://www.saasforecom.com/saasforecom-pricing/?preview_id=8169&preview_nonce=2eff823a8f&preview=true
589	https://www.saasforecom.com/saasforecom/sohelp/help/prestashop.htm
606	https://www.saasforecom.com/trial-settings-2-set-admin-company-details/
607	https://www.saasforecom.com/zendesk-request-form-test/
608	https://www.saasforecom.com/zendesk-request-form-test/?preview_id=15696&preview_nonce=83bf02d08a&preview=true

203 rows × 1 columns

Combine the GA and GSC URLs, dropping duplicates, ready for crawling:

```
crawl_urls = pd.concat([ga_raw_urls, all_gsc_urls]).drop_duplicates()
crawl_urls.to_csv('data/urls_to_crawl.csv')
crawl_urls
```

This results in the following:

	url
0	https://www.saasforecom.com/
26	https://www.saasforecom.com/cloud-erp-system-saas-erp-guide-wholesalers/
37	https://www.saasforecom.com/cloud-based-inventory-management/
39	https://www.saasforecom.com/help/
40	https://www.saasforecom.com/cloud-erp-solution/
...	...
978	https://www.saasforecom.com/documentation/tag/contacts/
979	https://www.saasforecom.com/b2b-ecommerce-solution/#elementor-toc__heading-anchor-7
980	https://www.saasforecom.com/help/sales-accounting/using-sales-invoice/#elementor-toc__heading-anchor-54
981	https://www.saasforecom.com/help/start/setup-document-template/#elementor-toc__heading-anchor-5
982	https://www.saasforecom.com/documentation/tag/tax-codes/

1077 rows × 1 columns

With the crawl completed, we're ready to import the data, clean the columns, and view the raw data:

```
audit_urls_raw = pd.read_csv('data/all_urls__excluding_uncrawled__
filtered_20210803163126.csv')
audit_urls_raw.columns = audit_urls_raw.columns.str.lower()
audit_urls_raw.columns = audit_urls_raw.columns.str.replace('/', '',
regex = False)
audit_urls_raw.columns = audit_urls_raw.columns.str.replace('.', '',
regex = False)
audit_urls_raw.columns = audit_urls_raw.columns.str.replace('% ', '',
regex = False)
audit_urls_raw.columns = audit_urls_raw.columns.str.replace('  ', ' ',
regex = False)
```

```
audit_urls_raw.columns = audit_urls_raw.columns.str.replace(' ', '_',
regex = False)
#audit_urls_raw['ctr'] = audit_urls_raw.ctr.str.replace('%', '',
regex = False)
#audit_urls_raw.drop_duplicates()
audit_urls_raw
```

This results in the following:

	url	crawl_depth	crawl_status	host	is_subdomain	scheme	crawl_source	first_parent_
0	https://www.saasforecom.com/	0	Success	www.saasforecom.com	No	https	Crawler	No
1	https://www.saasforecom.com/ecommerce-order-management-system/	1	Success	www.saasforecom.com	No	https	Crawler	https://www.saasforecom.co
2	https://www.saasforecom.com/software-for-warehouse-inventory-management/	1	Success	www.saasforecom.com	No	https	Crawler	https://www.saasforecom.co
3	https://www.saasforecom.com/drop-shipping-automation-software/	1	Success	www.saasforecom.com	No	https	Crawler	https://www.saasforecom.co
4	https://www.saasforecom.com/refer-and-earn/	1	Success	www.saasforecom.com	No	https	Crawler	https://www.saasforecom.co
...	...	...	...	...	...	...	...	
1190	https://www.saasforecom.com/help/order-management-software-guide/using-drop-ship-sales-orders/	Not Set	Success	www.saasforecom.com	No	https	Url List	No
1191	https://www.saasforecom.com/cloud-based-erp-software-customization/	Not Set	Redirect	www.saasforecom.com	No	https	Url List	No
1192	https://www.saasforecom.com/help/crm-for-wholesalers-guide/using-sales-quote/	Not Set	Success	www.saasforecom.com	No	https	Url List	No
1193	https://www.saasforecom.com/crm-for-distributors-wholesalers-ecommerce/	Not Set	Redirect	www.saasforecom.com	No	https	Url List	No
1194	https://www.saasforecom.com/b2b-ecommerce-solution/	Not Set	Success	www.saasforecom.com	No	https	Url List	No

1184 rows × 21 columns

We can see that most of the server status has not been extracted. This is likely to be a bug in the crawling software. The best thing to do is to take it up with the software vendor and recrawl with a longer timeout setting and at a slower pace to improve the numbers.

```
audit_urls_raw.groupby('final_redirect_url_status_code').size()

final_redirect_url_status_code
200          452
404            1
Not Set      731
dtype: int64
```

After our recrawl, 452 URLs is the best we could come up with. Next, we're ensuring any rows with duplicate URLs are dropped:

```
audit_urls_map = audit_urls_raw.drop_duplicates(subset = ['url']).
reset_index()
audit_urls_map.to_csv('exports/audit_urls_map.csv')
audit_urls_map
```

This results in the following:

	index	url	crawl_depth	crawl_status	host	is_subdomain	scheme	crawl_source	
0	0	https://www.saasforecom.com/	0	Success	www.saasforecom.com	No	https	Crawler	
1	1	https://www.saasforecom.com/ecommerce-order-management-system/	1	Success	www.saasforecom.com	No	https	Crawler	h
2	2	https://www.saasforecom.com/software-for-warehouse-inventory-management/	1	Success	www.saasforecom.com	No	https	Crawler	h
3	3	https://www.saasforecom.com/drop-shipping-automation-software/	1	Success	www.saasforecom.com	No	https	Crawler	h
4	4	https://www.saasforecom.com/refer-and-earn/	1	Success	www.saasforecom.com	No	https	Crawler	h
...	...	...	...	...	...	...	...	...	
917	1183	https://www.saasforecom.com/inventory-management-software-for-wichita-wholesalers/	Not Set	Redirect	www.saasforecom.com	No	https	Url List	
918	1186	https://www.saasforecom.com/documentation/related/prospects/	Not Set	Redirect	www.saasforecom.com	No	https	Url List	
919	1187	https://www.saasforecom.com/help/wms-console/wms-kits-assemblies/	Not Set	Not Found	www.saasforecom.com	No	https	Url List	
920	1188	https://www.saasforecom.com/documentation/banking/working-with-bank-accounts	Not Set	Redirect	www.saasforecom.com	No	https	Url List	
921	1189	https://www.saasforecom.com/documentation/banking/working-with-credit-cards/	Not Set	Redirect	www.saasforecom.com	No	https	Url List	

922 rows × 22 columns

The row count has now dropped from 1122 to 922 rows. Next, we'll find the final redirect URL so we can see where the URLs map to. Again, this seemingly unnecessary step is taken to overcome any glitches produced by the audit software.

Prepare the columns for content evaluation:

```
audit_urls = audit_urls_map[['url', 'redirect_url', 'final_redirect_url']]
ult_dest_url = []
```

This function will take a row, turn it into a list, and take the last value that isn't equal to "No Data" and stick the URL in the list ult_dest_url created earlier:

```
def find_ult_dest_url(row):
    data = row.values.tolist()
    data = [e for e in data if str(e) not in ('No Data')]
```

```
data = data[-1]
#print(data)
ult_dest_url.append(data)
```

The preceding function is applied by calling the following function to take the dataframe row by row, which is considered to be a less computationally intensive way to iterate over a dataframe, certainly faster than iterrows:

```
def apply_fudu(df):
  return df.apply(lambda row: find_ult_dest_url(row), axis=1)
```

Call the apply_fudu function that calls the find_ult_dest_url function:

```
apply_fudu(audit_urls)
```

The resulting list is now converted into a dataframe:

```
ult_dest_url_df = pd.DataFrame(ult_dest_url, columns=['ult_dest_url'])
ult_dest_url_df
```

This results in the following:

	ult_dest_url
0	https://www.saasforecom.com/
1	https://www.saasforecom.com/ecommerce-order-management-system/
2	https://www.saasforecom.com/software-for-warehouse-inventory-management/
3	https://www.saasforecom.com/drop-shipping-automation-software/
4	https://www.saasforecom.com/refer-and-earn/
...	...
917	http://www.saasforecom.com/cloud-based-inventory-management/
918	https://www.saasforecom.com/help/
919	https://www.saasforecom.com/help/wms-console/wms-kits-assemblies/
920	https://www.saasforecom.com/help/
921	https://www.saasforecom.com/help/

922 rows × 1 columns

Append the dataframe to the audit dataframe:

```
audit_urls_map_prep = audit_urls_map.join(ult_dest_url_df)
audit_urls_map_prep
```

This results in the following:

index		url	crawl_depth	crawl_status	host	is_subdomain	scheme	crawl_source	
0	0	https://www.saasforecom.com/	0	Success	www.saasforecom.com	No	https	Crawler	
1	1	https://www.saasforecom.com/ecommerce-order-management-system/	1	Success	www.saasforecom.com	No	https	Crawler	h
2	2	https://www.saasforecom.com/software-for-warehouse-inventory-management/	1	Success	www.saasforecom.com	No	https	Crawler	h
3	3	https://www.saasforecom.com/drop-shipping-automation-software/	1	Success	www.saasforecom.com	No	https	Crawler	h
4	4	https://www.saasforecom.com/refer-and-earn/	1	Success	www.saasforecom.com	No	https	Crawler	h
...	...	...	...	...	...	...	...	...	
917	1183	https://www.saasforecom.com/inventory-management-software-for-wichita-wholesalers/	Not Set	Redirect	www.saasforecom.com	No	https	Url List	
918	1186	https://www.saasforecom.com/documentation/related/prospects/	Not Set	Redirect	www.saasforecom.com	No	https	Url List	
919	1187	https://www.saasforecom.com/help/wms-console/wms-kits-assemblies/	Not Set	Not Found	www.saasforecom.com	No	https	Url List	
920	1188	https://www.saasforecom.com/documentation/banking/working-with-bank-accounts	Not Set	Redirect	www.saasforecom.com	No	https	Url List	
921	1189	https://www.saasforecom.com/documentation/banking/working-with-credit-cards/	Not Set	Redirect	www.saasforecom.com	No	https	Url List	

922 rows × 23 columns

With the ultimate destination URLs found, we need a simple way to test how similar they are. We can do this by measuring the string distance between the URL and the redirect URL. We'll use Sorensen-Dice which is fast and effective for SEO purposes:

```
audit_urls_map_prep['content_simi'] = audit_urls_map_prep.loc[:, ['url',
                                                        'ult_dest_url']].
apply(lambda x: sorensen_dice(*x), axis=1)

audit_urls_map = audit_urls_map_prep
audit_urls_map
```

This results in the following:

tus_code	redirect_url	redirect_url_status_code	urls_with_similar_content	unnamed: 20	ult_dest_url	content_simi
Not Set	No Data	Not Set	0	NaN	https://www.saasforecom.com/	1.000000
Not Set	No Data	Not Set	0	NaN	https://www.saasforecom.com/ecommerce-order-management-system/	1.000000
Not Set	No Data	Not Set	0	NaN	https://www.saasforecom.com/software-for-warehouse-inventory-management/	1.000000
Not Set	No Data	Not Set	0	NaN	https://www.saasforecom.com/drop-shipping-automation-software/	1.000000
Not Set	No Data	Not Set	0	NaN	https://www.saasforecom.com/refer-and-earn/	1.000000
...	...	...	...	...	...	...
Not Set	http://www.saasforecom.com/cloud-based-inventory-management/	Not Set	0	NaN	http://www.saasforecom.com/cloud-based-inventory-management/	0.788732
200	http://www.saasforecom.com/help/	301	0	NaN	https://www.saasforecom.com/help/	0.688172
Not Set	No Data	Not Set	0	NaN	https://www.saasforecom.com/help/wms-console/wms-kits-assemblies/	1.000000
200	http://www.saasforecom.com/help/	301	0	NaN	https://www.saasforecom.com/help/	0.568807
200	http://www.saasforecom.com/help/	301	0	NaN	https://www.saasforecom.com/help/	0.568807

Segmenting URLs

With all of these audit URLs, we'd want to make sense of them so we can discern trends by content type. Since we don't have a trained neural network at hand, we're going to use a crude yet useful method of grouping URLs by their URL address.

This method is not only fast, it's also cheap in that you won't require a million content documents to train an AI to categorize web documents by content type.

We'll start by extracting the URLs and ensuring they are unique before sticking them into a dataframe:

```
crawled_urls_unq = audit_urls_raw.url.drop_duplicates().to_frame()
crawled_urls_unq
```

This results in the following:

	url
0	https://www.saasforecom.com/
1	https://www.saasforecom.com/ecommerce-order-management-system/
2	https://www.saasforecom.com/software-for-warehouse-inventory-management/
3	https://www.saasforecom.com/drop-shipping-automation-software/
4	https://www.saasforecom.com/refer-and-earn/
...	...
1183	https://www.saasforecom.com/inventory-management-software-for-wichita-wholesalers/
1186	https://www.saasforecom.com/documentation/related/prospects/
1187	https://www.saasforecom.com/help/wms-console/wms-kits-assemblies/
1188	https://www.saasforecom.com/documentation/banking/working-with-bank-accounts
1189	https://www.saasforecom.com/documentation/banking/working-with-credit-cards/

922 rows × 1 columns

```
all_urls = pd.concat([crawl_urls, crawled_urls_unq])
all_urls = all_urls.drop_duplicates()
all_urls
```

This results in the following:

	url
0	https://www.saasforecom.com/
26	https://www.saasforecom.com/cloud-erp-system-saas-erp-guide-wholesalers/
37	https://www.saasforecom.com/cloud-based-inventory-management/
39	https://www.saasforecom.com/help/
40	https://www.saasforecom.com/cloud-erp-solution/
...	...
566	https://www.saasforecom.com/channel/UCi9Rkl_DituYZLciMtmNbfw
1084	https://www.saasforecom.com/cloud-erp-system-blog/?preview_id=14202&preview_nonce=73739bfaba&preview=true
1104	https://www.saasforecom.com/cloud-erp-system-saas-erp-guide-wholesalers/?preview_id=14202&preview_nonce=9d48a4694e&preview=true
1132	https://www.saasforecom.com/help-2-0/?preview_id=14818&preview_nonce=58c5969977&preview=true
1180	https://www.saasforecom.com/video-help-wall/?preview_id=14261&preview_nonce=fa90ab88ac&preview=true

1251 rows × 1 columns

This code cleans up the URLs ready for some text processing so we can start grouping the URLs. We'll start with removing the domain portion of the URL as that is constant throughout the URLs:

```
classified_start = all_urls[['url']]
classified_start['slug'] = classified_start.url.str.replace(root_url, '',
regex = True)
```

The home page can be immediately classified:

```
classified_start['slug'] = np.where(classified_start.slug == "/", "home",
classified_start.slug)
```

The following will deal with dates which won't add value to the classification:

```
classified_start['slug'] = classified_start.slug.str.replace("\
\d{4}\\-(0[1-9]|1[012])\\-(0[1-9]|[12][0-9]|3[01])",
                                                            '', regex = True)
```

Remove excessive spaces between words:

```
classified_start['slug'] = classified_start.slug.str.replace
("[^\w\s]", " ", regex = True)
```

Remove spaces at the beginning and end:

```
classified_start['slug'] = classified_start.slug.str.strip()
classified_start = classified_start.reset_index()
del classified_start['index']
classified_start.head(10)
```

This results in the following:

	url	slug
0	https://www.saasforecom.com/	home
1	https://www.saasforecom.com/cloud-erp-system-saas-erp-guide-wholesalers/	cloud erp system saas erp guide wholesalers
2	https://www.saasforecom.com/cloud-based-inventory-management/	cloud based inventory management
3	https://www.saasforecom.com/help/	help
4	https://www.saasforecom.com/cloud-erp-solution/	cloud erp solution
5	https://www.saasforecom.com/ecommerce-order-management-system/	ecommerce order management system
6	https://www.saasforecom.com/drop-shipping-automation-software/	drop shipping automation software
7	https://www.saasforecom.com/software-for-warehouse-inventory-management/	software for warehouse inventory management
8	https://www.saasforecom.com/benefits-of-cloud-erp/	benefits of cloud erp
9	https://www.saasforecom.com/crm-for-wholesale-distributors/	crm for wholesale distributors
10	https://www.saasforecom.com/cloud-erp-order-management-software-about-us/	cloud erp order management software about us

The result is the URL words without all of the characters, that is, the slug. We'll want to apply some numbers to get a sense of priority, so we'll use GSC traffic data to weight the slugs.

Get GSC traffic data:

```
url_clicks_gsc = all_gsc_raw[['top_pages', 'clicks']]
url_clicks_gsc = url_clicks_gsc.rename(columns = {'top_pages': 'url'})
url_clicks_gsc
```

This results in the following:

	url	clicks
0	https://www.saasforecom.com/	6911
1	https://www.saasforecom.com/documentation/products-services/working-with-items/non-stock-item/	428
2	https://www.saasforecom.com/documentation/sales/working-with-sales-orders/allocating-stock-to-saasforecoms/pick-pack-ship-process/	417
3	https://www.saasforecom.com/documentation/sales/working-with-sales-orders/allocating-stock-to-saasforecoms/	365
4	https://www.saasforecom.com/warehouse-management-system/	336
...	...	...
978	https://www.saasforecom.com/documentation/tag/contacts/	0
979	https://www.saasforecom.com/b2b-ecommerce-solution/#elementor-toc__heading-anchor-7	0
980	https://www.saasforecom.com/help/sales-accounting/using-sales-invoice/#elementor-toc__heading-anchor-54	0
981	https://www.saasforecom.com/help/start/setup-document-template/#elementor-toc__heading-anchor-5	0
982	https://www.saasforecom.com/documentation/tag/tax-codes/	0

983 rows × 2 columns

Then merge into the URL slug table earlier:

```
classified_stats = classified_start.merge(url_clicks_gsc, on = 'url',
how = 'left')
```

Remove URLs with no clicks. Choose your GSC date range wisely here. If you just went for 28 days, then there's the risk of seasonal bias as some content may not receive traffic at certain times of the year. Our recommendation is to select 16 months, the maximum possible for extraction from GSC.

```
classified_stats = classified_stats[classified_stats.clicks > 0]
classified_stats
```

This results in the following:

	url	slug	clicks
0	https://www.saasforecom.com/	home	6911.0
1	https://www.saasforecom.com/cloud-erp-system-saas-erp-guide-wholesalers/	cloud erp system saas erp guide wholesalers	37.0
2	https://www.saasforecom.com/cloud-based-inventory-management/	cloud based inventory management	49.0
3	https://www.saasforecom.com/help/	help	4.0
4	https://www.saasforecom.com/cloud-erp-solution/	cloud erp solution	49.0
...	...	...	...
1246	https://www.saasforecom.com/channel/UCi9Rkl_DituYZLciMtmNbfw	channel UCi9Rkl_DituYZLciMtmNbfw	0.0
1247	https://www.saasforecom.com/cloud-erp-system-blog/?preview_id=14202&preview_nonce=73739bfaba&preview=true	cloud erp system blog preview_id 14202 preview_nonce 73739bfaba preview true	0.0
1248	https://www.saasforecom.com/cloud-erp-system-saas-erp-guide-wholesalers/?preview_id=14202&preview_nonce=9d48a4694e&preview=true	cloud erp system saas erp guide wholesalers preview_id 14202 preview_nonce 9d48a4694e preview true	0.0
1249	https://www.saasforecom.com/help-2-0/?preview_id=14818&preview_nonce=58c5969977&preview=true	help 2 0 preview_id 14818 preview_nonce 58c5969977 preview true	0.0
1250	https://www.saasforecom.com/video-help-wall/?preview_id=14261&preview_nonce=fa90ab88ac&preview=true	video help wall preview_id 14261 preview_nonce fa90ab88ac preview true	0.0

1251 rows × 3 columns

We're going to explode the slug column into unigrams. That means taking the slug and expanding the column into several rows such that each word in the slug has its own row as one column:

```
bigrams = classified_stats['slug']
bigrams = bigrams.str.split(' ').explode().to_frame()
bigrams = bigrams.rename(columns = {'slug': 'ngram'})
bigrams.head(10)
```

This results in the following:

	ngram
0	home
1	cloud
1	erp
1	system
1	saas
1	erp
1	guide
1	wholesalers
2	cloud
2	based

With the slugs "exploded" into ngrams, this will be mapped to their original URL and traffic stats table:

```
bigrams_df = classified_stats.join(bigrams)

bigrams_df.head(20)
```

This results in the following:

	url	slug	clicks	ngram
0	https://www.saasforecom.com/	home	6911.0	home
1	https://www.saasforecom.com/cloud-erp-system-saas-erp-guide-wholesalers/	cloud erp system saas erp guide wholesalers	37.0	cloud
1	https://www.saasforecom.com/cloud-erp-system-saas-erp-guide-wholesalers/	cloud erp system saas erp guide wholesalers	37.0	erp
1	https://www.saasforecom.com/cloud-erp-system-saas-erp-guide-wholesalers/	cloud erp system saas erp guide wholesalers	37.0	system
1	https://www.saasforecom.com/cloud-erp-system-saas-erp-guide-wholesalers/	cloud erp system saas erp guide wholesalers	37.0	saas
1	https://www.saasforecom.com/cloud-erp-system-saas-erp-guide-wholesalers/	cloud erp system saas erp guide wholesalers	37.0	erp
1	https://www.saasforecom.com/cloud-erp-system-saas-erp-guide-wholesalers/	cloud erp system saas erp guide wholesalers	37.0	guide
1	https://www.saasforecom.com/cloud-erp-system-saas-erp-guide-wholesalers/	cloud erp system saas erp guide wholesalers	37.0	wholesalers
2	https://www.saasforecom.com/cloud-based-inventory-management/	cloud based inventory management	49.0	cloud
2	https://www.saasforecom.com/cloud-based-inventory-management/	cloud based inventory management	49.0	based
2	https://www.saasforecom.com/cloud-based-inventory-management/	cloud based inventory management	49.0	inventory
2	https://www.saasforecom.com/cloud-based-inventory-management/	cloud based inventory management	49.0	management
3	https://www.saasforecom.com/help/	help	4.0	help
4	https://www.saasforecom.com/cloud-erp-solution/	cloud erp solution	49.0	cloud
4	https://www.saasforecom.com/cloud-erp-solution/	cloud erp solution	49.0	erp
4	https://www.saasforecom.com/cloud-erp-solution/	cloud erp solution	49.0	solution
5	https://www.saasforecom.com/ecommerce-order-management-system/	ecommerce order management system	125.0	ecommerce
5	https://www.saasforecom.com/ecommerce-order-management-system/	ecommerce order management system	125.0	order
5	https://www.saasforecom.com/ecommerce-order-management-system/	ecommerce order management system	125.0	management
5	https://www.saasforecom.com/ecommerce-order-management-system/	ecommerce order management system	125.0	system

With the data merged, we'll want to remove some rows containing some stop words and other unhelpful words that could be used when creating group names.

A note of warning: The code is a bit repetitive on purpose to give you practice and build your muscle memory even if there are smarter ways to do the entire block in two lines – think list and '|'.join(list):

```
bigrams_df = bigrams_df[['url', 'ngram', 'clicks']]
bigrams_df = bigrams_df[~bigrams_df.ngram.str.contains(r'\band\b',
regex = True)]
bigrams_df = bigrams_df[~bigrams_df.ngram.str.contains(r'\bfor\b',
regex = True)]
bigrams_df = bigrams_df[~bigrams_df.ngram.str.contains(r'\bto\b',
regex = True)]
bigrams_df = bigrams_df[~bigrams_df.ngram.str.contains(r'\ba\b',
regex = True)]
```

```
bigrams_df = bigrams_df[~bigrams_df.ngram.str.contains(r'\ban\b',
regex = True)]
bigrams_df = bigrams_df[~bigrams_df.ngram.str.contains(r'\bin\b',
regex = True)]
bigrams_df = bigrams_df[~bigrams_df.ngram.str.contains(r'\bcom\b',
regex = True)]
bigrams_df = bigrams_df[~bigrams_df.ngram.str.contains(r'\bwww\b',
regex = True)]
bigrams_df = bigrams_df[~bigrams_df.ngram.str.contains(r'\bthe\b',
regex = True)]
bigrams_df = bigrams_df[~bigrams_df.ngram.str.contains(r'\busing\b',
regex = True)]
bigrams_df = bigrams_df[~bigrams_df.ngram.str.contains(r'\bwith\b',
regex = True)]
bigrams_df = bigrams_df[~bigrams_df.ngram.str.contains(r'\b(http|https)\b',
regex = True)]
bigrams_df['ngram'] = bigrams_df.ngram.str.strip()
bigrams_df = bigrams_df[~bigrams_df.ngram.isnull()]
bigrams_df
```

This results in the following:

	url	ngram	clicks
0	https://www.saasforecom.com/	home	6911.0
1	https://www.saasforecom.com/cloud-erp-system-saas-erp-guide-wholesalers/	cloud	37.0
1	https://www.saasforecom.com/cloud-erp-system-saas-erp-guide-wholesalers/	erp	37.0
1	https://www.saasforecom.com/cloud-erp-system-saas-erp-guide-wholesalers/	system	37.0
1	https://www.saasforecom.com/cloud-erp-system-saas-erp-guide-wholesalers/	saas	37.0
...	...	...	...
1250	https://www.saasforecom.com/video-help-wall/?preview_id=14261&preview_nonce=fa90ab88ac&preview=true	14261	0.0
1250	https://www.saasforecom.com/video-help-wall/?preview_id=14261&preview_nonce=fa90ab88ac&preview=true	preview_nonce	0.0
1250	https://www.saasforecom.com/video-help-wall/?preview_id=14261&preview_nonce=fa90ab88ac&preview=true	fa90ab88ac	0.0
1250	https://www.saasforecom.com/video-help-wall/?preview_id=14261&preview_nonce=fa90ab88ac&preview=true	preview	0.0
1250	https://www.saasforecom.com/video-help-wall/?preview_id=14261&preview_nonce=fa90ab88ac&preview=true	true	0.0

8242 rows × 3 columns

The table has ngrams with more sensible labels which can now be sum aggregated to pick the most common labels per URL:

```
bigram_stats = bigrams_df[['ngram', 'clicks']]
bigram_stats = bigram_stats[bigram_stats.ngram.str.contains(r'[\w\s]',
regex = True)]
ngram_ins = pd.DataFrame(bigram_stats.value_counts(subset=['ngram']),
columns = ['count'])
bigram_stats = bigram_stats.merge(ngram_ins, on = 'ngram', how = 'left')
```

The idea here is to create an index based on traffic and the amount of instances of the ngram label:

```
bigram_stats['g_score'] = bigram_stats['clicks'] * bigram_stats['count']
bigram_stats = bigram_stats.sort_values('g_score', ascending = False).
reset_index()
del bigram_stats['index']
bigram_stats.head(10)
```

This results in the following:

	ngram	clicks	count	g_score
0	management	336.0	331	111216.0
1	documentation	428.0	242	103576.0
2	documentation	417.0	242	100914.0
3	documentation	365.0	242	88330.0
4	sales	417.0	206	85902.0
5	sales	417.0	206	85902.0
6	management	233.0	331	77123.0
7	management	232.0	331	76792.0
8	documentation	311.0	242	75262.0
9	sales	365.0	206	75190.0
10	sales	365.0	206	75190.0

We now have a table with ngrams with their stats and their ultimate score. The following function will select the highest score per ngram:

```python
def filter_highest_stat(df, delcol, metric):
    del df[delcol]
    max_count = df[metric].max()
    df = df[df[metric] == max_count]
    df = df.iloc[0]
    return df
```

```python
ngram_stats_map = bigram_stats.groupby('ngram').apply(lambda x:
filter_highest_stat(x, 'ngram', 'g_score')).reset_index()
ngram_stats_map = ngram_stats_map.sort_values('g_score',
ascending = False).reset_index()
del ngram_stats_map['index']
ngram_stats_map.head(10)
```

This results in the following:

	ngram	clicks	count	g_score
0	management	336.0	331	111216.0
1	documentation	428.0	242	103576.0
2	sales	417.0	206	85902.0
3	working	428.0	146	62488.0
4	software	240.0	215	51600.0
5	inventory	233.0	165	38445.0
6	order	198.0	184	36432.0
7	stock	428.0	82	35096.0
8	system	336.0	95	31920.0
9	warehouse	336.0	95	31920.0
10	item	428.0	52	22256.0

The result is a prioritized table showing the most common ngrams that can be used to categorize URLs as segments.

Using the scores, we'll create two levels of segments, taking the most popular ngrams as labels while classifying the rest as "other." We're creating two levels so that we have a more high-level and a more detailed view to hand.

```
ngram_segments = ngram_stats_map[['ngram', 'g_score']]
ngram_segments['segment_one'] = np.where(ngram_segments.index < 11,
ngram_segments.ngram, 'other')
ngram_segments['segment_two'] = np.where(ngram_segments.index < 21,
ngram_segments.ngram, 'other')
ngram_segments.head(10)
```

This results in the following:

	ngram	g_score	segment_one	segment_two
0	management	111216.0	management	management
1	documentation	103576.0	documentation	documentation
2	sales	85902.0	sales	sales
3	working	62488.0	working	working
4	software	51600.0	software	software
5	inventory	38445.0	inventory	inventory
6	order	36432.0	order	order
7	stock	35096.0	stock	stock
8	system	31920.0	system	system
9	warehouse	31920.0	warehouse	warehouse
10	item	22256.0	item	item

We'll join the segment labels to the dataset so that all URLs are now classified by segment label:

```
# Join stats and then select highest
#bigram_stats
urls_grams_stats = bigrams_df.merge(ngram_segments, on = 'ngram',
how = 'left').sort_values(['url', 'g_score'], ascending = False)
urls_grams_stats
```

This results in the following:

	url	ngram	clicks	g_score	segment_one	segment_two
1242	https://www.saasforecom.com/zendesk-request-form-test/?preview_id=15696&preview_nonce=83bf02d08a&preview=true	zendesk	0.0	0.0	other	other
1243	https://www.saasforecom.com/zendesk-request-form-test/?preview_id=15696&preview_nonce=83bf02d08a&preview=true	request	0.0	0.0	other	other
1244	https://www.saasforecom.com/zendesk-request-form-test/?preview_id=15696&preview_nonce=83bf02d08a&preview=true	form	0.0	0.0	other	other
1245	https://www.saasforecom.com/zendesk-request-form-test/?preview_id=15696&preview_nonce=83bf02d08a&preview=true	test	0.0	0.0	other	other
1247	https://www.saasforecom.com/zendesk-request-form-test/?preview_id=15696&preview_nonce=83bf02d08a&preview=true	preview_id	0.0	0.0	other	other
...	...	...	...	...	...	...
7650	http://www.saasforecom.com/help/		0.0	NaN	NaN	NaN
7651	http://www.saasforecom.com/help/		0.0	NaN	NaN	NaN
7567	http://www.saasforecom.com/	saasforecom	0.0	2760.0	other	other
7565	http://www.saasforecom.com/		0.0	NaN	NaN	NaN
7566	http://www.saasforecom.com/		0.0	NaN	NaN	NaN

There are multiple rows per URL; however, we only want the top result, so we'll apply a function to filter for the row with the highest g_score:

```
urls_stats_grams_map = urls_grams_stats.groupby('url').apply(lambda x:
filter_highest_stat(x, 'url', 'g_score')).reset_index()

pd.set_option('display.max_colwidth', None)
urls_grams_map = urls_stats_grams_map
urls_grams_map = urls_grams_map.drop_duplicates()
del urls_grams_map['clicks']
del urls_grams_map['g_score']
#urls_grams_map.iloc[0, 'ngram'] = 'home'
urls_grams_map['subpath'] =
urls_grams_map.url.str.replace(r'(http|https)://www.saasforecom.com', '',
regex = True)
urls_grams_map
```

This results in the following:

	url	ngram	clicks	g_score	segment_one	segment_two
0	http://www.saasforecom.com/	saasforecom	0.0	2760.0	other	other
1	http://www.saasforecom.com/help/	help	0.0	16038.0	other	help
2	http://www.saasforecom.com/help/accounting/using-currencies-in-order-management/	management	0.0	111216.0	management	management
3	http://www.saasforecom.com/help/accounting/using-currencies/	help	0.0	16038.0	other	help
4	http://www.saasforecom.com/help/accounting/using-payment-terms/	help	0.0	16038.0	other	help
...	...	...	...	...	...	...
1246	https://www.saasforecom.com/whats-in-saasforecom/	saasforecom	0.0	2760.0	other	other
1247	https://www.saasforecom.com/woocommerce-inventory-management/	management	233.0	111216.0	management	management
1248	https://www.saasforecom.com/workflow-automation-for-inventory-and-order-management/	management	64.0	111216.0	management	management
1249	https://www.saasforecom.com/zendesk-request-form-test/	zendesk	0.0	0.0	other	other
1250	https://www.saasforecom.com/zendesk-request-form-test/?preview_id=15696&preview_nonce=83bf02d08a&preview=true	zendesk	0.0	0.0	other	other

1251 rows × 6 columns

All the preceding URLs have a unique row and are categorized by segment label. Let's summarize the data by segment to see the distribution of content:

```
urls_grams_map.groupby('segment_one').count().reset_index()
```

This results in the following:

	segment_one	url	ngram	segment_two	subpath
0	documentation	242	242	242	242
1	home	3	3	3	3
2	inventory	7	7	7	7
3	item	12	12	12	12
4	management	319	319	319	319
5	order	8	8	8	8
6	other	483	483	483	483
7	sales	101	101	101	101
8	software	46	46	46	46
9	stock	17	17	17	17
10	system	13	13	13	13

Most of the traffic is in the other classification, followed by management, documentation, and sales.

The next step is merging the performance data from GA with the segment labels and dropping duplicate URL combinations:

```
ga_segmented = ga_stats.merge(urls_grams_map, on = 'subpath', how = 'left')
ga_segmented = ga_segmented.drop_duplicates(subset=['date', 'url'],
keep='last')

ga_segmented = ga_segmented.drop_duplicates(subset=['subpath', 'date'],
keep='first')
```

Clean up the data such that null sessions are zero and new_sessions are treated as whole numbers (i.e., integers):

```
ga_segmented['new_sessions'] = np.where(ga_segmented.new_sessions.isnull(),
0, ga_segmented.new_sessions)
ga_segmented['new_sessions'] = ga_segmented['new_sessions'].astype(int)
ga_segmented
```

This results in the following:

	subpath	date	new_sessions	avg_session_duration	url
0	/	2020-12-08	35	1.240000e-07	https://www.saasforecom.com/
1	/	2020-12-18	24	1.900000e-07	https://www.saasforecom.com/
2	/	2020-12-07	23	2.830000e-07	https://www.saasforecom.com/
3	/	2020-11-30	23	2.710000e-07	https://www.saasforecom.com/
4	/	2020-12-15	27	1.520000e-07	https://www.saasforecom.com/
...	...	...	...	...	...
604	/submit-support-ticket/	2020-12-15	0	3.122000e-06	https://www.saasforecom.com/submit-support-ticket/
605	/submit-support-ticket/	2020-12-19	0	0.000000e+00	https://www.saasforecom.com/submit-support-ticket/
606	/trial-settings-2-set-admin-company-details/	2020-12-27	0	4.516000e-06	https://www.saasforecom.com/trial-settings-2-set-admin-company-details/
607	/zendesk-request-form-test/	2020-12-11	0	2.370000e-07	https://www.saasforecom.com/zendesk-request-form-test/
608	/zendesk-request-form-test/?preview_id=15696&preview_nonce=83bf02d08a&preview=true	2020-12-10	0	3.330000e-07	https://www.saasforecom.com/zendesk-request-form-test/?preview_id=15696&preview_nonce=83bf02d08a&preview=true

609 rows × 8 columns

The result is a dataset ready for time series analysis that can be broken down by segment.

Time Trends and Change Point Analysis

Time trends use time series data to help us understand and demonstrate to our colleagues the changes in search traffic over time. This includes

- Confirming the change point of traffic

- Seeing which types of content were impacted (thanks to the segmentation work done earlier)

There is a bit of a limitation in that it's quite difficult (though not impossible) to get time series data from Google Search Console (GSC). For Google Analytics (GA), getting time series data at a URL isolated to organic search is also very difficult.

Time series data can also be quite noisy by nature due to the way it cycles over the week such that there are peaks and troughs. To tease a trend, we'll need to dampen the noise which we will achieve by computing a moving average.

We start by grouping sessions by date:

```
time_trends = ga_segmented.groupby('date')['new_sessions'].sum().to_
frame().reset_index()
```

Then apply the rolling function to compute a seven-day average:

```
sess_trends_roll = time_trends.rolling(7, min_periods=1)
sess_trends_mean = sess_trends_roll.mean()
time_trends['avg_sess'] = sess_trends_mean
time_trends
```

This results in the following:

	date	new_sessions	avg_sess
7	2020-12-06	783	1103.625000
8	2020-12-07	1265	1121.555556
9	2020-12-08	1195	958.111111
10	2020-12-09	1201	996.555556
11	2020-12-10	1436	1012.777778
12	2020-12-11	671	937.777778
13	2020-12-12	2949	1199.666667
14	2020-12-13	583	1181.777778
15	2020-12-14	687	1196.666667
16	2020-12-15	1073	1228.888889
17	2020-12-16	432	1136.333333
18	2020-12-17	837	1096.555556
19	2020-12-18	1034	1078.000000
20	2020-12-19	750	1001.777778

Let's visualize:

```
pre_time_trends_plt = (
    ggplot(time_trends, aes(x = 'date', y = 'avg_sess', group = 1)) +
    geom_line(alpha = 0.6, colour = 'blue', size = 3) +
    labs(y = 'GA Sessions', x = 'Date') +
    scale_y_continuous() +
    scale_x_date() +
    theme(legend_position = 'right',
        axis_text_x=element_text(rotation=90, hjust=1)
        ))
Pre_time_trends_plt
```

The shift is quite evident in Figure 9-1 with the better half of the traffic trend switching over to the worse half at around the 20th of December 2020.

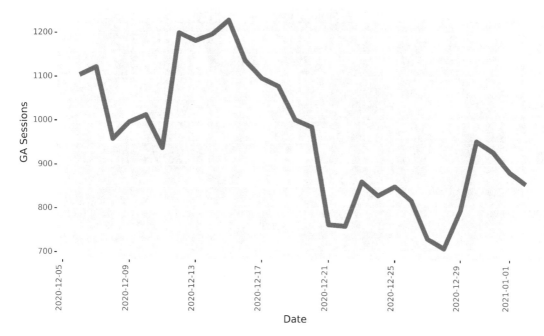

Figure 9-1. *Time series of analytics visits*

Using change point analysis, let's confirm this analytically using the ruptures package:

```
import ruptures as rpt
import matplotlib.pyplot as plt

points = np.array(overall_trends['avg_sess'])

model="rbf"
algo = rpt.Pelt(model=model).fit(points)
result = algo.predict(pen=6)
rpt.display(points, result, figsize=(10, 6))
plt.title('Change Point Detection using Pelt')
plt.show()
```

The change point analysis in Figure 9-2 confirms that on the 20th of December, there's a shift downward in traffic.

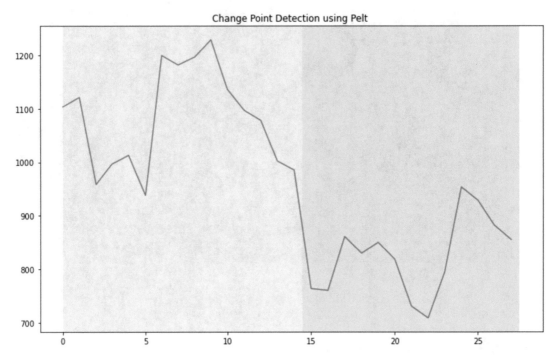

Figure 9-2. *Time series of analytics visits with estimated change point between before (blue shaded area) and after (red)*

Yes, it could be coinciding with the Christmas holidays, but unfortunately for this particular company, we don't have the data for the previous year to confirm how much of the downward change is attributable to seasonality vs. the new site relaunch migration.

Segmented Time Trends

With the change point confirmed, let's now see which content areas were affected. We'll start by performing a similar aggregation to calculate the rolling average:

```
segmented_trends = postmortem_df.groupby(['date', 'segment_two'])['new_
sessions'].sum().to_frame().reset_index()#.rolling(7).
sessseg_trends_roll = segmented_trends.rolling(8, min_periods=1)
```

```
sessseg_trends_mean = sessseg_trends_roll.mean()
segmented_trends['avg_sess'] = sessseg_trends_mean

segmented_trends
```

This results in the following:

	date	segment_one	new_sessions	avg_sess
7	2020-11-30	home	23	374.714286
8	2020-11-30	management	132	379.285714
9	2020-11-30	other	600	365.000000
10	2020-11-30	sales	0	293.571429
11	2020-11-30	software	100	179.285714
...	...	...	...	...
197	2021-01-01	stock	100	167.714286
198	2021-01-02	home	50	174.857143
199	2021-01-02	management	300	213.000000
200	2021-01-02	other	100	125.000000
201	2021-01-02	software	100	110.714286

195 rows × 4 columns

The data is in long format with the rolling averages calculated ready for visualization:

```
ga_seg_trends_plt = (
    ggplot(time_trends_segmented, aes(x = 'date', y = 'avg_sess',
                          group = 'segment_one', colour = 'segment_
                          one')) +
    geom_line(alpha = 0.7, size = 2) +
    labs(y = 'GA Sessions', x = 'Date') +
    scale_y_continuous() +
    scale_x_date() +
    theme(legend_position = 'right',
        axis_text_x=element_text(rotation=90, hjust=1)
      ))
```

```
ga_seg_trends_plt.save(filename = 'images/1_ga_seg_trends_plt.png',
height=5, width=15, units = 'in', dpi=1000)
ga_seg_trends_plt
```

No obvious trends are apparent in Figure 9-3, as most (if not all) of the content appear to move in the same direction over time. It's not as if a couple of segments decreased while others increased or were unchanged.

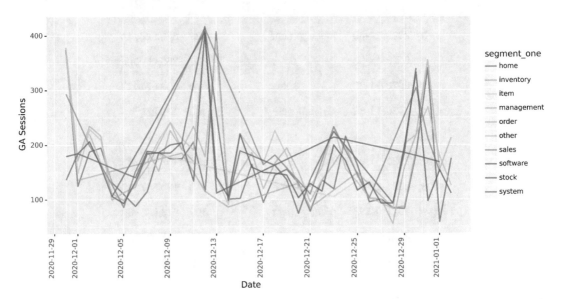

Figure 9-3. *Time series of analytics visits segmented by content type*

Analysis Impact

With the time trends dissected, we turn our attention to analyzing the before and after impact of the migration to hopefully generate recommendations or areas for further research.

We'll use GSC data at the page level which can be segmented by merging the map created earlier:

```
gsc_before = pd.read_csv('data/gsc_before.csv')
```

Clean the column names as usual:

```
gsc_before.columns = gsc_before.columns.str.lower()
gsc_before.columns = gsc_before.columns.str.replace('/', '', regex = False)
gsc_before.columns = gsc_before.columns.str.replace('.', '', regex = False)
gsc_before.columns = gsc_before.columns.str.replace('% ', '',
regex = False)
gsc_before.columns = gsc_before.columns.str.replace('  ', ' ',
regex = False)
gsc_before.columns = gsc_before.columns.str.replace(' ', '_',
regex = False)
gsc_before['ctr'] = gsc_before.ctr.str.replace('%', '', regex = False)
```

just so we know which phase of the migration this data refers to:

```
gsc_before['phase'] = 'before'
```

Rename the top_pages column before we merge the segment labels:

```
gsc_before = gsc_before.rename(columns = {'top_pages': 'url'})
gsc_before = gsc_before.merge(urls_grams_map, on = 'url', how = 'left')
gsc_before['count'] = 1
gsc_before['ngram'] = np.where(gsc_before['ngram'].isnull(), 'other', gsc_
before['ngram'])
gsc_before['segment_one'] = np.where(gsc_before['segment_one'].isnull(),
'other', gsc_before['segment_one'])
gsc_before['segment_two'] = np.where(gsc_before['segment_two'].isnull(),
'other', gsc_before['segment_two'])
gsc_before
```

This results in the following:

	url	clicks	impressions	ctr	position	phase	ngram	segment_one	segment_two	
0	https://www.saasforecom.com/	661	125889	0.53	35.84	before	home	home	home	
1	https://www.saasforecom.com/documentation/sales/working-with-sales-orders/allocating-stock-to-saasforecoms/pick-pack-ship-process/	104	19510	0.53	30.08	before	documentation	documentation	documentation	sa
2	https://www.saasforecom.com/documentation/sales/working-with-sales-orders/allocating-stock-to-saasforecoms/	89	2918	3.05	10.66	before	documentation	documentation	documentation	
3	https://www.saasforecom.com/documentation/products-services/working-with-items/stock-item/checking-stock-availability/	88	3668	2.4	13.62	before	documentation	documentation	documentation	
4	https://www.saasforecom.com/documentation/products-services/working-with-items/non-stock-item/	83	3704	2.24	8.89	before	documentation	documentation	documentation	se
...	...	...	...	...	...	...	...	...	...	
315	https://www.saasforecom.com/documentation/related/tax-codes/	0	1	0	213.00	before	documentation	documentation	documentation	
316	https://www.saasforecom.com/inventory-management-software-for-san-jose-wholesalers/	0	1	0	219.00	before	management	management	management	'
317	https://www.saasforecom.com/documentation/miscellaneous/working-with-sales-documents/creating-new-transactions/	0	1	0	254.00	before	documentation	documentation	documentation	/do
318	https://www.saasforecom.com/documentation/banking/working-with-credit-cards/entering-paying-a-credit-card/	0	1	0	292.00	before	documentation	documentation	documentation	/c
319	https://www.saasforecom.com/documentation/accounting/working-with-us-sales-tax/working-with-tax-groups/	0	1	0	294.00	before	documentation	documentation	documentation	

320 rows × 11 columns

So we have the before GSC data at the page level which is now segmented. The operations are repeated for the phase post migration, known as "after":

```
gsc_after = pd.read_csv('data/gsc_after.csv')
gsc_after.columns = gsc_after.columns.str.lower()
gsc_after.columns = gsc_after.columns.str.replace('/', '', regex = False)
gsc_after.columns = gsc_after.columns.str.replace('.', '', regex = False)
gsc_after.columns = gsc_after.columns.str.replace('% ', '', regex = False)
gsc_after.columns = gsc_after.columns.str.replace('  ', ' ', regex = False)
gsc_after.columns = gsc_after.columns.str.replace(' ', '_', regex = False)
gsc_after['ctr'] = gsc_after.ctr.str.replace('%', '', regex = False)
gsc_after['phase'] = 'after'
gsc_after = gsc_after.rename(columns = {'top_pages': 'url'})
gsc_after = gsc_after.merge(urls_grams_map, on = 'url', how = 'left')
gsc_after['count'] = 1
gsc_after['ngram'] = np.where(gsc_after['ngram'].isnull(), 'other',
gsc_after['ngram'])
gsc_after['segment_one'] = np.where(gsc_after['segment_one'].isnull(),
'other', gsc_after['segment_one'])
```

```
gsc_after['segment_two'] = np.where(gsc_after['segment_two'].isnull(),
'other', gsc_after['segment_two'])
gsc_after
```

This results in the following:

	url	clicks	impressions	ctr	position	phase	ngram	segment_one	segment_two	
0	https://www.saasforecom.com/	681	126029	0.54	31.75	after	home	home	home	home
1	https://www.saasforecom.com/drop-shipping-automation-software/	55	10880	0.51	44.07	after	software	software	software	/drop-automation-
2	https://www.saasforecom.com/marketplace-integration/etsy/	51	3083	1.65	37.38	after	integration	other	other	/mar integra
3	https://www.saasforecom.com/software-for-warehouse-inventory-management/	37	131793	0.03	49.64	after	management	management	management	/software-for-wa inventory-man
4	https://www.saasforecom.com/ecommerce-order-management-system/	31	30614	0.1	55.51	after	management	management	management	/ecommer managemen
...	...	...	...	...	...	...	...	...	...	
584	https://www.saasforecom.com/documentation/products-services/working-with-items/	0	1	0	133.00	after	documentation	documentation	documentation	/documentation/ services/work
585	https://www.saasforecom.com/order-management-software-for-denver-wholesalers/	0	1	0	134.00	after	management	management	management	/order-man software-fc wh
586	https://www.saasforecom.com/order-management-software-for-omaha-wholesalers/	0	1	0	135.00	after	management	management	management	/order-man software-fc wh
587	https://www.saasforecom.com/order-management-software-for-portland-wholesalers/	0	1	0	137.00	after	management	management	management	/order-man software-for- wh
588	https://www.saasforecom.com/help/start/setup-document-template/#elementor-toc__heading-anchor-5	0	1	0	157.00	after	help	other	help	/help/st d template/#e toc__heading

589 rows × 11 columns

With both datasets imported and cleaned, we're ready to start analyzing using aggregations, starting with weighted average rank positions by phase.

The weighted average rank position function (wavg_rank_imps) takes two arguments (position and impressions) and returns the calculation result using the column name "wavg_rank":

```
def wavg_rank_imps(x):
    names = {'wavg_rank': (x['position'] * x['impressions']).sum()/
    (x['impressions']).sum()}
    return pd.Series(names, index=['wavg_rank']).round(1)
```

We'll make a copy of the "before" dataset before applying the function:

```
gsc_before_agg = gsc_before
gsc_before_wavg = gsc_before_agg.groupby('phase').apply(wavg_rank_imps).
reset_index()
```

In addition to the weighted average ranking positions, we're also interested in the total number of URLs and the total number of clicks (organic search traffic):

```
gsc_before_sum = gsc_before_agg.groupby('phase').agg({'count': 'sum',
                                                      'clicks': 'sum'}).
                                                      reset_index()

gsc_before_stats = gsc_before_wavg.merge(gsc_before_sum, on = ['phase'],
how = 'left')
```

The index is a ratio of clicks to count that forms our index to give us some sense of proportion:

```
gsc_before_stats['index'] = gsc_before_stats['clicks']/gsc_before_
stats['wavg_rank']
gsc_before_stats.sort_values('index', ascending = False)
```

This results in the following:

	phase	wavg_rank	count	clicks	index
0	before	44.8	375	600	13.392857

That's the stats before the migration. Now let's look at the stats after the migration, applying the same methods used earlier to data post migration:

```
gsc_after_agg = gsc_after
gsc_after_wavg = gsc_after_agg.groupby('phase').apply(wavg_rank_imps).
reset_index()
gsc_after_sum = gsc_after_agg.groupby('phase').agg({'count': 'sum',
                                                    'clicks': 'sum'}).
                                                    reset_index()

gsc_after_stats = gsc_after_wavg.merge(gsc_after_sum, on = ['phase'],
how = 'left')
gsc_after_stats['index'] = gsc_after_stats['clicks']/gsc_after_
stats['wavg_rank']
gsc_after_stats.sort_values('index', ascending = False)
```

This results in the following:

	phase	wavg_rank	count	clicks	index
0	after	45.0	302	205	4.555556

With both datasets aggregated, we can concatenate them into a single table to compare directly:

```
pd.concat([gsc_before_stats, gsc_after_stats])
```

This results in the following:

	phase	wavg_rank	count	clicks	index
0	before	44.8	375	600	13.392857
0	after	45.0	302	205	4.555556

So the average rank doesn't appear to have changed that much, which implies the dramatic change could be more seasonal. However, as we'll see later, averages can often mask what's really happening.

The amount of pages receiving traffic has decreased by roughly –20%, which is telling as that appears to be migration related.

We'll start visualizing some data to help us investigate deeper:

```
overall_clicks_plt = (
    ggplot(pd.concat([gsc_before_stats, gsc_after_stats]),
           aes(x = 'reorder(phase, -clicks)', y = 'clicks' ,
           fill = 'phase')) +
    geom_bar(stat = 'identity', alpha = 0.6, position = 'dodge') +
    position=position_stack(vjust=0.01)) +
    labs(y = 'GSC Clicks', x = 'phase') +

    theme(legend_position = 'right',
        )
)

overall_clicks_plt.save(filename = 'images/2_overall_clicks_plt.png',
height=5, width=10, units = 'in', dpi=1000)
overall_clicks_plt
```

Clicks are down in Figure 9-4, which we obviously know, by about –67%.

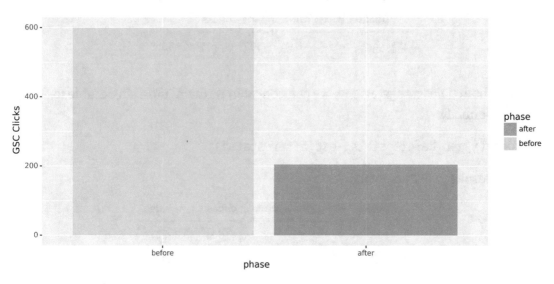

Figure 9-4. *Column chart of before and after Google Search Console (GSC) clicks*

Let's break it down at the segment level.

We'll start by computing segment rank averages and total clicks and derive an index of visibility based on the ratio of clicks to rank:

```
gsc_before_seg_agg = gsc_before
gsc_before_seg_wavg = gsc_before_seg_agg.groupby(['segment_two', 'phase']).
apply(wavg_rank_imps).reset_index()
gsc_before_seg_sum = gsc_before_seg_agg.groupby(['segment_two', 'phase']).
agg({'count': 'sum', 'clicks': 'sum'}).reset_index()

gsc_before_seg_stats = gsc_before_seg_wavg.merge(gsc_before_seg_sum,
on = ['segment_two', 'phase'], how = 'left')
gsc_before_seg_stats['index'] = gsc_before_seg_stats['clicks']/
gsc_before_seg_stats['wavg_rank']
gsc_before_seg_stats.sort_values('index', ascending = False)
```

This results in the following:

	segment_two	phase	wavg_rank	count	clicks	index
3	home	before	31.3	1	364	11.629393
9	other	before	37.5	51	55	1.466667
7	management	before	51.2	107	71	1.386719
11	software	before	43.6	38	50	1.146789
1	documentation	before	34.3	96	36	1.049563
14	wholesalers	before	10.4	1	6	0.576923
0	crm	before	31.7	6	11	0.347003
12	stock	before	5.1	5	1	0.196078
2	help	before	19.3	36	2	0.103627
13	system	before	69.3	2	3	0.043290
4	inventory	before	67.6	1	1	0.014793
5	item	before	13.4	2	0	0.000000
6	items	before	5.3	2	0	0.000000
8	order	before	22.8	2	0	0.000000
10	sales	before	19.8	25	0	0.000000

The preceding segment breakdown shows much of the content is in the "management" classification, followed by "documentation." We'll repeat the aggregations for the postmigration data:

```
gsc_after_seg_agg = gsc_after
gsc_after_seg_wavg = gsc_after_seg_agg.groupby(['segment_two', 'phase']).
apply(wavg_rank_imps).reset_index()
gsc_after_seg_sum = gsc_after_seg_agg.groupby(['segment_two', 'phase']).
agg({'count': 'sum', 'clicks': 'sum'}).reset_index()

gsc_after_seg_stats = gsc_after_seg_wavg.merge(gsc_after_seg_sum,
on = ['segment_two', 'phase'], how = 'left')
```

```
gsc_after_seg_stats['index'] = gsc_after_seg_stats['clicks']/gsc_after_seg_
stats['wavg_rank']
gsc_after_seg_stats.sort_values('index', ascending = False)
```

This results in the following:

	segment_two	phase	wavg_rank	count	clicks	index
3	home	after	31.6	2	140	4.430380
11	software	after	44.8	37	29	0.647321
7	management	after	51.5	111	17	0.330097
9	other	after	38.7	33	10	0.258398
0	crm	after	31.4	7	7	0.222930
12	stock	after	8.0	2	1	0.125000
1	documentation	after	33.1	50	1	0.030211
2	help	after	13.1	31	0	0.000000
4	inventory	after	72.1	3	0	0.000000
5	item	after	17.1	2	0	0.000000
6	items	after	5.0	1	0	0.000000
8	order	after	7.7	2	0	0.000000
10	sales	after	34.1	16	0	0.000000
13	system	after	66.4	2	0	0.000000
14	wholesalers	after	29.2	3	0	0.000000

Curiously, "management" has 10% more URLs ranking than premigration with no real change in ranking. "Documentation" has lost virtually all of its clicks and half of its URLs.

To visualize this, the dataframes will need to be concatenated in long format to feed the graphics code:

```
gsc_long_seg_stats = pd.concat([gsc_before_seg_stats, gsc_after_seg_stats])
gsc_long_seg_stats['phase'] = gsc_long_seg_stats['phase'].
astype('category')
```

```
gsc_long_seg_stats['phase'].cat.reorder_categories(['before', 'after'],
inplace=True)

segment_clicks_plt = (
    ggplot(gsc_long_seg_stats,
            aes(x = 'reorder(segment_two, -clicks)', y = 'clicks' , fill =
            'phase')) +
    geom_bar(stat = 'identity', alpha = 0.6, position = 'dodge') +
    position=position_stack(vjust=0.01)) +
    labs(y = 'GSC Clicks', x = '') +
    theme(legend_position = 'right',
        axis_text_x=element_text(rotation=90, hjust=1)
        )
)

segment_clicks_plt.save(filename = 'images/2_segment_clicks_plt.png',
height=5, width=10, units = 'in', dpi=1000)
segment_clicks_plt
```

As shown in Figure 9-5, most of the click losses appear to have happened at the home page.

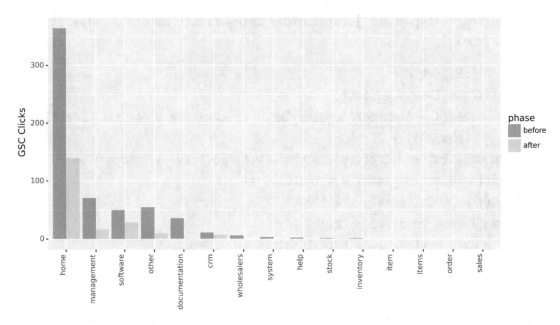

Figure 9-5. *Column chart of before and after Google Search Console (GSC) clicks by content segment*

What about the number of URLs receiving traffic from Google?

```
segment_urls_plt = (
    ggplot(gsc_long_seg_stats,
           aes(x = 'reorder(segment_two, -count)', y = 'count' ,
           fill = 'phase')) +
    geom_bar(stat = 'identity', alpha = 0.6, position = 'dodge') +
    position=position_stack(vjust=0.01)) +
    labs(y = 'GSC URL Count', x = '') +
    theme(legend_position = 'right',
        axis_text_x=element_text(rotation=90, hjust=1)
        )
)

segment_urls_plt.save(filename = 'images/2_segment_urls_plt.png', height=5,
width=10, units = 'in', dpi=1000)
segment_urls_plt
```

So there are more management URLs receiving traffic post migration (Figure 9-6).

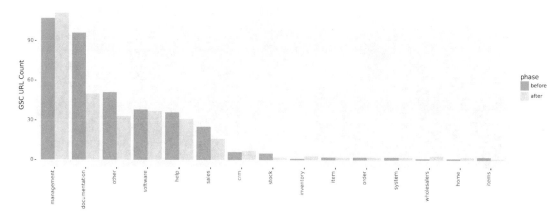

Figure 9-6. *Column chart of before and after Google Search Console (GSC) URL counts by content segment*

However, there is much less in "documentation" and "other" and a bit less in "help," "sales," and "stock." What about Google rank positions?

```
segment_rank_plt = (
    ggplot(gsc_long_seg_stats,
           aes(x = 'reorder(segment_two, -wavg_rank)', y = 'wavg_rank' ,
           fill = 'phase')) +
    geom_bar(stat = 'identity', alpha = 0.6, position = 'dodge') +
    #geom_text(dd_factor_df, aes(label = 'serps_name'), position=position_
    stack(vjust=0.01)) +
    labs(y = 'GSC Clicks', x = '') +
    scale_y_reverse() +
    #theme_classic() +
    theme(legend_position = 'right',
          axis_text_x=element_text(rotation=90, hjust=1)
          )
)

segment_rank_plt.save(filename = 'images/2_segment_rank_plt.png', height=5,
width=15, units = 'in', dpi=1000)
segment_rank_plt
```

Rankings fell for the inventory, sales, wholesalers, and stock classifications (Figure 9-7).

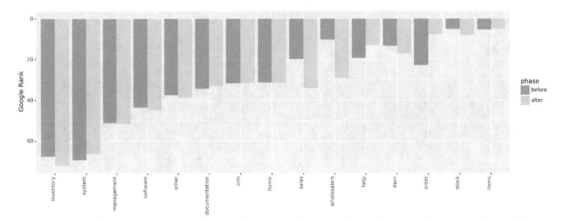

Figure 9-7. *Column chart of before and after Google Search Console (GSC) rank position averages by content segment*

So there is some correlation between the losses in traffic and rankings. As a general conclusion, some of the downshift in organic performance, as initially suspected, is a mixture of seasonality and site migration.

Diagnostics

So we see that rankings fell for inventory and others, but why?

To understand what went wrong, we're now going to merge performance data with crawl data to help us diagnose what went wrong. We'll also append the segment names so we can diagnose by content area.

Select the clicks and rank columns we want before merging:

```
gsc_before_diag = gsc_before[['url', 'clicks', 'segment_two', 'position']]
gsc_after_diag = gsc_after[['url', 'clicks', 'segment_two', 'position']]
```

Now merge to create a new dataframe gsc_ba_diag so we can compare performance at the URL level before and after.

We use an outer join to capture all URLs before and after the migration. If we did a left join (equivalent of a vlookup in Excel), Pandas would assume an inner join, which means we'd miss out on any URLs that had no data post migration.

```
gsc_ba_diag = gsc_before_diag.merge(gsc_after_diag, on = ['url', 'segment_
two'], how = 'outer')
```

Because the dataframes of the before and after share the same column names, Pandas interprets this as unintended and correctly assumes these columns are different and therefore adds the suffixes _x and _y. So we're renaming them to be more user-friendly:

```
gsc_ba_diag = gsc_ba_diag.rename(columns = {'segment_two': 'segment',
                                            'clicks_x': 'clicks_before',
                                            'clicks_y': 'clicks_after',
                                            'position_x': 'rank_before',
                                            'position_y': 'rank_after' })
```

After joining, we'd expect to see some rows where they have null clicks before or (more likely) after the migration. So we're cleaning up the data to replace "not a number" (NaNs) values with 100 for rankings and 0 for clicks:

```
gsc_ba_diag['rank_before'] = np.where(gsc_ba_diag['rank_before'].isnull(),
100, gsc_ba_diag['rank_before'])

gsc_ba_diag['rank_after'] = np.where(gsc_ba_diag['rank_after'].isnull(),
100, gsc_ba_diag['rank_after'])

gsc_ba_diag['clicks_before'] = np.where(gsc_ba_diag['clicks_before'].
isnull(), 0,  gsc_ba_diag['clicks_before'])

gsc_ba_diag['clicks_after'] = np.where(gsc_ba_diag['clicks_after'].
isnull(), 0, gsc_ba_diag['clicks_after'])
```

With the data in wide format and the null values cleaned up, we can compute the differences in clicks and rankings before and after, which we will now do:

```
gsc_ba_diag['rank_delta'] = gsc_ba_diag['rank_before'] - gsc_ba_
diag['rank_after']
gsc_ba_diag['clicks_delta'] = gsc_ba_diag['clicks_after'] - gsc_ba_
diag['clicks_before']

gsc_ba_diag
```

This results in the following:

	url	clicks_before	segment	rank_before	clicks_after	rank_after	rank_delta	clicks_delta
0	https://www.saasforecom.com/	364.0	home	31.29	140.0	31.57	-0.28	-224.0
1	https://www.saasforecom.com/drop-shipping-automation-software/	35.0	software	42.32	11.0	40.84	1.48	-24.0
2	https://www.saasforecom.com/software-for-warehouse-inventory-management/	29.0	management	49.05	9.0	51.37	-2.32	-20.0
3	https://www.saasforecom.com/marketplace-integration/etsy/	26.0	other	34.04	0.0	100.00	-65.96	-26.0
4	https://www.saasforecom.com/ecommerce-order-management-system/	21.0	management	54.27	4.0	56.30	-2.03	-17.0
...	...	...	...	...	...	...	...	...
459	https://www.saasforecom.com/documentation/related/tax-codes/	0.0	documentation	100.00	0.0	84.00	16.00	0.0
460	https://www.saasforecom.com/inventory-management-software-for-san-jose-wholesalers/	0.0	management	100.00	0.0	87.00	13.00	0.0
461	https://www.saasforecom.com/documentation/tag/special-orders/	0.0	documentation	100.00	0.0	89.00	11.00	0.0
462	https://www.saasforecom.com/help/accounts-receivable/using-payment-terms/	0.0	help	100.00	0.0	89.00	11.00	0.0
463	https://www.saasforecom.com/inventory-management-software-for-el-paso-wholesalers/	0.0	management	100.00	0.0	93.00	7.00	0.0

464 rows × 8 columns

The performance deltas are now in place, so we can merge the crawl data with performance data into a new dataframe aptly named "perf_crawl."

Since we have all the URLs we want and there's a lot of unwanted URLs in the crawl data, we'll take the desired URLs (perf_crawl) and join the crawl data specified in the merge function, which will be set to "left." This is equivalent to an Excel vlookup, which will only join the desired crawl URLs.

```
perf_crawl = gsc_ba_diag.merge(audit_urls_map, on = 'url', how = 'left')

perf_crawl = perf_crawl[['url', 'segment', 'clicks_after', 'clicks_before',
'clicks_delta', 'crawl_depth',
                'host', 'crawl_source', 'http_status_code',
                'indexable_status', 'canonical_url',
                'canonical_status', 'redirect_url', 'redirect_url_
                status_code',
                'final_redirect_url', 'final_redirect_url_status_
                code', 'urls_with_similar_content',
                'ult_dest_url', 'content_simi']]

perf_crawl
```

This results in the following:

	url	segment	clicks_after	clicks_before	clicks_delta	crawl_depth	host	crawl_sourc
0	https://www.saasforecom.com/	home	140.0	364.0	-224.0	0	www.saasforecom.com	Crawl
1	https://www.saasforecom.com/drop-shipping-automation-software/	software	11.0	35.0	-24.0	1	www.saasforecom.com	Crawl
2	https://www.saasforecom.com/software-for-warehouse-inventory-management/	management	9.0	29.0	-20.0	1	www.saasforecom.com	Crawl
3	https://www.saasforecom.com/marketplace-integration/etsy/	other	0.0	26.0	-26.0	Not Set	www.saasforecom.com	Url Li
4	https://www.saasforecom.com/ecommerce-order-management-system/	management	4.0	21.0	-17.0	1	www.saasforecom.com	Crawl
...	...	...	...	...	...	...	...	...
459	https://www.saasforecom.com/documentation/related/tax-codes/	documentation	0.0	0.0	0.0	Not Set	www.saasforecom.com	Url Li
460	https://www.saasforecom.com/inventory-management-software-for-san-jose-wholesalers/	management	0.0	0.0	0.0	Not Set	www.saasforecom.com	Url Li
461	https://www.saasforecom.com/documentation/tag/special-orders/	documentation	0.0	0.0	0.0	Not Set	www.saasforecom.com	Url Li
462	https://www.saasforecom.com/help/accounts-receivable/using-payment-terms/	help	0.0	0.0	0.0	2	www.saasforecom.com	Crawl
463	https://www.saasforecom.com/inventory-management-software-for-el-paso-wholesalers/	management	0.0	0.0	0.0	Not Set	www.saasforecom.com	Url Li

464 rows × 19 columns

More fun awaits us as we now get to diagnose the URLs. To do this, we're going to use a set of conditions in the data, such that when they are met, they will be given a diagnosis value. This is where your SEO experience comes in, because your ability to spot patterns dictates the conditions you will set as follows:

```
perf_diags = perf_crawl.copy()
```

Create a list of our conditions:

```
modifier_conds = [
    (perf_crawl['http_status_code'] == '200') & (perf_crawl['crawl_source']
    != 'Crawler'),
    (perf_crawl['redirect_url_status_code'] == '301'),
    (perf_crawl['http_status_code'].isnull()),
    perf_crawl['http_status_code'].isin(['400', '403', '404']),
    (perf_diags['canonical_status'] != 'Missing') & (perf_diags['indexable_
    status'] == 'Noindex'),
    perf_diags['content_simi'] < 1
]
```

Create a list of the values we want to assign for each condition:

```
segment_values = ['outside_ia', 'redirect_chain', 'lost_content', 'error',
'robots_conflict', 'lost_content']
```

Create a new column and use np.select to assign values to it using our lists as arguments:

```
perf_diags['diagnosis'] = np.select(modifier_conds, segment_values,
default = 'other')
perf_diags['diagnosis'] = np.where((perf_diags['diagnosis'] == 'redirect_
chain') & (perf_diags['content_simi'] < 1),
                                    'lost_content', perf_diags['diagnosis'])
```

```
perf_diags
```

This results in the following:

ode	final_redirect_url	final_redirect_url_status_code	urls_with_similar_content	ult_dest_url	content_simi	diagnosis
Set	No Data	Not Set	0.0	https://www.saasforecom.com/	1.000000	other
Set	No Data	Not Set	0.0	https://www.saasforecom.com/drop-shipping-automation-software/	1.000000	other
Set	No Data	Not Set	0.0	https://www.saasforecom.com/software-for-warehouse-inventory-management/	1.000000	other
Set	No Data	Not Set	0.0	http://www.saasforecom.com/ecommerce-order-management-system/	0.813559	lost_content
Set	No Data	Not Set	0.0	https://www.saasforecom.com/ecommerce-order-management-system/	1.000000	other
...	...	...	...	...	...	...
301	https://www.saasforecom.com/help/	200	0.0	https://www.saasforecom.com/help/	0.666667	lost_content
Set	No Data	Not Set	0.0	http://www.saasforecom.com/cloud-based-inventory-management/	0.769231	lost_content
301	https://www.saasforecom.com/help/	200	0.0	https://www.saasforecom.com/help/	0.680851	lost_content
Set	No Data	Not Set	0.0	https://www.saasforecom.com/help/accounts-receivable/using-payment-terms/	1.000000	other
Set	No Data	Not Set	0.0	http://www.saasforecom.com/cloud-based-inventory-management/	0.774648	lost_content

A new column "diagnosis" has been added based on the rules we just created, helping us to make sense, at the URL level, what has happened.

With each URL labeled, we can start to quantify the diagnosis:

```
diagnosis_clicks = perf_diags.groupby('diagnosis').agg({'clicks_delta':
'sum'}).reset_index()
diagnosis_urls = perf_diags.groupby('diagnosis').agg({'url': 'count'}).
reset_index()
```

```
diagnosis_stats = diagnosis_clicks.merge(diagnosis_urls, on = 'diagnosis',
how = 'left')
```

```
diagnosis_stats['clicks_pURL'] = (diagnosis_stats['clicks_delta'] /
diagnosis_stats['url']).round(2)
```

```
diagnosis_stats
```

This results in the following:

	diagnosis	clicks_delta	url	clicks_pURL
0	error	-14.0	34	-0.41
1	lost_content	-84.0	373	-0.23
2	other	-297.0	50	-5.94
3	outside_ia	0.0	7	0.00

According to the analysis, around 25% of the total loss of clicks is down to error codes (HTTP server status 4XX) and lost content (URLs redirected to a parent folder).

Most of the URLs affected are those 373 redirected which is most of the website.

Other (with no URLs) implies the traffic loss would be seasonal and/or an indirect effect of the migration errors.

If you want to share what you found, you could visualize this for your colleagues using the following code:

```
diagnosis_plt = (
    ggplot(diagnosis_stats,
           aes(x = 'reorder(diagnosis, -clicks_delta)', y = 'clicks_
           delta')) +
    geom_bar(stat = 'identity', alpha = 0.6, position = 'dodge', fill =
    'blue') +
position=position_stack(vjust=0.01)) +
    labs(y = 'GSC Clicks Impact', x = '') +
    coord_flip() +
    theme(legend_position = 'right',
        axis_text_x=element_text(rotation=90, hjust=1)
        )
)
```

```
diagnosis_plt.save(filename = 'images/3_diagnosis_plt.png', height=5,
width=10, units = 'in', dpi=1000)
diagnosis_plt
```

"Other" (probably seasonality) was the major reason for the click losses (Figure 9-8), followed by lost_content (i.e., URLs redirected).

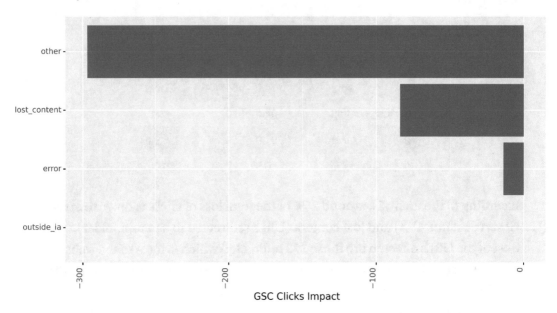

Figure 9-8. *Bar chart of Google Search Console (GSC) click impact by tech SEO diagnosis*

But what was the number of URLs impacted?

```
diagnosis_count_dat = perf_diags.groupby('diagnosis').agg({'url':
'count'}).reset_index()
print(diagnosis_count_dat)

diagnosis_urlcount_plt = (
    ggplot(diagnosis_count_dat,
            aes(x = 'reorder(diagnosis, url)', y = 'url')) +
    geom_bar(stat = 'identity', alpha = 0.6, position = 'dodge',
    fill = 'blue') +
position=position_stack(vjust=0.01)) +
    labs(y = 'URL Count', x = '') +
    coord_flip() +
```

```
    theme(legend_position = 'right',
        axis_text_x=element_text(rotation=90, hjust=1)
        )
)

diagnosis_urlcount_plt.save(filename = 'images/3_diagnosis_urlcount_plt.
png', height=5, width=15, units = 'in', dpi=1000)
diagnosis_urlcount_plt
```

Despite "Other" losing the most clicks, it was "lost content" that impacted the most URLs (Figure 9-9).

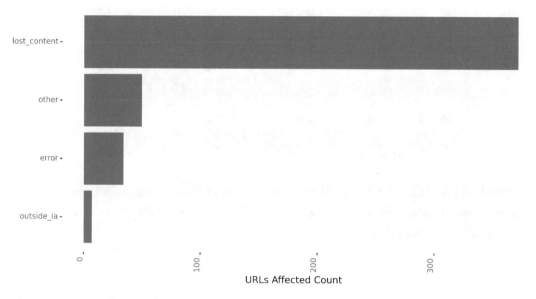

Figure 9-9. *Column chart of Google Search Console (GSC) URLs affected counts by tech SEO diagnosis*

That's the overview done; let's break it down by content type. We'll use the content segment labels to get click impact stats:

```
diagnosis_seg_clicks = perf_diags.groupby(['diagnosis', 'segment']).
agg({'clicks_delta': 'sum'}).reset_index()
diagnosis_seg_urls = perf_diags.groupby(['diagnosis', 'segment']).
agg({'url': 'count'}).reset_index()
```

```
diagnosis_seg_stats = diagnosis_seg_clicks.merge(diagnosis_seg_urls,
                                        on = ['diagnosis',
                                        'segment'], how = 'left')
diagnosis_seg_stats['clicks_p_url'] = (diagnosis_seg_stats['clicks_delta']
/ diagnosis_seg_stats['url']).round(2)
diagnosis_seg_stats.sort_values('clicks_p_url')
```

This results in the following:

	diagnosis	segment	clicks_delta	url	clicks_p_url
23	other	home	-224.0	1	-224.00
28	other	software	-22.0	2	-11.00
20	lost_content	wholesalers	-6.0	1	-6.00
21	other	crm	-4.0	1	-4.00
25	other	management	-46.0	14	-3.29
30	other	system	-2.0	1	-2.00
3	error	other	-13.0	8	-1.62

So not only is "Other" the biggest reason for the click losses, most of it impacted the home page. This would be consistent with the idea of seasonality, that is, a characteristically quiet December.

While tables are useful, we'll make use of data visualization to see the overall picture more easily:

```
diagnosis_seg_clicks_plt = (
    ggplot(diagnosis_seg_stats,
           aes(x = 'diagnosis', y = 'segment', fill = 'clicks_delta')) +
    geom_tile(stat = 'identity', alpha = 0.6) + position=position_
stack(vjust=0.00)) +
    labs(y = '', x = '') +
    theme_classic() +
    theme(legend_position = 'right')
)
```

```
diagnosis_seg_clicks_plt.save(filename = 'images/5_diagnosis_seg_clicks_
plt.png', height=5, width=10, units = 'in', dpi=1000)
diagnosis_seg_clicks_plt
```

The home page followed by "management content" is the most affected within "other" (Figure 9-10).

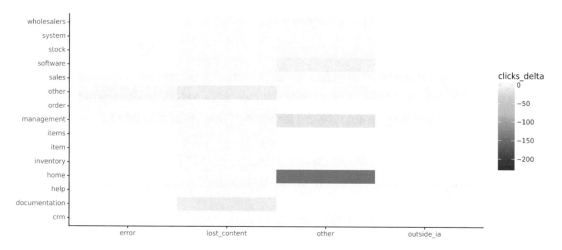

Figure 9-10. *Heatmap chart of clicks delta by content type and SEO diagnosis*

In terms of lost content, these are mostly "documentation" and "other." This is quite useful for deciding where to focus our attention.

Although "other" as a reason isn't overly helpful for fixing a site post migration, we can still explain where some of the migration errors occurred, start labeling URLs for recommended actions, and visualize. This is what we're doing next.

Road Map

We'll start with our dataframe "perf_diags" and copy it into "perf_recs" before creating the recommendations based on the errors found:

```
perf_recs = perf_diags
```

The aptly named diag_conds is a list of diagnoses based on the value of the diagnosis column in the perf_recs dataframe. The np.select function (shortly later on) will draw from this list to assign a recommendation.

```
diag_conds = [
    perf_recs['diagnosis'] == 'outside_ia',
    perf_recs['diagnosis'] == 'redirect_chain',
    perf_recs['diagnosis'] == 'error',
    perf_recs['diagnosis'] == 'robots_conflict',
    perf_recs['diagnosis'] == 'lost_content',
    perf_recs['diagnosis'] == 'other'
]
```

rec_values is a list of recommendations to go with the preceding diagnosis list. At this point, it's assumed that you have done the detective work to know what recommendations you're putting forward for each of the preceding labels.

The recommendation list items are ordered to match the order of the diag_conds list. For example, if the diagnosis cell value is "lost_content," then the recommendation is to "create_integrate," which means to create the content redirected, unredirect, and reintegrate into the website.

```
rec_values = ['integrate', 'disintermediate', 'fix/remove links to error
URL', 'remove canonical', 'create_integrate', 'no further action']
```

With the lists in place, we can now match them when we create a new column and use np.select to assign values to it using our lists as arguments:

```
perf_recs['recommendation'] = np.select(diag_conds, rec_values,
default = 'na')
perf_recs = perf_recs.sort_values('clicks_delta')

perf_recs.to_csv('exports/' + hostname + '_migration_data_1.csv')

perf_recs
```

This results in the following:

final_redirect_url	final_redirect_url_status_code	urls_with_similar_content	ult_dest_url	content_simi	diagnosis	recommendation
No Data	Not Set	0.0	https://www.saasforecom.com/	1.000000	other	no further action
No Data	Not Set	0.0	http://www.saasforecom.com/ecommerce-order-management-system/	0.813559	lost_content	create_integrate
No Data	Not Set	0.0	https://www.saasforecom.com/drop-shipping-automation-software/	1.000000	other	no further action
No Data	Not Set	0.0	https://www.saasforecom.com/software-for-warehouse-inventory-management/	1.000000	other	no further action
No Data	Not Set	0.0	https://www.saasforecom.com/ecommerce-order-management-system/	1.000000	other	no further action
...	...	...	...	...	...	...
No Data	Not Set	0.0	https://www.saasforecom.com/b2b-ecommerce-solution/	1.000000	other	no further action
No Data	Not Set	0.0	https://www.saasforecom.com/saasforecom-pricing/	1.000000	error	fix/remove links to error URL
sforecom.com/custom-erp-for-wholesalers/	200	0.0	https://www.saasforecom.com/custom-erp-for-wholesalers/	0.852459	lost_content	create_integrate
No Data	Not Set	0.0	http://www.saasforecom.com/software-for-warehouse-inventory-management/	0.844444	lost_content	create_integrate
No Data	Not Set	0.0	https://www.saasforecom.com/accounting-software-for-wholesale-distribution/	1.000000	other	no further action

You'll now see the perf_recs dataframe updated with a new column to match the diagnosis.

Of course, we'll now want to quantify all of this for our presentation decks to our colleagues, using the hopefully familiar groupby() function:

```
recs_clicksurl = perf_recs.groupby('recommendation')['clicks_delta'].
agg(['sum', 'count']).reset_index()
recs_clicksurl['recovery_clicks_url'] = np.abs(recs_clicksurl['sum'] /
recs_clicksurl['count'])
```

We're taking the absolute as we want to put a positive slant on the presentation of the numbers:

```
recs_clicksurl['sum'] = np.abs(recs_clicksurl['sum'])

recs_clicksurl
```

This results in the following:

	recommendation	sum	count	recovery_clicks_url
0	create_integrate	84.0	373	0.225201
1	fix/remove links to error URL	14.0	34	0.411765
2	integrate	0.0	7	0.000000
3	no further action	297.0	50	5.940000

The preceding table shows the recommendation with clicks to be recovered (sum), URL count (count), and the potential recovery clicks per URL. Although it may seem strange to recover 297 clicks per month through "no further action," some may well be recovered by fixing the other issues.

Time to visualize:

```
recs_clicks_plt = (
    ggplot(recs_clicksurl,
            aes(x = 'reorder(recommendation, sum)', y = 'sum')) +
    geom_bar(stat = 'identity', alpha = 0.6, position = 'dodge',
    fill = 'blue') +
position=position_stack(vjust=0.01)) +
    labs(y = 'Recovery Clicks Available', x = '') +
    coord_flip() +
    theme(legend_position = 'right',
        axis_text_x=element_text(rotation=90, hjust=1)
        )
)

recs_clicks_plt.save(filename = 'images/8_recs_clicks_plt.png',
                    height=5, width=10, units = 'in', dpi=1000)
recs_clicks_plt
```

Figure 9-11 visualizes the recommendations.

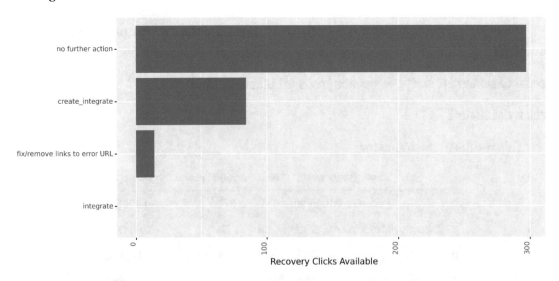

Figure 9-11. *Bar chart of estimated recovery clicks available by tech SEO diagnosis*

Summary

This chapter covered site migration mapping so that you could set the structure of your new site and semiautomate the formation of your migration URLs. Some of the techniques used are as follows:

- String manipulation

- Iterating through dataframe rows by converting these into a list

- Using NLP to compare URL strings

While these techniques were applied to speed up the processing of data for a site migration, they can easily be applied to other use cases. In the next chapter, we will show how algorithm updates can be better understood using data.

CHAPTER 10

Google Updates

Just as death and taxes are the certainties of life, algorithm updates are a certainty for any SEO career. That's right, Google frequently introduces changes to its ranking algorithm, which means your website (and many others) may experience fluctuations in rankings and, by extension, traffic. These changes may be positive or negative, and in some cases, you'll discern no impact at all.

To compound matters, Google in particular gives rather vague information as to what the algorithm updates are about and how business and SEO professionals should respond. Naturally, the lack of prescriptive advice from Google other than delivering "a great user experience" and "creating compelling content" means SEOs must find answers using various analysis tools. Fortunately, for the SEO

- Google is a system of algorithms. That means that the changes in ranking factors are likely to be consistent and predictable and not at the whim of a human. These changes are likely to have been tested beforehand.

- The outcomes of Google's algorithm changes are in the public domain by virtue of the Search Engine Results Pages (SERPs), which means that there is data available for analysis, even if it is against the Google Webmaster Guidelines.

- Even without the SERPs, Google Search Console is a valuable data source for understanding the nature of Google's updates.

In this chapter, we will cover algorithm updates analysis which is to analyze the difference in search results before and after the algorithm update event at different levels:

- Domains
- Result types

A. Voniatis, *Data-Driven SEO with Python*, https://doi.org/10.1007/978-1-4842-9175-7_10

- Cannibalization

- Keywords

- Within client tracked queries (target)

- Segmented SERPs

Algo Updates

The general approach here is to compare performance between the before and after phases of the Google algorithm update. In this case, we'll focus on a newly listed webinar company known as ON24. ON24 suffered from the December 2019 core update.

   With some analysis and visualization, we can get an idea of what's going on with the update. As well as the usual libraries, we'll be importing SERPs data from getSTAT (an enterprise-level rank tracking platform, available at getstat.com):

```
import re
import time
import random
import pandas as pd
import numpy as np
import datetime
import re
import time
import requests
import json
from datetime import timedelta
from glob import glob
import os
from textdistance import sorensen_dice
from plotnine import *
import matplotlib.pyplot as plt
from pandas.api.types import is_string_dtype
from pandas.api.types import is_numeric_dtype
import uritools

pd.set_option('display.max_colwidth', None)
```

```
%matplotlib inline

root_domain = 'on24.com'
hostdomain = 'www.on24.com'
hostname = 'on24'
full_domain = 'https://www.on24.com'
target_name = 'ON24'

getstat_raw.head()
```

The following is the printout of the getSTAT data:

	Keyword	Market	Location	Device	Global Monthly Search Volume	Regional Monthly Search Volume	Rank	Result Types for Nov 19, 2020	Protocol for Nov 19, 2020	Ranking URL on Nov 19, 2020	Result Types for Dec 17, 2020	Protocol for Dec 17, 2020	Ranking URL on Dec 17 202(
0	6sense webinars	US-en	NaN	desktop	0	0	1	organic	https	hub.6sense.com/upcoming-events	organic	https	hub.6sense.com/upcoming event:
1	6sense webinars	US-en	NaN	desktop	0	0	2	organic	https	hub.6sense.com/on-demand-events/webinars	organic	https	hub.6sense.com/on demand-event:
2	6sense webinars	US-en	NaN	desktop	0	0	3	organic	https	hub.6sense.com/webinars/	organic	https	hub.6sense.com/on demand-events/webinar:
3	6sense webinars	US-en	NaN	desktop	0	0	4	organic	https	hub.6sense.com/on-demand-events	organic	https	6sense.com/prg-in-action
4	6sense webinars	US-en	NaN	desktop	0	0	5	organic	https	hub.6sense.com/webinars/analyst-webinars	organic	https	6sense.com/breakthrough

To make the column names more data-friendly, we'll do some cleaning:

```
getstat_cleancols = getstat_raw
```

Convert to lowercase:

```
getstat_cleancols.columns = [x.lower() for x in getstat_cleancols.columns]
```

Given ON24 is a global brand and using a single website to capture all English language searches worldwide, we're using the global monthly search volume instead of the usual regional (country level) numbers. Hence, we're renaming the global volumes as the search volume:

```
getstat_cleancols = getstat_cleancols.rename(columns = {'global monthly
search volume': 'search_volume'})
```

We filter out rows for brand searches as we would expect ON24 to rank well for its own brand and we're more interested in the general core update:

```
getstat_cleancols = getstat_cleancols[~getstat_cleancols['keyword'].str.
contains('24')]
getstat_cleancols
```

The columns are now in lowercase with some columns renamed:

	keyword	market	location	device	search_volume	regional monthly search volume	rank	result types for nov 19, 2020	protocol for nov 19, 2020	ranking url on nov 19, 2020	result type o 2(
0	6sense webinars	US-en	NaN	desktop	0	0	1	organic	https	hub.6sense.com/upcoming-events	orga
1	6sense webinars	US-en	NaN	desktop	0	0	2	organic	https	hub.6sense.com/on-demand-events/webinars	orga
2	6sense webinars	US-en	NaN	desktop	0	0	3	organic	https	hub.6sense.com/webinars/	orga
3	6sense webinars	US-en	NaN	desktop	0	0	4	organic	https	hub.6sense.com/on-demand-events	orga
4	6sense webinars	US-en	NaN	desktop	0	0	5	organic	https	hub.6sense.com/webinars/analyst-webinars	orga
...	...	...	...	...	...	...	...	...		...	
27635	zoom webinars	US-en	NaN	smartphone	135000	40500	16	image / organic	https	impact.extension.org/zoom/	ima orga
27636	zoom webinars	US-en	NaN	smartphone	135000	40500	17	image / organic	https	its.umich.edu/communication/videoconferencing/zoom/meetings-vs-webinars	ima orga
27637	zoom webinars	US-en	NaN	smartphone	135000	40500	18	image / organic	https	www.owllabs.com/blog/zoom-webinar	ima orga
27638	zoom webinars	US-en	NaN	smartphone	135000	40500	19	image / organic	https	is.oregonstate.edu/zoom/webinar-licenses	ima orga
27639	zoom webinars	US-en	NaN	smartphone	135000	40500	20	image / organic	https	www.sandiego.edu/its/support/software/zoom-webinar.php	ima orga

27560 rows × 13 columns

To make the calculations easier, we'll split the dataframe column-wise into before and after. The splits will be aggregated and then compared to each other.

We'll start with the before dataframe, selecting the before columns:

```
getstat_before = getstat_cleancols[['keyword', 'market', 'location',
'device', 'search_volume', 'rank',
                              'result types for nov 19, 2020',
                              'protocol for nov 19, 2020',
                              'ranking url on nov 19, 2020']]
```

We build the full URL:

```
getstat_before['url'] = getstat_before['protocol for nov 19, 2020'] +
'://' + getstat_before['ranking url on nov 19, 2020']
```

Change the values of the URL column such that if there are any blanks (null values), then replace it with ' ' as opposed to a NaN (not a number). This helps avoid any errors when aggregating later on.

```
getstat_before['url'] = np.where(getstat_before['url'].isnull(), '',
getstat_before['url'])
```

We'll derive site names using the urisplit function (embedded inside a list comprehension) to extract the domain name. This will be useful to summarize performance at the site level.

```
getstat_before['site'] = [uritools.urisplit(x).authority if uritools.
isuri(x) else x for x in getstat_before['url']]
```

We initialize a list named strip_subdomains to help strip out string components of the URL:

```
strip_subdomains = ['hub\.', 'blog\.', 'www\.', 'impact\.', 'harvard\.',
'its\.', 'is\.', 'support\.']
```

We change the site field to replace any strip_subdomains strings found in the site column and replace with nothing:

```
getstat_before['site'] = getstat_before['site'].str.replace('|'.join(strip_
subdomains), '')
```

We set a new column phase to "before":

```
getstat_before['phase'] = 'before'
```

Stratifying the ranking position data helps us perform more detailed aggregations so that we can break down performance into Google's top 3, page 1, etc. This uses np.where which is the Python equivalent of Excel's if function:

```
getstat_before['rank_profile'] = np.where(getstat_before['rank'] < 11,
'page_1', 'page_2')
getstat_before['rank_profile'] = np.where(getstat_before['rank'] < 3,
'top_3',
```

```
                getstat_before['rank_profile'])
```

Here, we'll rename some columns as we don't need the month year in the column title:

```
getstat_before = getstat_before.rename(columns = {'result types for nov 19,
2020': 'snippets'})
```

Column selection is not absolutely necessary, but it does help clean up the dataframe and remind us of what we're working with:

```
getstat_before = getstat_before[['keyword', 'market', 'phase', 'device',
'search_volume', 'rank',
                              'url', 'site', 'snippets',
                              'rank_profile']]
```

We'll set zero search volumes to one so that we don't get "divide by zero errors" later on when deriving calculations:

```
getstat_before['search_volume'] = np.where(getstat_before['search_volume']
== 0, 1, getstat_before['search_volume'])
```

Initialize a new column count which also comes in handy for aggregations:

```
getstat_before['count'] = 1
```

Sometimes, you'll want to dissect the SERPs by head, middle, and long tail. To make this possible, we'll initialize a column called "token_count" which counts the amount of gaps between the words (and add 1) to extract the query word count in the "keyword" column:

```
getstat_before['token_count'] = getstat_before['keyword'].str.
count(' ') + 1
```

Thanks to the word count, we use the np.select() function to classify the query length:

```
before_length_conds = [
    getstat_before['token_count'] == 1,
    getstat_before['token_count'] == 2,
```

```
       getstat_before['token_count'] > 2]

length_vals = ['head', 'middle', 'long']

getstat_before['token_size'] = np.select(before_length_conds, length_vals)

getstat_before
```

Here are the before dataset with additional features to make the analysis more useful.

:e	search_volume	rank	url	site	snippets	rank_profile	count	token_count	token_size
)p	1	1	https://hub.6sense.com/upcoming-events	6sense.com	organic	top_3	1	2	middle
)p	1	2	https://hub.6sense.com/on-demand-events/webinars	6sense.com	organic	top_3	1	2	middle
)p	1	3	https://hub.6sense.com/webinars/	6sense.com	organic	page_1	1	2	middle
)p	1	4	https://hub.6sense.com/on-demand-events	6sense.com	organic	page_1	1	2	middle
)p	1	5	https://hub.6sense.com/webinars/analyst-webinars	6sense.com	organic	page_1	1	2	middle
...	...	...	...	...	...	...	...	...	...
1e	135000	16	https://impact.extension.org/zoom/	extension.org	image / organic	page_2	1	2	middle
1e	135000	17	https://its.umich.edu/communication/videoconferencing/zoom/meetings-vs-webinars	umich.edu	image / organic	page_2	1	2	middle
1e	135000	18	https://www.owllabs.com/blog/zoom-webinar	owllabs.com	image / organic	page_2	1	2	middle
1e	135000	19	https://is.oregonstate.edu/zoom/webinar-licenses	oregonstate.edu	image / organic	page_2	1	2	middle
1e	135000	20	https://www.sandiego.edu/its/support/software/zoom-webinar.php	sandiego.edu	image / organic	page_2	1	2	middle

Let's repeat the data transformation steps for the after dataset:

```
getstat_after = getstat_cleancols[['keyword', 'market', 'location',
'device', 'search_volume', 'rank',
                            'result types for dec 17, 2020',
                            'protocol for dec 17, 2020',
                            'ranking url on dec 17, 2020']]
getstat_after['url'] = getstat_after['protocol for dec 17, 2020'] + '://' +
getstat_after['ranking url on dec 17, 2020']
getstat_after['url'] = np.where(getstat_after['url'].isnull(), '', getstat_
after['url'])

getstat_after['site'] = [uritools.urisplit(x).authority if uritools.
isuri(x) else x for x in getstat_after['url']]
strip_subdomains = ['hub\.', 'blog\.', 'www\.', 'impact\.', 'harvard\.',
'its\.', 'is\.', 'support\.']
```

```python
getstat_after['site'] = getstat_after['site'].str.replace('|'.join(strip_
subdomains), '')

getstat_after['phase'] = 'after'

getstat_after = getstat_after.rename(columns = {'result types for dec 17,
2020': 'snippets'})

getstat_after = getstat_after[['keyword', 'market', 'phase', 'device',
'search_volume', 'rank', 'url', 'site', 'snippets']]

getstat_after['search_volume'] = np.where(getstat_after['search_volume'] ==
0, 1, getstat_after['search_volume'])

getstat_after['count'] = 1

getstat_after['rank_profile'] = np.where(getstat_after['rank'] < 11,
'page_1', 'page_2')
getstat_after['rank_profile'] = np.where(getstat_after['rank'] < 3,
'top_3', getstat_after['rank_profile'])

getstat_after['token_count'] = getstat_after['keyword'].str.count(' ') + 1

after_length_conds = [
    getstat_after['token_count'] == 1,
    getstat_after['token_count'] == 2,
    getstat_after['token_count'] > 2,
]

getstat_after['token_size'] = np.select(after_length_conds, length_vals)

getstat_after
```

getstat_after is the after dataset transformation which is now complete, allowing us to proceed to the next step of deduplicating our data:

:e	search_volume	rank	url	site	snippets	count	rank_profile	token_count	token_size
)p	1	1	https://hub.6sense.com/upcoming-events	6sense.com	organic	1	top_3	2	middle
)p	1	2	https://hub.6sense.com/on-demand-events	6sense.com	organic	1	top_3	2	middle
)p	1	3	https://hub.6sense.com/on-demand-events/webinars	6sense.com	organic	1	page_1	2	middle
)p	1	4	https://6sense.com/prg-in-action/	6sense.com	organic	1	page_1	2	middle
)p	1	5	https://6sense.com/breakthrough/	6sense.com	organic	1	page_1	2	middle
...	...	...	...	...	...	...	...	...	...
ie	135000	16	https://www.owllabs.com/blog/zoom-webinar	owllabs.com	image / organic	1	page_2	2	middle
ie	135000	17	https://its.umich.edu/communication/videoconferencing/zoom/meetings-vs-webinars	umich.edu	image / organic	1	page_2	2	middle
ie	135000	18	https://is.oregonstate.edu/zoom/webinar-licenses	oregonstate.edu	image / organic	1	page_2	2	middle
ie	135000	19	https://www.bu.edu/tech/services/cccs/conf/online/zoom/zoom-webinars/	bu.edu	image / organic	1	page_2	2	middle
ie	135000	20	https://harvard.service-now.com/ithelp?id=kb_article&sys_id=a996d6dddbe78c1496ab5682ca9619cf	service-now.com	image / organic	1	page_2	2	middle

Dedupe

The reason for deduplication is that the search engines often rank multiple URLs from the same SERPs. This is fine if you want to evaluate SERPs share or rates of cannibalization (i.e., multiple URLs from the same domain competing for the same ranking and ultimately constraining the maximum ranking achieved). However, in our use case of just seeing which sites come first, in what rank order, and how often, deduplication is key.

Using the transformed datasets, we will group by site, selecting and keeping the highest ranked URL in the unique (deduplicated) dataset:

```
getstat_bef_unique = getstat_before.sort_values('rank').groupby(['site',
'device', 'keyword']).first()
getstat_bef_unique = getstat_bef_unique.reset_index()
getstat_bef_unique = getstat_bef_unique[getstat_bef_unique['site'] != '']
getstat_bef_unique = getstat_bef_unique.sort_values(['keyword', 'device',
'rank'])
```

```
getstat_bef_unique = getstat_bef_unique[['keyword', 'market', 'phase',
'device', 'search_volume',
    'rank', 'url', 'site', 'snippets', 'rank_profile', 'count', 'token_
    count','token_size']]
```

```
getstat_bef_unique
```

This results in the following:

	keyword	market	phase	device	search_volume	rank	url	site
177	6sense webinars	US-en	before	desktop	1	1	https://hub.6sense.com/upcoming-events	6sense.com
5647	6sense webinars	US-en	before	desktop	1	16	https://www.drift.com/webinars/not-abm/	drift.com
16455	6sense webinars	US-en	before	desktop	1	17	https://www.playbigger.com/media/coffee-talk-webinar-w/play-bigger-and-6sense	playbigger.com
17496	6sense webinars	US-en	before	desktop	1	18	https://resources.pedowitzgroup.com/webinar-slides/techtalk-slides-account-based-marketing-6sense	resources.pedowitzgroup.com
2447	6sense webinars	US-en	before	desktop	1	19	https://www.brighttalk.com/webcast/12753/270461/webinar-feat-forrester-taking-the-pulse-of-b2b-predictive-marketing-analytics	brighttalk.com
...	...	...	...	...	...	...		...
6994	zoom webinars	US-en	before	smartphone	135000	16	https://impact.extension.org/zoom/	extension.org
21418	zoom webinars	US-en	before	smartphone	135000	17	https://its.umich.edu/communication/videoconferencing/zoom/meetings-vs-webinars	umich.edu
15927	zoom webinars	US-en	before	smartphone	135000	18	https://www.owllabs.com/blog/zoom-webinar	owllabs.com
15835	zoom webinars	US-en	before	smartphone	135000	19	https://is.oregonstate.edu/zoom/webinar-licenses	oregonstate.edu
18023	zoom webinars	US-en	before	smartphone	135000	20	https://www.sandiego.edu/its/support/software/zoom-webinar.php	sandiego.edu

23677 rows × 13 columns

The dataset has been reduced noticeably from 27,000 to 23,600 rows. We'll repeat the same operation for the after dataset:

```
getstat_aft_unique = getstat_after.sort_values('rank').groupby(['site',
'device', 'keyword']).first()
getstat_aft_unique = getstat_aft_unique.reset_index()
```

```
getstat_aft_unique = getstat_aft_unique[getstat_aft_unique['site'] != '']
getstat_aft_unique = getstat_aft_unique.sort_values(['keyword', 'device',
'rank'])
```

```
getstat_aft_unique = getstat_aft_unique[['keyword', 'market', 'phase',
'device', 'search_volume',
    'rank', 'url', 'site', 'snippets', 'rank_profile', 'count',
    'token_count','token_size']]
```

```
getstat_aft_unique
```

This results in the following:

	keyword	market	phase	device	search_volume	rank	url	site	snippets	r
117	6sense webinars	US-en	after	desktop	1	1	https://hub.6sense.com/upcoming-events	6sense.com	organic	
120	6sense webinars	US-en	after	smartphone	1	1	https://hub.6sense.com/upcoming-events	6sense.com	organic	
9276	6sense webinars	US-en	after	smartphone	1	7	http://www.google.com/	google.com	related searches	
160	abm	US-en	after	desktop	110000	1	https://www.abm.com/	abm.com	organic / sitelinks	
8818	abm	US-en	after	desktop	110000	2	http://www.google.com/	google.com	people also ask	
...	...	...	...	...	...	...	...	...	...	
16806	zoom webinars	US-en	after	smartphone	135000	16	https://www.owllabs.com/blog/zoom-webinar	owllabs.com	image / organic	
22606	zoom webinars	US-en	after	smartphone	135000	17	https://its.umich.edu/communication/videoconferencing/zoom/meetings-vs-webinars	umich.edu	image / organic	
16717	zoom webinars	US-en	after	smartphone	135000	18	https://is.oregonstate.edu/zoom/webinar-licenses	oregonstate.edu	image / organic	
2541	zoom webinars	US-en	after	smartphone	135000	19	https://www.bu.edu/tech/services/cccs/conf/online/zoom/zoom-webinars/	bu.edu	image / organic	
19371	zoom webinars	US-en	after	smartphone	135000	20	https://harvard.service-now.com/ithelp?id=kb_article&sys_id=a996d6dddbe78c1496ab5682ca9619cf	service-now.com	image / organic	

24972 rows × 13 columns

With both datasets deduplicated, we can start performing aggregations from different viewpoints and generate insights.

Domains

One of the most common questions of any algo update is which sites gained and which ones lost. We will start by filtering for those in the top 10 to calculate the "reach" and sum these by site:

```
before_unq_reach = getstat_bef_unique
before_unq_reach = before_unq_reach[before_unq_reach['rank'] < 11 ]
before_unq_reach = before_unq_reach.groupby(['site']).agg({'count':
'sum'}).reset_index()
```

Rename count as reach:

```
before_unq_reach = before_unq_reach.rename(columns = {'count': 'reach'})
before_unq_reach = before_unq_reach[['site', 'reach']]
```

Swap null values for zero:

```
before_unq_reach['reach'] = np.where(before_unq_reach['reach'].isnull(), 0,
before_unq_reach['reach'])
before_unq_reach.sort_values('reach', ascending = False).head(10)
```

Unsurprisingly, Google has the most keyword presence of any site. After that, it's HubSpot, then ON24, our site of interest. Note that this is before the Google update.

	site	reach
879	google.com	914
962	hubspot.com	323
1429	on24.com	221
881	gotomeeting.com	153
334	capterra.com	148
822	g2.com	135
2224	wordstream.com	114
1294	medium.com	92
1217	m.youtube.com	88
1253	marketo.com	88

We'll repeat the domain reach aggregation for after the update:

```
after_unq_reach = getstat_aft_unique
after_unq_reach = after_unq_reach[after_unq_reach['rank'] < 11 ]
after_unq_reach = after_unq_reach.groupby(['site']).agg({'count': 'sum'}).
reset_index()
after_unq_reach = after_unq_reach.rename(columns = {'count': 'reach'})
after_unq_reach['reach'] = np.where(after_unq_reach['reach'].isnull(), 0,
after_unq_reach['reach'])

after_unq_reach = after_unq_reach[['site', 'reach']]
after_unq_reach.sort_values('reach', ascending = False).head(10)
```

This results in the following:

	site	reach
879	google.com	1011
969	hubspot.com	304
1437	on24.com	222
318	capterra.com	156
826	g2.com	148
881	gotomeeting.com	148
1302	medium.com	118
2264	wordstream.com	110
1231	m.youtube.com	96
2313	zoom.us	95

Google is an even bigger winner post its own update. HubSpot has lost out slightly, and ON24 is virtually unchanged. Or so it appears on the surface as we'll see later on when we get deeper into the analysis.

Rather than eyeballing two separate dataframes, we'll join them together for a side-by-side comparison:

```
compare_reach_loser = before_unq_reach.merge(after_unq_reach, on =
['site'], how = 'outer')
```

Rename the columns to be more user-friendly:

```
compare_reach_loser = compare_reach_loser.rename(columns = {'reach_x':
'before_reach', 'reach_y': 'after_reach'})
compare_reach_loser['before_reach'] = np.where(compare_reach_loser['before_
reach'].isnull(),
        0, compare_reach_loser['before_reach'])
```

Swap null values with zero to prevent errors for the next step:

```
compare_reach_loser['after_reach'] = np.where(compare_reach_loser['after_
reach'].isnull(),
        0, compare_reach_loser['after_reach'])
```

Create new columns to quantify the difference in reach between before and after:

```
compare_reach_loser['delta_reach'] = compare_reach_loser['after_reach'] -
compare_reach_loser['before_reach']
compare_reach_loser = compare_reach_loser.sort_values('delta_reach')
compare_reach_loser = compare_reach_loser[['site', 'before_reach', 'after_
reach', 'delta_reach']]
compare_reach_loser.head(10)
```

This results in the following:

	site	before_reach	after_reach	delta_reach
2227	workcast.com	49.0	0.0	-49.0
962	hubspot.com	323.0	304.0	-19.0
1538	podcastinsights.com	41.0	28.0	-13.0
1253	marketo.com	88.0	76.0	-12.0
730	eventbrite.com	36.0	25.0	-11.0
1504	pega.com	13.0	3.0	-10.0
1649	resources.engagio.com	7.0	0.0	-7.0
408	clickmeeting.com	58.0	51.0	-7.0
1007	inc.com	27.0	20.0	-7.0
1059	inxpo.com	23.0	16.0	-7.0

The biggest loser by far appears to be WorkCast, a major player in the webinar software space, followed by HubSpot. As you'll realize, having the tables aggregated separately and then joined makes the comparison much easier. Let's repeat to find the winners:

```
compare_reach_winners = compare_reach_loser.sort_values('delta_reach',
ascending = False)
compare_reach_winners.head(10)
```

This results in the following:

	site	before_reach	after_reach	delta_reach
879	google.com	914.0	1011.0	97.0
1023	info.workcast.com	34.0	74.0	40.0
1171	liferay.com	19.0	46.0	27.0
1294	medium.com	92.0	118.0	26.0
1607	qualtrics.com	47.0	70.0	23.0
1910	superoffice.com	38.0	61.0	23.0
1785	sitecore.com	17.0	38.0	21.0
424	cmswire.com	25.0	45.0	20.0
794	forbes.com	48.0	67.0	19.0
536	cvent.com	12.0	28.0	16.0

Interesting, so WorkCast lost, yet its subdomain gained. A few publishers like Medium and blogs from indirect B2B software competitors also gain. Intuitively, this looks like blogs and guides have been favored.

Time to visualize. We'll convert to long format which is the data structure of choice for data visualization graphing packages (think pivot tables):

```
compare_reach_losers_long = compare_reach_loser[['site', 'before_
reach','after_reach']].head(28)
compare_reach_losers_long = compare_reach_losers_long.melt(id_vars =
['site'], var_name='Phase', value_name='Reach')
compare_reach_losers_long['Phase'] = compare_reach_losers_long['Phase'].
str.replace('_reach', '')
compare_reach_losers_long['Phase'] = compare_reach_losers_long['Phase'].
astype('category')
compare_reach_losers_long['Phase'] = compare_reach_losers_long['Phase'].
cat.reorder_categories(['before', 'after'])

stop_doms = ['en.wikipedia.org', 'google.com', 'youtube.com',
'lexisnexis.com']
compare_reach_losers_long = compare_reach_losers_long[~compare_reach_
losers_long['site'].isin(stop_doms)]

compare_reach_losers_long.head(10)
```

This results in the following:

	site	Phase	Reach
0	workcast.com	before	49.0
1	hubspot.com	before	323.0
2	podcastinsights.com	before	41.0
3	marketo.com	before	88.0
4	eventbrite.com	before	36.0
5	pega.com	before	13.0
6	resources.engagio.com	before	7.0
7	clickmeeting.com	before	58.0
8	inc.com	before	27.0
9	inxpo.com	before	23.0

```
#VIZ
compare_reach_losers_plt = (
    ggplot(compare_reach_losers_long, aes(x = 'reorder(site, Reach)', y =
    'Reach', fill = 'Phase')) +
    geom_bar(stat = 'identity', position = 'dodge', alpha = 0.8) +
    #geom_text(dd_factor_df, aes(label = 'market_name'), position=position_
    stack(vjust=0.01)) +
    labs(y = 'Reach', x = ' ') +
    #scale_y_reverse() +
    coord_flip() +
    theme(legend_position = 'right', axis_text_x=element_text(rotation=0,
    hjust=1, size = 12)) +
    facet_wrap('device')
)

compare_reach_plt.save(filename = 'images/1_compare_reach_losers_plt.png',
height=5, width=10, units = 'in', dpi=1000)
compare_reach_plt
```

It seems not all websites had a consistent presence across both device search result types (Figure 10-1).

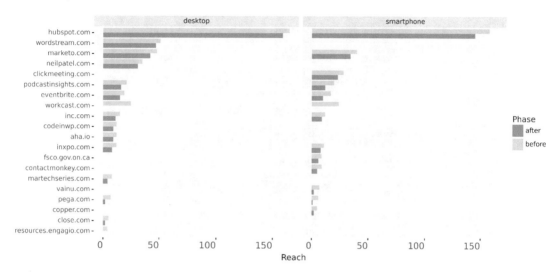

Figure 10-1. *Website top 10 ranking counts (reach) before and after by browser device*

Reach Stratified

Reach is helpful, but as always the devil is in the detail, and no doubt you and your colleagues will want to drill down further by rank strata, that is, rankings in the top 3 positions or perhaps only rankings on page 1 of Google, etc. Let's aggregate only this time with reach strata starting with the before dataset:

```
before_unq_reachstrata = getstat_bef_unique
before_unq_reachstrata = before_unq_reachstrata.groupby(['site', 'rank_
profile']).agg({'count': 'sum'}).reset_index()
before_unq_reachstrata = before_unq_reachstrata.rename(columns = {'count':
'reach'})
before_unq_reachstrata = before_unq_reachstrata[['site', 'rank_profile',
'reach']]
before_unq_reachstrata.sort_values('reach', ascending = False).head(10)
```

This results in the following:

	site	rank_profile	reach
2426	google.com	page_1	475
2428	google.com	top_3	439
2651	hubspot.com	page_1	243
3950	on24.com	page_1	170
938	capterra.com	page_1	117
3951	on24.com	page_2	114
2433	gotomeeting.com	page_1	113
3543	medium.com	page_2	110
2652	hubspot.com	page_2	105
2434	gotomeeting.com	page_2	101

Now we have an ordered dataframe by reach, this time split by rank_profile, thus stratifying the reach metric. For example, we see HubSpot has twice as many keywords on page 1 of Google search results compared to page 2, whereas with ON24, it's more or less equal.

Repeat the operation for the after dataset:

```
after_unq_reachstrata = getstat_aft_unique
after_unq_reachstrata = after_unq_reachstrata.groupby(['site', 'rank_
profile']).agg({'count': 'sum'}).reset_index()
after_unq_reachstrata = after_unq_reachstrata.rename(columns = {'count':
'reach'})
after_unq_reachstrata = after_unq_reachstrata[['site', 'rank_profile',
'reach']]
after_unq_reachstrata.sort_values('reach', ascending = False).head(10)
```

This results in the following:

	site	rank_profile	reach
2465	google.com	page_1	518
2467	google.com	top_3	493
2714	hubspot.com	page_1	240
4005	on24.com	page_1	149
2715	hubspot.com	page_2	125
916	capterra.com	page_1	124
3602	medium.com	page_2	108
2471	gotomeeting.com	page_1	108
4006	on24.com	page_2	105
3601	medium.com	page_1	101

As you can imagine, it's less easy to see who won and lost by eyeballing the separate dataframes, so we will merge as usual:

```
compare_strata_loser = before_unq_reachstrata.merge(after_unq_reachstrata,
on = ['site', 'rank_profile'], how = 'outer')
compare_strata_loser = compare_strata_loser.rename(columns = {'reach_x':
'before_reach', 'reach_y': 'after_reach'})

compare_strata_loser['before_reach'] = np.where(compare_strata_
loser['before_reach'].isnull(), 0, compare_strata_loser['before_reach'])
compare_strata_loser['after_reach'] = np.where(compare_strata_loser['after_
reach'].isnull(), 0, compare_strata_loser['after_reach'])
compare_strata_loser['delta_reach'] = compare_strata_loser['after_reach'] -
compare_strata_loser['before_reach']

compare_strata_loser = compare_strata_loser.sort_values('delta_reach')
compare_strata_loser.head(10)
```

This results in the following:

	site	rank_profile	before_reach	after_reach	delta_reach
6083	workcast.com	page_1	30.0	0.0	-30.0
3950	on24.com	page_1	170.0	149.0	-21.0
6084	workcast.com	page_2	21.0	1.0	-20.0
6085	workcast.com	top_3	19.0	0.0	-19.0
688	bloggingwizard.com	page_2	27.0	9.0	-18.0
2653	hubspot.com	top_3	80.0	64.0	-16.0
3443	marketo.com	page_2	87.0	74.0	-13.0
2434	gotomeeting.com	page_2	101.0	90.0	-11.0
6184	youtube.com	page_2	85.0	74.0	-11.0
2776	inc.com	page_2	26.0	16.0	-10.0

This dataframe merge makes things much clearer as we can now see ON24 lost most of its rankings on page 1, whereas WorkCast has lost everywhere.

We'll turn our attention to the reach winners stratified by rank profile:

```
compare_strata_winners = before_unq_reachstrata.merge(after_unq_
reachstrata, on = ['site', 'rank_profile'], how = 'outer')
compare_strata_winners = compare_strata_winners.rename(columns =
{'reach_x': 'before_reach', 'reach_y': 'after_reach'})

compare_strata_winners['before_reach'] = np.where(compare_strata_
winners['before_reach'].isnull(),
        0, compare_strata_winners['before_reach'])
compare_strata_winners['after_reach'] = np.where(compare_strata_
winners['after_reach'].isnull(),
        0, compare_strata_winners['after_reach'])
compare_strata_winners['delta_reach'] = compare_strata_winners['after_
reach'] - compare_strata_winners['before_reach']

compare_strata_winners = compare_strata_winners.sort_values('delta_reach',
ascending = False)
```

```
compare_strata_winners = compare_strata_winners[['site', 'rank_
profile','before_reach', 'after_reach', 'delta_reach']]
compare_strata_winners = compare_strata_winners[compare_strata_
winners['delta_reach'] > 0]
```

```
compare_strata_winners.head(10)
```

This results in the following:

	site	rank_profile	before_reach	after_reach	delta_reach
2428	google.com	top_3	439.0	493.0	54.0
2426	google.com	page_1	475.0	518.0	43.0
2813	info.workcast.com	page_2	30.0	59.0	29.0
4415	qualtrics.com	page_1	36.0	63.0	27.0
3237	liferay.com	page_1	14.0	39.0	25.0
1631	digital.com	page_2	6.0	31.0	25.0
3952	on24.com	top_3	51.0	73.0	22.0
3542	medium.com	page_1	80.0	101.0	21.0
1170	cmswire.com	page_1	19.0	40.0	21.0
2814	info.workcast.com	top_3	2.0	23.0	21.0

Although WorkCast's info subdomain gained 40 positions overall, their main site lost 69 positions, so it's a net loss. Time to visualize, we'll take the top 28 sites using the head() function:

```
compare_strata_losers_long = compare_strata_loser[['site', 'rank_profile',
'before_reach','after_reach']].head(28)
```

The melt() function helps reshape the data from wide format (as per the preceding dataframe) to long format (where the column names are now in a single column as rows):

```
compare_strata_losers_long = compare_strata_losers_long.melt(id_vars =
['site', 'rank_profile'], var_name='Phase', value_name='Reach')
compare_strata_losers_long['Phase'] = compare_strata_losers_long['Phase'].
str.replace('_reach', '')
```

The astype() function allows us to instruct Pandas to treat a data column as a different data type. In this case, we're asking Pandas to treat Phase as a category as this is a discrete variable which then allows us to order these categories:

```
compare_strata_losers_long['Phase'] = compare_strata_losers_long['Phase'].
astype('category')
```

With Phase now set as a category, we can now set the order:

```
compare_strata_losers_long['Phase'] = compare_strata_losers_long['Phase'].
cat.reorder_categories(['before', 'after'])
```

The same applies to rank profile. Top 3 is obviously better than page 1, which is better than page 2.

```
compare_strata_losers_long['rank_profile'] = compare_strata_losers_
long['rank_profile'].astype('category')
compare_strata_losers_long['rank_profile'] = compare_strata_losers_
long['rank_profile'].cat.reorder_categories(['top_3', 'page_1', 'page_2'])
```

The stop_doms list is used to weed out domains from our analysis that the audience wouldn't be interested in:

```
stop_doms = ['en.wikipedia.org', 'google.com', 'youtube.com']
```

With the stop_doms list, we can filter the dataframe of these undesirable domain names by negating any sites that are in (using the isin() function) the stop_doms list:

```
compare_strata_losers_long = compare_strata_losers_long[~compare_strata_
losers_long['site'].isin(stop_doms)]
```

```
compare_strata_losers_long.head(10)
```

This results in the following:

	site	rank_profile	Phase	Reach
0	workcast.com	page_1	before	30.0
1	on24.com	page_1	before	170.0
2	workcast.com	page_2	before	21.0
3	workcast.com	top_3	before	19.0
4	bloggingwizard.com	page_2	before	27.0
5	hubspot.com	top_3	before	80.0
6	marketo.com	page_2	before	87.0
7	gotomeeting.com	page_2	before	101.0
9	inc.com	page_2	before	26.0
10	founderjar.com	page_2	before	10.0

The data is now in long format with the Phase extracted from the before_reach and after_reach columns and pushed into a column called Phase. The values of the two columns sit under a new single column Reach. Let's visualize:

```
compare_strata_losers_plt = (
    ggplot(compare_strata_losers_long, aes(x = 'reorder(site, Reach)',
    y = 'Reach', fill = 'rank_profile')) +
    geom_bar(stat = 'identity', position = 'fill', alpha = 0.8) +
    position=position_stack(vjust=0.00)) +
    labs(y = 'Reach', x = ' ') +
    coord_flip() +
    theme(legend_position = 'right', axis_text_x=element_text(rotation=0,
    hjust=1, size = 12)) +
    facet_wrap('Phase')
)

compare_strata_losers_plt.save(filename = 'images/1_compare_strata_losers_
plt.png', height=5, width=10, units = 'in', dpi=1000)
compare_strata_losers_plt
```

We see the proportions of keywords in their rank profile, which are much easier to see thanks to the fixed lengths (Figure 10-2).

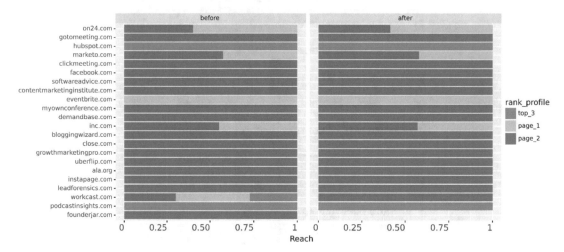

Figure 10-2. *Website Google rank proportions (reach) by top 3, page 1, and page 2 before and after*

For example, WorkCast had a mixture of top 3 and page 1 rankings which are now all on page 2. FounderJar had page 2 listings, which are now nowhere to be found.

The fixed lengths are set in the geom_bar() function using the parameter position set to "fill." Despite following best practice data visualization as shown earlier, you may have to acquiesce to your business audience who may want multilength bars as well as proportions (even if it's much harder to infer from the chart). So instead of the position set to fill, we will set it to "stack":

```
compare_strata_losers_plt = (
    ggplot(compare_strata_losers_long, aes(x = 'reorder(site, Reach)',
    y = 'Reach', fill = 'rank_profile')) +
    geom_bar(stat = 'identity', position = 'stack', alpha = 0.8) +
    position=position_stack(vjust=0.01)) +
    labs(y = 'Reach', x = ' ') +
    coord_flip() +
    theme(legend_position = 'right', axis_text_x=element_text(rotation=0,
    hjust=1, size = 12)) +
    facet_wrap('Phase')
)
```

```
compare_strata_losers_plt.save(filename = 'images/1_compare_strata_losers_
stack_plt.png', height=5, width=10, units = 'in', dpi=1000)
compare_strata_losers_plt
```

Admittedly, in cases like ON24 where in the fixed bar length chart above (Figure 10-2), the differences were not as obvious (Figure 10-3).

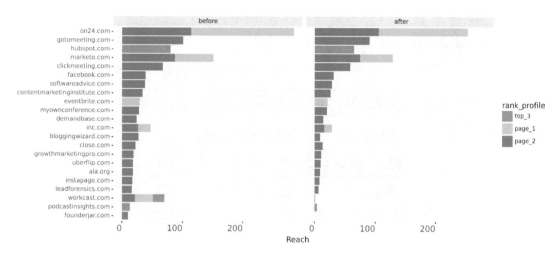

Figure 10-3. *Website Google rank counts (reach) by top 3, page 1, and page 2 before and after*

In contrast, with the free length bars, we can see that ON24 lost at least 10% of their reach.

Rankings

While reach is nice, as a single metric on its own it is not enough. If you consider the overall value of your organic presence as a function of price and volume, then reach is the volume (which we have just addressed). And now we must come to the price, which in the organic value paradigm is ranking positions.

We'll aggregate rankings by site for both before and after the core update, starting with the before dataset:

```
before_unq_ranks = getstat_bef_unique
```

Unlike reach where we took the sum of keyword search results, in this case, we're taking the average (also known as the mean):

```
before_unq_ranks = before_unq_ranks.groupby(['site']).agg({'rank':
'mean'}).reset_index()
before_unq_ranks = before_unq_ranks[['site', 'rank']]
before_unq_ranks.sort_values('rank').head(10)
```

This results in the following:

	site	rank
3156	qlik.com	1.0
2154	jungleworks.com	1.0
3280	revenuegrowthsolutions.com	1.0
2148	journals.elsevier.com	1.0
3705	strategyassociation.org	1.0
928	convene.com	1.0
3761	talentlyft.com	1.0
3882	themarketingmentors.com	1.0
2759	nngroup.com	1.0
2094	it.rutgers.edu	1.0

The table shows the average rank by site. As you may infer, the rank per se is quite meaningless because

- Some keywords have higher search volumes than others.

- The average rank is not zero inflated for keywords the sites don't rank for. For example, qlik.com's average rank of 1 may be just on one keyword.

Instead of going through the motions, repeating code to calculate and visualize the rankings for the after dataset and then comparing, we'll move on to a search volume weighted average ranking.

WAVG Search Volume

This time, we will weight the average ranking position by search volume:

```
before_unq_svranks = getstat_bef_unique
```

Define the function that takes the dataframe and uses the rank column. The weighted average is calculated by multiplying the rank by the search volume and then dividing by the total weight (being the search volume sum):

```
def wavg_rank_sv(x):
    names = {'wavg_rank': (x['rank'] * x['search_volume']).sum()/
    (x['search_volume']).sum()}
    return pd.Series(names, index=['wavg_rank']).round(1)
```

With the function in place, we'll now use the apply() function to apply the wavg_rank() function just defined earlier:

```
before_unq_svranks = before_unq_svranks.groupby(['site']).apply(wavg_rank_
sv).reset_index()
before_unq_svranks.sort_values('wavg_rank').head(10)
```

This results in the following:

	site	wavg_rank
2506	mediamotiononline.com	1.0
2508	mediapost.com	1.0
4357	www-356.ibm.com	1.0
3246	resources.gitcom	1.0
2479	masterpositioning.com	1.0
1201	docs.chorus.ai	1.0
2437	marketingoutfield.com	1.0
664	casecoach.com	1.0
84	acrpnet.org	1.0
3882	themarketingmentors.com	1.0

We can see already that the list of sites have changed due to the search volume weighting. Even though the weighted average rankings don't add much value from a business insight perspective, this is an improvement. However, what we really need is the full picture being the overall visibility.

Visibility

The visibility will be our index metric for evaluating the value of a site's organic search presence taking both reach and ranking into account.

Merge the search volume weighted rank data with reach:

```
before_unq_visi = before_unq_svranks.merge(before_unq_reach, on = 'site',
how = 'left')
```

Clean the columns of null values:

```
before_unq_visi['reach'] = np.where(before_unq_visi['reach'].isnull(), 0,
before_unq_visi['reach'])
before_unq_visi['wavg_rank'] = np.where(before_unq_visi['wavg_rank'].
isnull(), 100, before_unq_visi['wavg_rank'])
```

Computing the visibility index is derived by dividing the reach by the weighted average rank. That's because the smaller the weighted average rank number, the more visible the site is, hence why rank is the divisor. In contrast, the reach is the numerator because the higher the number, the higher your visibility.

```
before_unq_visi['visibility'] = before_unq_visi['reach'] / before_unq_
visi['wavg_rank']
before_unq_visi = before_unq_visi.sort_values('visibility',
ascending = False)
before_unq_visi
```

This results in the following:

	site	wavg_rank	reach	visibility
1707	google.com	1.8	914.0	507.777778
1864	hubspot.com	6.6	323.0	48.939394
3388	scholar.google.com	1.0	33.0	33.000000
1710	gotomeeting.com	6.4	153.0	23.906250
2822	on24.com	11.9	221.0	18.571429
...	...	...	...	...
2005	inq.dk	20.0	0.0	0.000000
2006	inreachce.com	16.5	0.0	0.000000
2007	inriver.com	19.0	0.0	0.000000
2009	insideview.com	12.2	0.0	0.000000
4431	zuora.com	18.0	0.0	0.000000

4432 rows × 4 columns

The results are looking a lot more sensible and reflect what we would expect to see in the webinar software space. We can also see that gotomeeting.com, despite having less reach, has a higher visibility score by virtue of ranking higher on more sought-after search terms. We can thus conclude the visibility score works.

Compute the same for the after dataset:

```
after_unq_visi = after_unq_svranks.merge(after_unq_reach, on = 'site', how
= 'left')
after_unq_visi['reach'] = np.where(after_unq_visi['reach'].isnull(), 0,
after_unq_visi['reach'])
after_unq_visi['wavg_rank'] = np.where(after_unq_visi['wavg_rank'].
isnull(), 100, after_unq_visi['wavg_rank'])

after_unq_visi['visibility'] = after_unq_visi['reach'] / after_unq_
visi['wavg_rank']
after_unq_visi = after_unq_visi.sort_values('visibility', ascending
= False)
after_unq_visi
```

This results in the following:

	site	wavg_rank	reach	visibility
1741	google.com	1.6	1011.0	631.875000
1912	hubspot.com	6.0	304.0	50.666667
3425	scholar.google.com	1.0	31.0	31.000000
1743	gotomeeting.com	5.1	148.0	29.019608
653	capterra.com	8.2	156.0	19.024390
...	...	...	...	...
2013	infinitee.com	16.0	0.0	0.000000
2014	infinitiresearch.com	14.0	0.0	0.000000
2015	influence.bloglovin.com	17.5	0.0	0.000000
2018	info.at-event.com	17.3	0.0	0.000000
4519	zuora.com	18.0	0.0	0.000000

4520 rows × 4 columns

GoToMeeting has gained in visibility, and ON24 is no longer in the top 5.

Join the tables to compare before and after directly in a single dataframe:

```
compare_visi_losers = before_unq_visi.merge(after_unq_visi, on = ['site'],
how = 'outer')
compare_visi_losers = compare_visi_losers.rename(columns = {'wavg_rank_x':
'before_rank', 'wavg_rank_y': 'after_rank',
                    'reach_x': 'before_reach', 'reach_y': 'after_reach',
                    'visibility_x': 'before_visi', 'visibility_y':
                    'after_visi'
                })
```

```
compare_visi_losers['before_visi'] = np.where(compare_visi_losers['before_
visi'].isnull(), 0, compare_visi_losers['before_visi'])
compare_visi_losers['after_visi'] = np.where(compare_visi_losers['after_
visi'].isnull(), 0, compare_visi_losers['after_visi'])
compare_visi_losers['delta_visi'] = compare_visi_losers['after_visi'] -
compare_visi_losers['before_visi']
```

```
compare_visi_losers = compare_visi_losers.sort_values('delta_visi')

compare_visi_losers.head(10)
```

This results in the following:

	site	before_rank	before_reach	before_visi	after_rank	after_reach	after_visi	delta_visi
25	workcast.com	7.1	49.0	6.901408	20.0	0.0	0.000000	-6.901408
4	on24.com	11.9	221.0	18.571429	17.8	222.0	12.471910	-6.099518
12	entrepreneur.com	1.4	15.0	10.714286	3.2	15.0	4.687500	-6.026786
16	podcastinsights.com	4.3	41.0	9.534884	7.6	28.0	3.684211	-5.850673
15	eventbrite.com	3.6	36.0	10.000000	4.1	25.0	6.097561	-3.902439
57	trainingcheck.com	1.0	4.0	4.000000	6.5	3.0	0.461538	-3.538462
22	en.rockcontent.com	5.3	40.0	7.547170	12.0	49.0	4.083333	-3.463836
64	mysql.com	1.1	4.0	3.636364	16.7	4.0	0.239521	-3.396843
56	heinemann.com	1.0	4.0	4.000000	6.0	4.0	0.666667	-3.333333
8	wordstream.com	8.7	114.0	13.103448	11.0	110.0	10.000000	-3.103448

The comparison view is much clearer, and ON24 and WorkCast are the biggest losers of the 2019 core update from Google.

Let's see the winners:

```
compare_visi_winners = before_unq_visi.merge(after_unq_visi, on = ['site'],
how = 'outer')
compare_visi_winners = compare_visi_winners.rename(columns = {'wavg_
rank_x': 'before_rank', 'wavg_rank_y': 'after_rank',
                    'reach_x': 'before_reach', 'reach_y': 'after_reach',
                    'visibility_x': 'before_visi', 'visibility_y':
                    'after_visi'
                    })

compare_visi_winners['before_visi'] = np.where(compare_visi_
winners['before_visi'].isnull(), 0, compare_visi_winners['before_visi'])
compare_visi_winners['after_visi'] = np.where(compare_visi_winners['after_
visi'].isnull(), 0, compare_visi_winners['after_visi'])
compare_visi_winners['delta_visi'] = compare_visi_winners['after_visi'] -
compare_visi_winners['before_visi']
```

```
compare_visi_winners = compare_visi_winners.sort_values('delta_visi',
ascending   = False)

compare_visi_winners.head(10)
```

This results in the following:

	site	before_rank	before_reach	before_visi	after_rank	after_reach	after_visi	delta_visi
0	google.com	1.8	914.0	507.777778	1.6	1011.0	631.875000	124.097222
28	liferay.com	2.9	19.0	6.551724	3.6	46.0	12.777778	6.226054
60	pcmag.com	9.6	37.0	3.854167	3.3	33.0	10.000000	6.145833
457	tallyfy.com	6.1	5.0	0.819672	1.2	8.0	6.666667	5.846995
71	toolshero.com	2.1	7.0	3.333333	1.0	9.0	9.000000	5.666667
784	netsuite.com	12.4	6.0	0.483871	1.0	6.0	6.000000	5.516129
3	gotomeeting.com	6.4	153.0	23.906250	5.1	148.0	29.019608	5.113358
18	medium.com	10.9	92.0	8.440367	9.4	118.0	12.553191	4.112825
26	qualtrics.com	7.0	47.0	6.714286	6.5	70.0	10.769231	4.054945
66	info.workcast.com	9.6	34.0	3.541667	10.1	74.0	7.326733	3.785066

The biggest winners are publishers which include nonindustry players like PCMag and Medium.

Here's some code to convert to long format for visualization:

```
compare_visi_losers_long = compare_visi_losers[['site', 'before_
visi','after_visi']].head(12)
compare_visi_losers_long = compare_visi_losers_long.melt(id_vars =
['site'], var_name='Phase', value_name='Visi')
compare_visi_losers_long['Phase'] = compare_visi_losers_long['Phase'].str.
replace('_visi', '')

compare_visi_losers_long['Phase'] = compare_visi_losers_long['Phase'].
astype('category')
compare_visi_losers_long['Phase'] = compare_visi_losers_long['Phase'].cat.
reorder_categories(['before', 'after'])

stop_doms = ['en.wikipedia.org', 'google.com', 'youtube.com']
compare_visi_losers_long = compare_visi_losers_long[~compare_visi_losers_
long['site'].isin(stop_doms)]

compare_visi_losers_long.head(10)
```

This results in the following:

	site	Phase	Visi
0	workcast.com	before	6.901408
1	on24.com	before	18.571429
2	entrepreneur.com	before	10.714286
3	podcastinsights.com	before	9.534884
4	eventbrite.com	before	10.000000
5	trainingcheck.com	before	4.000000
6	en.rockcontent.com	before	7.547170
7	mysql.com	before	3.636364
8	heinemann.com	before	4.000000
9	wordstream.com	before	13.103448

The preceding data is in long format. This will now feed the following graphics code:

```
compare_visi_losers_plt = (
    ggplot(compare_visi_losers_long, aes(x = 'reorder(site, Visi)',
    y = 'Visi', fill = 'Phase')) +
    geom_bar(stat = 'identity', position = 'dodge', alpha = 0.8) +
    position=position_stack(vjust=0.01)) +
    labs(y = 'Visiblity', x = ' ') +
    coord_flip() +
    theme(legend_position = 'right', axis_text_x=element_text(rotation=0,
    hjust=1, size = 12)) +
    facet_wrap('Phase')
)

compare_visi_losers_plt.save(filename = 'images/1_compare_visi_losers_plt.
png', height=5, width=10, units = 'in', dpi=1000)
compare_visi_losers_plt
```

The separate panels are achieved by using the facet_wrap() function where we instruct plotnine (the graphics package) to separate panels by Phase as a parameter (Figure 10-4).

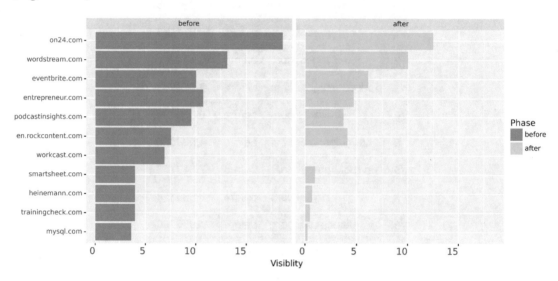

Figure 10-4. *Website Google visibility scores before and after*

Let's see the winners:

```
compare_visi_winners_long = compare_visi_winners[['site', 'before_
visi','after_visi']].head(12)
compare_visi_winners_long = compare_visi_winners_long.melt(id_vars =
['site'], var_name='Phase', value_name='Visi')
compare_visi_winners_long['Phase'] = compare_visi_winners_long['Phase'].
str.replace('_visi', '')

compare_visi_winners_long['Phase'] = compare_visi_winners_long['Phase'].
astype('category')
compare_visi_winners_long['Phase'] = compare_visi_winners_long['Phase'].
cat.reorder_categories(['before', 'after'])

stop_doms = ['en.wikipedia.org', 'google.com', 'youtube.com',
'lexisnexis.com']
compare_visi_winners_long = compare_visi_winners_long[~compare_visi_
winners_long['site'].isin(stop_doms)]

compare_visi_winners_long.head(10)
```

This results in the following:

	site	Phase	Visi
1	liferay.com	before	6.551724
2	pcmag.com	before	3.854167
3	tallyfy.com	before	0.819672
4	toolshero.com	before	3.333333
5	netsuite.com	before	0.483871
6	gotomeeting.com	before	23.906250
7	medium.com	before	8.440367
8	qualtrics.com	before	6.714286
9	info.workcast.com	before	3.541667
10	marketinginsidergroup.com	before	3.939394

```
compare_visi_winners_plt = (
    ggplot(compare_visi_winners_long, aes(x = 'reorder(site, Visi)', y =
    'Visi', fill = 'Phase')) +
    geom_bar(stat = 'identity', position = 'dodge', alpha = 0.8) +
 position=position_stack(vjust=0.01)) +
    labs(y = 'Rank', x = ' ') +
    coord_flip() +
    theme(legend_position = 'right', axis_text_x=element_text(rotation=0,
    hjust=1, size = 12)) +
    facet_wrap('Phase')
)

compare_visi_winners_plt.save(filename = 'images/1_compare_visi_winners_
plt.png', height=5, width=10, units = 'in', dpi=1000)
compare_visi_winners_plt
```

This time, we're not using the facet_wrap() function which puts both before and after bars on the same panel (Figure 10-5).

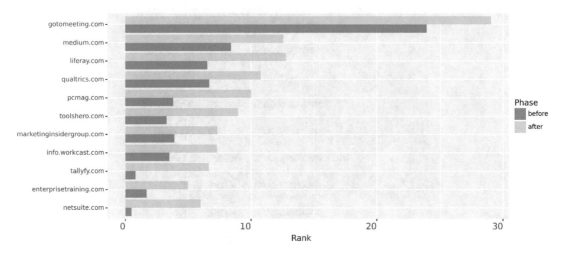

Figure 10-5. *Website Google rank average before and after*

This makes it easier to compare directly and even get a better sense of the difference for each site before and after.

Result Types

With the overall performance in hand, we'll drill down further, starting with result types. By result types, we mean the format in which the ranking is displayed. This could be

- Regular organic (think the usual ten blue links)

- Video

- Image

- News

- People Also Ask

As usual, we'll perform aggregations on both before and after datasets. Only this time, we'll group by the snippets column:

```
before_unq_snippets = getstat_bef_unique
```

We're aggregating by counting the number of keyword search results the snippet appears in, which is a form of reach. Most snippets rank in the top 5 positions of the Search Engine Results Pages, so we won't bother with snippet rankings.

```
before_unq_snippets = before_unq_snippets.groupby(['snippets']).
agg({'count': 'sum'}).reset_index()
before_unq_snippets = before_unq_snippets[['snippets', 'count']]
before_unq_snippets = before_unq_snippets.rename(columns = {'count':
'reach'})
before_unq_snippets.sort_values('reach', ascending = False).head(10)
```

This results in the following:

	snippets	reach
30	organic	14221
23	image / organic	5612
8	amp / organic	1062
32	people also ask	760
27	interesting finds	478
19	carousel / videos	278
21	faq / organic	276
13	answers / paragraph	197
11	answers / list	158
6	amp / interesting finds	146

Organic predictably has the most reach followed by images and AMP (accelerated mobile pages).

Repeat the process for the after dataset:

```
after_unq_snippets = getstat_aft_unique
after_unq_snippets = after_unq_snippets.groupby(['snippets']).agg({'count':
'sum'}).reset_index()
after_unq_snippets = after_unq_snippets[['snippets', 'count']]
after_unq_snippets = after_unq_snippets.rename(columns = {'count': 'reach'})
after_unq_snippets.sort_values('reach', ascending = False).head(10)
```

This results in the following:

	snippets	reach
31	organic	14606
24	image / organic	6129
6	amp / organic	1122
33	people also ask	862
28	interesting finds	564
22	faq / organic	352
20	carousel / videos	303
11	answers / paragraph	206
9	answers / list	178
4	amp / interesting finds	150

Organic has gone down implying that there could be more diversification of search results. Join the datasets to facilitate an easier comparison:

```
compare_snippets = before_unq_snippets.merge(after_unq_snippets, on =
['snippets'], how = 'outer')
compare_snippets = compare_snippets_losers.rename(columns = {'reach_x':
'before_reach', 'reach_y': 'after_reach'})
compare_snippets['before_reach'] = np.where(compare_snippets['before_
reach'].isnull(), 0, compare_snippets['before_reach'])
compare_snippets['after_reach'] = np.where(compare_snippets['after_reach'].
isnull(), 0, compare_snippets['after_reach'])
compare_snippets['delta_reach'] = compare_snippets['after_reach'] -
compare_snippets['before_reach']

compare_snippets_losers = compare_snippets.sort_values('delta_reach')
compare_snippets_losers.head(10)
```

This results in the following:

	snippets	before_reach	after_reach	delta_reach
31	organic / sitelinks	33.0	18.0	-15.0
33	placesv3 / ratings	30.0	17.0	-13.0
38	videos	44.0	38.0	-6.0
35	related searches	98.0	93.0	-5.0
1	accordion / answers / knowledge graph / paragraph	6.0	2.0	-4.0
4	amp / carousel / news	2.0	0.0	-2.0
12	answers / list / related searches	4.0	2.0	-2.0
15	carousel / knowledge graph / videos	26.0	24.0	-2.0
3	accordion / answers / paragraph	1.0	0.0	-1.0
20	events	22.0	21.0	-1.0

The table confirms that organic sitelinks' listings have fallen, followed by places, videos, and related searches. What does this mean? It means that Google is diversifying its results but not in the way of videos or local business results. Also, the fall in sitelinks implies the searches are less navigational, which possibly means more opportunity to rank for search phrases that were previously the preserve of certain brands.

```
compare_snippets_winners = compare_snippets.sort_values('delta_reach',
ascending = False)
compare_snippets_winners.head(10)
```

This results in the following:

	snippets	before_reach	after_reach	delta_reach
23	image / organic	5612.0	6129.0	517.0
30	organic	14259.0	14636.0	377.0
32	people also ask	762.0	865.0	103.0
27	interesting finds	478.0	564.0	86.0
21	faq / organic	276.0	353.0	77.0
8	amp / organic	1062.0	1122.0	60.0
19	carousel / videos	278.0	303.0	25.0
11	answers / list	158.0	178.0	20.0
37	unknown	2.0	19.0	17.0
24	image / ratings / organic	41.0	56.0	15.0

Comparing the winners, we see that images and pure organic have increased as has People Also Ask. So the high-level takeaway here is that the content should be more FAQ driven and tagged with schema markup. There should also be more use of images in the content. Let's visualize by reformatting the data and feeding it into plotnine:

```
each']].head(10)
compare_snippets_losers_long = compare_snippets_losers_long.melt(id_vars =
['snippets'], var_name='Phase', value_name='Reach')
compare_snippets_losers_long['Phase'] = compare_snippets_losers_
long['Phase'].str.replace('_reach', '')

compare_snippets_losers_long['Phase'] = compare_snippets_losers_
long['Phase'].astype('category')
compare_snippets_losers_long['Phase'] = compare_snippets_losers_
long['Phase'].cat.reorder_categories(['after', 'before'])
compare_snippets_losers_long = compare_snippets_losers_long[compare_
snippets_losers_long['snippets'] != 'organic']

compare_snippets_losers_long.head(10)
```

This results in the following:

	snippets	Phase	Reach
0	organic / sitelinks	before	33.0
1	placesv3 / ratings	before	30.0
2	videos	before	44.0
3	related searches	before	98.0
4	accordion / answers / knowledge graph / paragraph	before	6.0
5	amp / carousel / news	before	2.0
6	answers / list / related searches	before	4.0
7	carousel / knowledge graph / videos	before	26.0
8	accordion / answers / paragraph	before	1.0
9	events	before	22.0

```
compare_snippets_losers_plt = (
    ggplot(compare_snippets_losers_long, aes(x = 'reorder(snippets,
    Reach)', y = 'Reach', fill = 'Phase')) +
    geom_bar(stat = 'identity', position = 'dodge', alpha = 0.8) +
    #geom_text(dd_factor_df, aes(label = 'market_name'), position=position_
    stack(vjust=0.01)) +
    labs(y = 'Visiblity', x = ' ') +
    #scale_y_reverse() +
    coord_flip() +
    theme(legend_position = 'right', axis_text_x=element_text(rotation=0,
    hjust=1, size = 12))
)

compare_snippets_losers_plt.save(filename = 'images/2_compare_snippets_
losers_plt.png', height=5, width=10, units = 'in', dpi=1000)
compare_snippets_losers_plt
```

The great thing about charts like Figure 10-6 is that you get an instant sense of proportion.

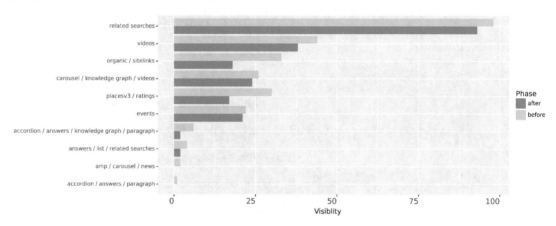

Figure 10-6. *Google visibility by result type before and after*

It's much easier to spot that there are more carousel videos than organic sitelinks post update.

```
compare_snippets_winners_long = compare_snippets_winners[['snippets',
'before_reach','after_reach']].head(10)
compare_snippets_winners_long = compare_snippets_winners_long.melt(id_vars
= ['snippets'], var_name='Phase', value_name='Reach')
compare_snippets_winners_long['Phase'] = compare_snippets_winners_
long['Phase'].str.replace('_reach', '')

compare_snippets_winners_long['Phase'] = compare_snippets_winners_
long['Phase'].astype('category')
compare_snippets_winners_long['Phase'] = compare_snippets_winners_
long['Phase'].cat.reorder_categories(['after', 'before'])
compare_snippets_winners_long = compare_snippets_winners_long[compare_
snippets_winners_long['snippets'] != 'organic']

compare_snippets_winners_long.head(10)
```

This results in the following:

	snippets	Phase	Reach
0	image / organic	before	5612.0
2	people also ask	before	762.0
3	interesting finds	before	478.0
4	faq / organic	before	276.0
5	amp / organic	before	1062.0
6	carousel / videos	before	278.0
7	answers / list	before	158.0
8	unknown	before	2.0
9	image / ratings / organic	before	41.0
10	image / organic	after	6129.0

```
compare_snippets_winners_plt = (
    ggplot(compare_snippets_winners_long, aes(x = 'reorder(snippets,
    Reach)', y = 'Reach', fill = 'Phase')) +
    geom_bar(stat = 'identity', position = 'dodge', alpha = 0.8)
    +position=position_stack(vjust=0.01)) +
    labs(y = 'Rank', x = ' ') +
    coord_flip() +
    theme(legend_position = 'right', axis_text_x=element_text(rotation=0,
    hjust=1, size = 12))
)

compare_snippets_winners_plt.save(filename = 'images/1_compare_snippets_
winners_plt.png', height=5, width=10, units = 'in', dpi=1000)
compare_snippets_winners_plt
```

Other than more of each snippet or result type, the increases across all types look relatively the same (Figure 10-7).

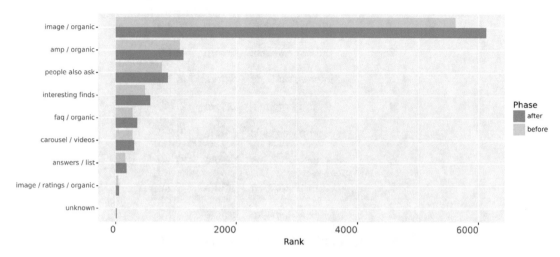

Figure 10-7. *Google's top 10 count reach by result type before and after*

Cannibalization

With performance determined, our attention turns to the potential drivers of performance, such as cannibals.

Cannibals occur when there are instances of sites with multiple URLs ranking in the search results for a single keyword.

We'll start by using the duplicated SERPs datasets and counting the number of URLs from the same site per keyword. This will involve a groupby() function on the keyword and site:

```
cannibals_before_agg = getstat_before.groupby(['keyword', 'site']).
agg({'count': 'sum'}).reset_index()
```

At this stage, we want to isolate the SERPs rows that are cannibalized. That means URLs that have other URLs from the same site appearing in the same keyword results.

```
cannibals_before_agg = cannibals_before_agg[cannibals_before_
agg['count'] > 1]
```

We reset the count to 1 so we can perform further sum aggregations:

```
cannibals_before_agg['count'] = 1
```

Next, we aggregate by keyword to count the number of cannibalized URLs in the SERPs data:

```
cannibals_before_agg = getstat_before[getstat_before['device'] ==
'desktop']
cannibals_before_agg = cannibals_before_agg.groupby(['keyword']).
agg({'count': 'sum'}).reset_index()
cannibals_before_agg = cannibals_before_agg.rename(columns = {'count':
'cannibals'})
cannibals_before_agg
```

This results in the following:

	keyword	cannibals
0	6sense webinars	1
1	abm	3
2	abm benchmarks	3
3	abm best practices	1
4	abm case studies	2
...	...	...
435	why is digital customer experience important	1
436	why user experience matters in digital marketing	1
437	zapier webinar integration	1
438	zapier webinars	1
439	zoom webinars	1

440 rows × 2 columns

You could argue that these numbers contain one URL per site that are not strictly cannibals. However, this looser calculation is simple and does a robust enough job to get a sense of the cannibalization trend.

Let's see how cannibalized the SERPs were following the update:

```
cannibals_after_agg = getstat_after.groupby(['keyword', 'site']).
agg({'count': 'sum'}).reset_index()
cannibals_after_agg = cannibals_after_agg[cannibals_after_agg['count'] > 1]
cannibals_after_agg['count'] = 1
cannibals_after_agg = cannibals_after_agg.groupby(['keyword']).
agg({'count': 'sum'}).reset_index()
cannibals_after_agg = cannibals_after_agg.rename(columns = {'count':
'cannibals'})
cannibals_after_agg
```

This results in the following:

	keyword	cannibals
0	6sense webinars	1
1	abm	3
2	abm benchmarks	3
3	abm best practices	1
4	abm case studies	2
...	...	...
432	webinars for dummies	1
433	what is a webinar	1
434	zapier webinar integration	1
435	zapier webinars	1
436	zoom webinars	1

437 rows × 2 columns

The preceding preview hints that not much has changed; however, this is hard to tell by looking at one table. So let's merge them together and get a side-by-side comparison:

```
compare_cannibals = cannibals_before_agg.merge(cannibals_after_agg, on =
'keyword', how = 'left')
```

```
compare_cannibals = compare_cannibals.rename(columns = {'cannibals_x':
'before_cannibals', 'cannibals_y': 'after_cannibals',
                })

compare_cannibals['before_cannibals'] = np.where(compare_cannibals['before_
cannibals'].isnull(),
        0, compare_cannibals['before_cannibals'])
compare_cannibals['after_cannibals'] = np.where(compare_cannibals['after_
cannibals'].isnull(),
        0, compare_cannibals['after_cannibals'])

compare_cannibals['delta_cannibals'] = compare_cannibals['after_
cannibals'] - compare_cannibals['before_cannibals']

compare_cannibals = compare_cannibals.sort_values('delta_cannibals')
compare_cannibals
```

This results in the following:

	keyword	before_cannibals	after_cannibals	delta_cannibals
255	live streaming software	4	1.0	-3.0
336	recorded webinars	3	0.0	-3.0
396	web conferencing attendance	2	0.0	-2.0
411	webcast guidelines	2	0.0	-2.0
353	salesforce webinar integration	3	1.0	-2.0
...	...	...	...	...
214	hubspot webinar integration	3	5.0	2.0
386	training tools	2	4.0	2.0
184	enterprise training platform	2	4.0	2.0
37	always on experiences software	1	3.0	2.0
78	cmo q and a	1	4.0	3.0

440 rows × 4 columns

The table shows at the keyword level that there are less cannibals for "webcast guidelines" but more for "enterprise training platform." But what was the overall trend?

```
cannibals_trend = compare_cannibals
cannibals_trend['project'] = target_name
cannibals_trend = cannibals_trend.groupby('project').agg({'before_
cannibals': 'sum',
                     'after_cannibals': 'sum',
                     'delta_cannibals': 'sum'}).reset_index()
cannibals_trend
```

This results in the following:

	project	before_cannibals	after_cannibals	delta_cannibals
0	ON24	664	575.0	-89.0

So there were less cannibals overall by just over 13%, following the core update, as we would expect.

Let's convert to format before graphing the top cannibals for both SERPs that gained and lost cannibals:

```
compare_cannibals_less = compare_cannibals[['keyword', 'before_cannibals',
'after_cannibals']].head(10)
compare_cannibals_less = compare_cannibals_less.melt(id_vars = ['keyword'],
             var_name = 'Phase', value_name = 'cannibals')

compare_cannibals_less['Phase'] = compare_cannibals_less['Phase'].str.
replace('_cannibals', '')

compare_cannibals_less['Phase'] = compare_cannibals_less['Phase'].
astype('category')
compare_cannibals_less['Phase'] = compare_cannibals_less['Phase'].cat.
reorder_categories(['after', 'before'])

compare_cannibals_less
```

This results in the following:

	keyword	Phase	cannibals
0	live streaming software	before	4.0
1	recorded webinars	before	3.0
2	web conferencing attendance	before	2.0
3	webcast guidelines	before	2.0
4	salesforce webinar integration	before	3.0
5	salesforce webinars	before	3.0
6	certification platform	before	2.0
7	brand leadership guide	before	4.0
8	act on webinars	before	4.0
9	enterprise certification benchmarks	before	3.0
10	live streaming software	after	1.0
11	recorded webinars	after	0.0

```
compare_cannibals_less_plt = (
    ggplot(compare_cannibals_less, aes(x = 'keyword', y = 'cannibals',
                fill = 'Phase')) +
    geom_bar(stat = 'identity', position = 'dodge', alpha = 0.8) +
    #geom_text(dd_factor_df, aes(label = 'market_name'), position=position_
    stack(vjust=0.01)) +
    labs(y = '# Cannibals in SERP', x = ' ') +
    #scale_y_reverse() +
    coord_flip() +
    theme(legend_position = 'right', axis_text_x=element_text(rotation=0,
    hjust=1, size = 12))
)

compare_cannibals_less_plt.save(filename = 'images/5_compare_cannibals_
less_plt.png', height=5, width=10, units = 'in', dpi=1000)
compare_cannibals_less_plt
```

Figure 10-8 shows keywords that lost their cannibalizing URLs or had less cannibals.

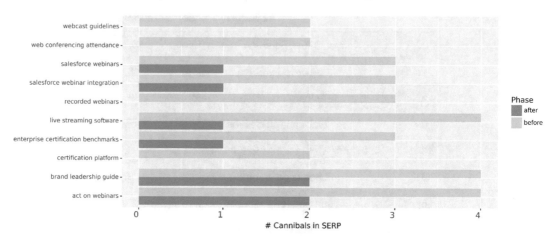

Figure 10-8. *Cannibalized SERP result instance counts by keyword*

The most dramatic loss appears to be "live streaming software" going from 4 to 1. All of the phrases appear to be quite generic apart from the term "act on webinars" which appears to be a brand term for act-on.com.

```
compare_cannibals_more = compare_cannibals[['keyword', 'before_cannibals',
'after_cannibals']].tail(10)
compare_cannibals_more = compare_cannibals_more.melt(id_vars = ['keyword'],
            var_name = 'Phase', value_name = 'cannibals')
```

```
compare_cannibals_more['Phase'] = compare_cannibals_more['Phase'].str.
replace('_cannibals', '')
```

```
compare_cannibals_more['Phase'] = compare_cannibals_more['Phase'].
astype('category')
compare_cannibals_more['Phase'] = compare_cannibals_more['Phase'].cat.
reorder_categories(['after', 'before'])
```

```
compare_cannibals_more
```

This results in the following:

	keyword	Phase	cannibals
0	sugarcrm webinars	before	1.0
1	continuing education platform	before	1.0
2	web conferencing registrations	before	1.0
3	brand leadership case studies	before	1.0
4	certification strategy	before	2.0
5	hubspot webinar integration	before	3.0
6	training tools	before	2.0
7	enterprise training platform	before	2.0
8	always on experiences software	before	1.0
9	cmo q and a	before	1.0
10	sugarcrm webinars	after	3.0
11	continuing education platform	after	3.0
12	web conferencing registrations	after	3.0

```
compare_cannibals_more_plt = (
    ggplot(compare_cannibals_more, aes(x = 'keyword', y = 'cannibals',
                fill = 'Phase')) +
    geom_bar(stat = 'identity', position = 'dodge', alpha = 0.8) +
    labs(y = '# Cannibals in SERP', x = ' ') +
    coord_flip() +
    theme(legend_position = 'right', axis_text_x=element_text(rotation=0,
    hjust=1, size = 12))
)

compare_cannibals_more_plt.save(filename = 'images/5_compare_cannibals_
more_plt.png', height=5, width=10, units = 'in', dpi=1000)
compare_cannibals_more_plt
```

Nothing obvious appears as to why these keywords received more cannibals when compared with the keywords that lost their cannibals, as both had a mixture of generic and brand hybrid keywords (Figure 10-9).

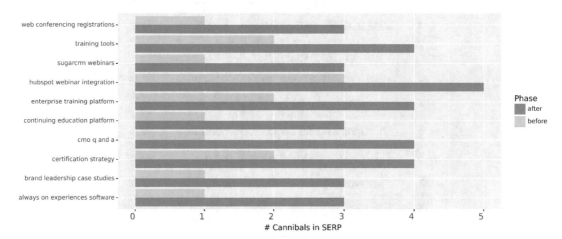

Figure 10-9. *Cannibalized SERP result instance counts by keyword, before and after*

Keywords

Let's establish a general trend before moving the analysis toward the site in question.

Token Length

Perhaps there are keyword patterns such as token length that could explain the gains and losses following the core update. We'll try token length which measures the number of keywords in a search query.

Metrics such as search volume before and after are not available in the getSTAT before and after the update. We're interested to see how many unique sites were present for each token length for a general trend view.

We'll analyze the SERPs for desktop devices; however, the code can easily be adapted for other devices such as mobiles:

```
tokensite_before = getstat_bef_unique[getstat_bef_unique['device'] ==
'desktop']
tokensite_after = getstat_aft_unique[getstat_aft_unique['device'] ==
'desktop']
tokensite_after.sort_values(['keyword', 'rank'])
```

	keyword	market	phase	device	search_volume	rank	url	site	snip
117	6sense webinars	US-en	after	desktop	1	1	https://hub.6sense.com/upcoming-events	6sense.com	org
160	abm	US-en	after	desktop	110000	1	https://www.abm.com/	abm.com	orga site
8818	abm	US-en	after	desktop	110000	2	http://www.google.com/	google.com	pe alsc
13591	abm	US-en	after	desktop	110000	3	https://locations.abm.com/ok/tulsa/facility-services-tulsa-ok-1334.html	locations.abm.com	plac / ra
6685	abm	US-en	after	desktop	110000	6	https://en.wikipedia.org/wiki/Account-based_marketing	en.wikipedia.org	org
...	...	...	...	...	...	...	...	...	...
8369	zoom webinars	US-en	after	desktop	135000	16	https://www.gend.co/zoom-webinars	gend.co	org
2346	zoom webinars	US-en	after	desktop	135000	17	https://www.brandeis.edu/its/services/communication/zoom/zoom_webinars.html	brandeedu	org
8126	zoom webinars	US-en	after	desktop	135000	18	https://www.g2.com/products/zoom-video-webinar/reviews	g2.com	org
23714	zoom webinars	US-en	after	desktop	135000	19	https://webinarninja.com/zoom-webinars/	webinarninja.com	org
12253	zoom webinars	US-en	after	desktop	135000	20	https://it.cornell.edu/zoom/whats-zoom-meeting-whats-zoom-webinar	it.cornell.edu	org

12909 rows × 13 columns

The first step is to aggregate both datasets by token size and phase for both before and after. We only want the top 10 sites; hence, the filter rank is less than 11. We start by aggregating at the keyword level within the token size and phase to sum the number of sites. Then aggregate again by token size and phase to get the overall number of sites ranking in the top 10 for the token size.

The two-step aggregation was made necessary because of the filtering for the top 10 sites within the keyword; otherwise, we would have aggregated within the phase and token size in one line.

```
tokensite_before_reach = tokensite_before[tokensite_before['rank'] <
11].groupby(['token_size', 'keyword', 'phase']).agg({'count': 'sum'}).
reset_index()
tokensite_before_reach = tokensite_before_reach.groupby(['token_size',
'phase']).agg({'count': 'sum'}).reset_index()
```

```
tokensite_before_agg = tokensite_before_reach.rename(columns = {'count':
'site_count'})
```

```
tokensite_before_agg
```

This results in the following:

	token_size	phase	site_count
0	head	before	35
1	long	before	4552
2	middle	before	1460

The two-step aggregation approach is repeated for the after dataset:

```
targetsite_after_token = targetsite_after.groupby(['token_size', 'phase']).
agg({'count': 'sum'}).reset_index()
targetsite_after_sv = targetsite_after.groupby(['token_size', 'phase']).
agg({'search_volume': 'sum'}).reset_index()
targetsite_after_agg = targetsite_after_token.merge(targetsite_after_sv, on
= ['token_size', 'phase'], how = 'left')
targetsite_after_agg = targetsite_after_agg.rename(columns = {'count':
'reach'})
targetsite_after_agg
```

This results in the following:

	token_size	phase	site_count
0	head	after	35
1	long	after	4848
2	middle	after	1521

With both phases aggregated by site count, we'll merge these for a side-by-side comparison:

```
tokensite_token_deltas = tokensite_before_agg.merge(tokensite_after_agg, on
= ['token_size'], how = 'left')
```

```
tokensite_token_deltas['sites_delta'] = (tokensite_token_deltas['site_
count_y'] - tokensite_token_deltas['site_count_x'])
```

Cast the token size as a category data type so that we can order these for the table and the graphs later:

```
tokensite_token_deltas['token_size'] = tokensite_token_deltas['token_
size'].astype('category')
tokensite_token_deltas['token_size'] = tokensite_token_deltas['token_
size'].cat.reorder_categories(['head', 'middle', 'long'])
```

```
tokensite_token_deltas = tokensite_token_deltas.sort_values('token_size')
```

```
tokensite_token_deltas
```

This results in the following:

	token_size	phase_x	site_count_x	phase_y	site_count_y	sites_delta
0	head	before	35	after	35	0
2	middle	before	1460	after	1521	61
1	long	before	4552	after	4848	296

So the table is sorted by token_size rather than in alphabetical order thanks to converting the data type from a string to category. Most of the changes have been in the long tail and middle body in that there are more sites in the top 10 than before, whereas the head terms didn't change much by volume. This may be a push by Google to diversify the search results and cut down on site dominance and cannibals.

Let's visualize:

```
targetsite_token_viz = pd.concat([targetsite_before_agg, targetsite_
after_agg])
```

```
targetsite_token_viz
```

This results in the following:

	token_size	phase	reach	search_volume
0	head	before	187	107595000
1	long	before	17624	8081056
2	middle	before	5866	50344090
0	head	after	182	105454500
1	long	after	18718	8311735
2	middle	after	6072	51223382

```
targetsite_token_sites_plt = (
    ggplot(tokensite_token_viz,
           aes(x = 'token_size', y = 'site_count', fill = 'phase')) +
    geom_bar(stat = 'identity', position = 'dodge', alpha = 0.8) +
    position=position_stack(vjust=0.01)) +
    labs(y = 'Unique Site Count', x = 'Query Length') +
    coord_flip() +
    theme(legend_position = 'right',
          axis_text_y =element_text(rotation=0, hjust=1, size = 12),
          legend_title = element_blank()
          )
)
```

```
targetsite_token_sites_plt.save(filename = 'images/8_targetsite_token_
sites_plt.png', height=5, width=10, units = 'in', dpi=1000)
targetsite_token_sites_plt
```

So that's the general trend graphed for our PowerPoint deck (Figure 10-10). The question is which sites gained and lost?

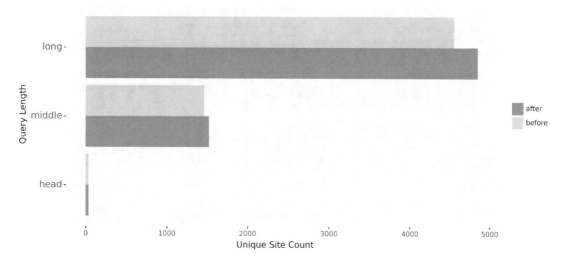

Figure 10-10. *Count of unique sites by query length*

Token Length Deep Dive

With the sense of the general trend in hand, we'll get into the details to see how sites were affected within the long tail.

As before, we'll focus on desktop results, in addition to filtering for the long tail:

```
longs_before = tokensite_before[tokensite_before['token_size'] == 'long']
longs_after = tokensite_after[tokensite_after['token_size'] == 'long']
longs_after
```

This results in the following:

	keyword	market	phase	device	search_volume	rank	url	site	snippets	rank_profile
21968	abm best practices	US-en	after	desktop	90	1	https://blog.topohq.com/account-based-marketing-11-tactics-to-drive-your-abm-process/	topohq.com	answers / list	top_3
516	abm best practices	US-en	after	desktop	90	2	https://adage.com/article/btob/practices-account-based-marketing-b-b/300361	adage.com	answers / list	top_3
19685	abm best practices	US-en	after	desktop	90	3	https://skaled.com/abm-best-practices/	skaled.com	organic	page_1
12895	abm best practices	US-en	after	desktop	90	4	https://www.leadspace.com/enterprise-level-account-based-marketing/	leadspace.com	organic	page_1
10843	abm best practices	US-en	after	desktop	90	5	https://blog.hubspot.com/marketing/account-based-marketing-guide	hubspot.com	organic	page_1
...	...	...	...	...	...	...	...	...	...	...
5675	why user experience matters in digital marketing	US-en	after	desktop	10	20	https://digitalmarketinginstitute.com/blog/why-user-experience-is-key-to-digital-marketing-success	digitalmarketinginstitute.com	organic	page_2
24638	zapier webinar integration	US-en	after	desktop	10	1	https://zapier.com/apps/gotowebinar/integrations	zapier.com	organic	top_3
13450	zapier webinar integration	US-en	after	desktop	10	18	https://livestorm.co/webinar-integrations/zapier/	livestorm.co	organic	page_2
16283	zapier webinar integration	US-en	after	desktop	10	19	https://www.on24.com/zapier/	on24.com	organic	page_2
13540	zapier webinar integration	US-en	after	desktop	10	20	https://www.livewebinar.com/integrations/zapier-integration	livewebinar.com	organic	page_2

9704 rows × 13 columns

The data is now filtered for the desktop and long tail, making it ready for analysis using aggregation:

```
longs_before_reach = longs_before.groupby('site').agg({'count': 'sum'}).
reset_index()
longs_before_rank = longs_before.groupby('site').apply(wavg_rank_sv).
reset_index()
longs_before_agg = longs_before_reach.merge(longs_before_rank, on = 'site',
how = 'left')
longs_before_agg['visi'] = longs_before_agg['count'] / longs_before_
agg['wavg_rank']
longs_before_agg['phase'] = 'before'
longs_before_agg = longs_before_agg.sort_values('count', ascending = False)
longs_before_agg.head()
```

This results in the following:

	site	count	wavg_rank	visi	phase
1239	google.com	298	3.1	96.129032	before
2319	qualtrics.com	42	1.2	35.000000	before
1365	hubspot.com	179	5.7	31.403509	before
3257	youtube.com	100	7.2	13.888889	before
1859	medium.com	84	7.3	11.506849	before

So far, we see that Qualtrics ranked around 1.2 (on average) on 112 long-tail keywords on desktop searches. We'll repeat the aggregation for the after data:

```
longs_after_reach = longs_after.groupby('site').agg({'count': 'sum'}).
reset_index()
longs_after_rank = longs_after.groupby('site').apply(wavg_rank_sv).
reset_index()
longs_after_agg = longs_after_reach.merge(longs_after_rank, on = 'site',
how = 'left')
longs_after_agg['visi'] = longs_after_agg['count'] / longs_after_
agg['wavg_rank']
longs_after_agg['phase'] = 'after'
longs_after_agg = longs_after_agg.sort_values('visi', ascending = False)
longs_after_agg.head()
```

This results in the following:

	site	count	wavg_rank	visi	phase
1271	google.com	349	2.5	139.600000	after
1408	hubspot.com	180	4.5	40.000000	after
2600	sitecore.com	33	1.9	17.368421	after
2344	qualtrics.com	58	4.1	14.146341	after
3336	youtube.com	91	7.3	12.465753	after

Following the core update, HubSpot and Sitecore have moved ahead of Qualtrics within the top 5 in the long tail. Medium has moved out of the top 5. Let's make this comparison easier:

```
compare_longs = longs_before_agg.merge(longs_after_agg, on = ['site'],
how = 'outer')
compare_longs = compare_longs.rename(columns = {'count_x': 'before_reach',
'count_y': 'after_reach',
        'wavg_rank_x': 'before_rank', 'wavg_rank_y': 'after_rank',
        'visi_x': 'before_visi', 'visi_y': 'after_visi',
       })

compare_longs['before_reach'] = np.where(compare_longs['before_reach'].
isnull(),
      0, compare_longs['before_reach'])
compare_longs['after_reach'] = np.where(compare_longs['after_reach'].
isnull(),
      0, compare_longs['after_reach'])

compare_longs['before_rank'] = np.where(compare_longs['before_rank'].
isnull(),
      100, compare_longs['before_rank'])
compare_longs['after_rank'] = np.where(compare_longs['after_rank'].
isnull(),
      100, compare_longs['after_rank'])

compare_longs['before_visi'] = np.where(compare_longs['before_visi'].
isnull(),
      0, compare_longs['before_visi'])
compare_longs['after_visi'] = np.where(compare_longs['after_visi'].
isnull(),
      0, compare_longs['after_visi'])

compare_longs['delta_reach'] = compare_longs['after_reach'] - compare_
longs['before_reach']
compare_longs['delta_rank'] = compare_longs['before_rank'] - compare_
longs['after_rank']
```

```
compare_longs['delta_visi'] = compare_longs['after_visi'] - compare_
longs['before_visi']
```

```
compare_longs.sort_values('delta_visi').head(12)
```

This results in the following:

	site	before_reach	before_rank	before_visi	phase_x	after_reach	after_rank	after_visi	phase_y	delta_reach	delta_rank	delta_visi
1	qualtrics.com	42.0	1.2	35.000000	before	58.0	4.1	14.146341	after	16.0	-2.9	-20.853659
15	podcastinsights.com	19.0	2.9	6.551724	before	17.0	4.5	3.777778	after	-2.0	-1.6	-2.773946
59	workcast.com	22.0	8.5	2.588235	before	1.0	20.0	0.050000	after	-21.0	-11.5	-2.538235
7	gotomeeting.com	76.0	8.0	9.500000	before	72.0	10.0	7.200000	after	-4.0	-2.0	-2.300000
32	profitwell.com	6.0	1.4	4.285714	before	3.0	1.4	2.142857	after	-3.0	0.0	-2.142857
33	fsco.gov.on.ca	16.0	3.9	4.102564	before	12.0	5.6	2.142857	after	-4.0	-1.7	-1.959707
57	mightynetworks.com	6.0	2.3	2.608696	before	5.0	6.1	0.819672	after	-1.0	-3.8	-1.789024
54	glisser.com	5.0	1.9	2.631579	before	7.0	7.6	0.921053	after	2.0	-5.7	-1.710526
86	zest.is	2.0	1.0	2.000000	before	2.0	5.5	0.363636	after	0.0	-4.5	-1.636364
13	ventureharbour.com	35.0	4.8	7.291667	before	35.0	6.1	5.737705	after	0.0	-1.3	-1.553962
19	trainingmag.com	8.0	1.5	5.333333	before	10.0	2.6	3.846154	after	2.0	-1.1	-1.487179
3	youtube.com	100.0	7.2	13.888889	before	91.0	7.3	12.465753	after	-9.0	-0.1	-1.423135

As confirmed, Qualtrics lost the most in the long tail. Let's visualize, starting with the losers:

```
longs_reach_losers_long = compare_longs.sort_values('delta_visi')
longs_reach_losers_long = longs_reach_losers_long[['site', 'before_visi',
'after_visi']]
longs_reach_losers_long = longs_reach_losers_long[~longs_reach_losers_
long['site'].isin(['google.co.uk', 'youtube.com'])]
longs_reach_losers_long = longs_reach_losers_long.head(10)
```

```
longs_reach_losers_long = longs_reach_losers_long.melt(id_vars = 'site',
var_name = 'phase', value_name = 'visi')
longs_reach_losers_long['phase'] = longs_reach_losers_long['phase'].str.
replace('_visi', '')
```

```
longs_reach_losers_long
```

This results in the following:

	site	phase	visi
0	qualtrics.com	before	35.000000
1	podcastinsights.com	before	6.551724
2	workcast.com	before	2.588235
3	gotomeeting.com	before	9.500000
4	profitwell.com	before	4.285714
5	fsco.gov.on.ca	before	4.102564
6	mightynetworks.com	before	2.608696
7	glisser.com	before	2.631579
8	zest.is	before	2.000000
9	ventureharbour.com	before	7.291667
10	qualtrics.com	after	14.146341
11	podcastinsights.com	after	3.777778
12	workcast.com	after	0.050000

```
longs_reach_losers_plt = (
    ggplot(longs_reach_losers_long,
           aes(x = 'reorder(site, visi)', y = 'visi', fill = 'phase')) +
    geom_bar(stat = 'identity', position = 'dodge', alpha = 0.8) +
    #geom_text(dd_factor_df, aes(label = 'market_name'), position=position_
    stack(vjust=0.01)) +
    labs(y = 'Visibility', x = '') +
    #scale_y_log10() +
    coord_flip() +
    theme(legend_position = 'right',
          axis_text_y =element_text(rotation=0, hjust=1, size = 12),
          legend_title = element_blank()
          )
)
```

```
longs_reach_losers_plt.save(filename = 'images/10_longs_visi_losers_plt.
png', height=5, width=10, units = 'in', dpi=1000)
longs_reach_losers_plt
```

Qualtrics and major competing brand GoToMeeting (Figure 10-11) were notably among the top losers following the Google update.

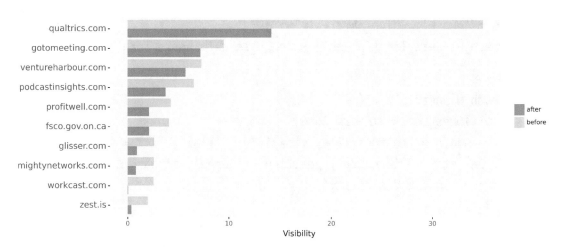

Figure 10-11. *Visibility by website before and after*

For the winners

```
longs_reach_winners_long = compare_longs.sort_values('delta_visi')
longs_reach_winners_long = longs_reach_winners_long[['site', 'before_visi',
'after_visi']]
```

We'll also remove Google and YouTube as Google may have biased their owned properties in search results following the algorithm update:

```
longs_reach_winners_long = longs_reach_winners_long[~longs_reach_winners_
long['site'].isin(['google.co.uk', 'google.com', 'youtube.com'])]
```

Taking the tail as opposed to the head allows us to select the winners as the table was ordered in ascending order of importance from lost visibility all the way down to sites that gained the most visibility:

```
longs_reach_winners_long = longs_reach_winners_long.tail(10)
```

```
longs_reach_winners_long = longs_reach_winners_long.melt(id_vars = 'site',
var_name = 'phase', value_name = 'visi')
longs_reach_winners_long['phase'] = longs_reach_winners_long['phase'].str.
replace('_visi', '')
```

```
longs_reach_winners_plt = (
    ggplot(longs_reach_winners_long,
           aes(x = 'reorder(site, visi)', y = 'visi', fill = 'phase')) +
    geom_bar(stat = 'identity', position = 'dodge', alpha = 0.8) +
    labs(y = 'Google Visibility', x = '') +
    coord_flip() +
    theme(legend_position = 'right',
          axis_text_y =element_text(rotation=0, hjust=1, size = 12),
          legend_title = element_blank()
          )
)
```

```
longs_reach_winners_plt.save(filename = 'images/10_longs_visi_winners_plt.
png', height=5, width=10, units = 'in', dpi=1000)
longs_reach_winners_plt
```

In the long-tail space, HubSpot and Sitecore are the clear winners (Figure 10-12).

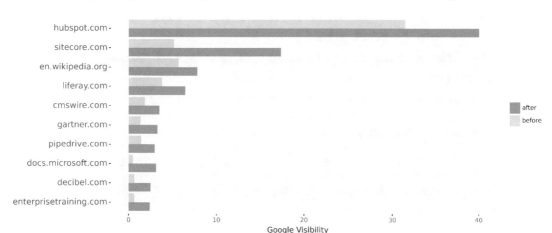

Figure 10-12. *Website Google visibility gainers, before and after*

This may be as a result of their numerous, well-produced, content-rich guides.

Target Level

With the general trends established, it's time to get into the details. Naturally, SEO practitioners and marketers want to know the performance by keywords and pages in terms of top gainers and losers. We'll split the analysis between keywords and pages.

Keywords

To achieve this, we'll filter for the target site "ON24" for both before and after the core update:

```
before_site = getstat_bef_unique[getstat_bef_unique['site'] == root_domain]
```

The weighted average rank doesn't apply here because we're aggregating at a keyword level where there is only value for a given keyword:

```
before_site_ranks = before_site.groupby(['keyword', 'search_volume']).
agg({'rank': 'mean'}).reset_index()
before_site_ranks = before_site_ranks.sort_values('search_volume',
ascending = False).head(10)
```

```
after_site = getstat_aft_unique[getstat_aft_unique['site'] == root_domain]
after_site_ranks = after_site.groupby(['keyword', 'search_volume']).
agg({'rank': 'mean'}).reset_index()
after_site_ranks = after_site_ranks.sort_values('search_volume', ascending
= False).head(10)
after_site_ranks
```

This results in the following:

	keyword	search_volume	wavg_rank
152	webinar	1000000	20.0
165	what is a webinar	60500	14.0
140	webcast	40500	9.0
100	live webinar	18100	18.0
102	live webinars	18100	14.0
161	webinar registrations	5400	9.0
93	live demo	4400	2.5
160	webinar marketing	2900	9.0
158	webinar events	2900	14.0
78	how do webinars work	2400	3.5

With the two datasets in hand, we'll merge them to get a side-by-side comparison:

```
compare_site_ranks = before_site_ranks. merge(after_site_ranks, on =
['keyword', 'search_volume'],
         how = 'outer')
compare_site_ranks = compare_site_ranks.rename(columns = {'rank_x':
'before_rank', 'rank_y': 'after_rank'})
compare_site_ranks['before_rank'] = np.where(compare_site_ranks['before_
rank'].isnull(), 100, compare_site_ranks['before_rank'])
compare_site_ranks['after_rank'] = np.where(compare_site_ranks['after_
rank'].isnull(), 100, compare_site_ranks['after_rank'])
```

```
compare_site_ranks['delta_rank'] = compare_site_ranks['before_rank'] -
compare_site_ranks['after_rank']
compare_site_ranks
```

This results in the following:

	keyword	search_volume	before_rank	after_rank	delta_rank
0	webinar	1000000	12.0	20.0	-8.0
1	what is a webinar	60500	9.5	14.0	-4.5
2	webcast	40500	15.0	9.0	6.0
3	online webinar	14800	17.0	100.0	-83.0
4	webinar registrations	5400	7.5	9.0	-1.5
5	live demo	4400	5.0	2.5	2.5
6	webinar marketing	2900	9.0	9.0	0.0
7	how do webinars work	2400	4.0	3.5	0.5
8	how does a webinar work	2400	2.5	100.0	-97.5
9	salesforce webinars	1900	15.0	100.0	-85.0
10	live webinar	18100	100.0	18.0	82.0
11	live webinars	18100	100.0	14.0	86.0
12	webinar events	2900	100.0	14.0	86.0

The biggest losing keyword was webinar, followed by "what is a webinar."
Let's visualize:

```
compare_site_ranks_long = compare_site_ranks[['keyword', 'before_rank',
'after_rank']]
compare_site_ranks_long = compare_site_ranks_long.melt(id_vars = 'keyword',
var_name = 'Phase', value_name = 'rank')
compare_site_ranks_long['Phase'] = compare_site_ranks_long['Phase'].str.
replace('_rank', '')

compare_site_ranks_long
```

```
compare_keywords_rank_plt = (
    ggplot(compare_site_ranks_long, aes(x = 'keyword', y = 'rank',
                    fill = 'Phase')) +
    geom_bar(stat = 'identity', position = 'dodge', alpha = 0.8) +
    labs(y = 'Google Rank', x = ' ') +
    scale_y_reverse() +
    coord_flip() +
    theme(legend_position = 'right', axis_text_x=element_text(rotation=0,
    hjust=1, size = 12))
)

compare_keywords_rank_plt.save(filename = 'images/6_compare_keywords_rank_
plt.png', height=5, width=10, units = 'in', dpi=1000)
compare_keywords_rank_plt
```

"Salesforce webinars" and "online webinars" really fell by the wayside going from the top 10 to nowhere (Figure 10-13).

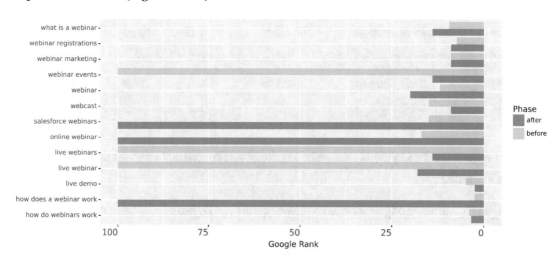

Figure 10-13. *Average rank positions by keyword for ON24 before and after*

By contrast, "webinar events" and "live webinar" gained. Knowing this information would help us prioritize keywords to analyze further to recover traffic back. For example, the SEO in charge of ON24 might want to analyze the top 20 ranking competitors for "webinar" to generate recovery recommendations.

Pages

Use the target keyword dataset which has been prefiltered to include the target site ON24:

```
targetURLs_before_reach = before_site.groupby(['url', 'phase']).
agg({'count': 'sum'}).reset_index()
targetURLs_before_sv = before_site.groupby(['url', 'phase']).agg({'search_
volume': 'mean'}).reset_index()
targetURLs_before_rank = before_site.groupby(['url', 'phase']).apply(wavg_
rank_sv).reset_index()
targetURLs_before_agg = targetURLs_before_reach.merge(targetURLs_before_sv,
on = ['url', 'phase'], how = 'left')
targetURLs_before_agg = targetURLs_before_agg.merge(targetURLs_before_rank,
on = ['url', 'phase'], how = 'left')
targetURLs_before_agg = targetURLs_before_agg.rename(columns = {'count':
'reach'})
targetURLs_before_agg['visi'] = (targetURLs_before_agg['search_volume'] /
targetURLs_before_agg['wavg_rank']).round(2)
targetURLs_before_agg
```

This results in the following:

	url	phase	reach	search_volume	wavg_rank	visi
0	https://www.on24.com/	before	12	125.750000	14.8	8.50
1	https://www.on24.com/act-on/	before	4	10.500000	9.7	1.08
2	https://www.on24.com/always-on-engagement-hub/	before	16	1.000000	3.2	0.31
3	https://www.on24.com/blog/12-amazing-tips-to-boost-webinar-registration-and-attendance/	before	7	772.285714	8.0	96.54
4	https://www.on24.com/blog/4-quick-and-easy-webinar-formats-you-can-use-right-now/	before	2	40.000000	17.0	2.35
...	...	...	...	...	...	...
89	https://www.on24.com/sql-server/	before	4	10.500000	13.0	0.81
90	https://www.on24.com/sugarcrm/	before	4	5.500000	4.9	1.12
91	https://www.on24.com/tableau-3/	before	4	240.500000	17.0	14.15
92	https://www.on24.com/webinar-benchmarks/	before	2	1.000000	8.5	0.12
93	https://www.on24.com/zapier/	before	5	30.400000	9.3	3.27

94 rows × 6 columns

```
targetURLs_after_reach = after_site.groupby(['url', 'phase']).agg({'count':
'sum'}).reset_index()
targetURLs_after_sv = after_site.groupby(['url', 'phase']).agg({'search_
volume': 'mean'}).reset_index()
targetURLs_after_rank = after_site.groupby(['url', 'phase']).apply(wavg_
rank_sv).reset_index()
targetURLs_after_agg = targetURLs_after_reach.merge(targetURLs_after_sv, on
= ['url', 'phase'], how = 'left')
targetURLs_after_agg = targetURLs_after_agg.merge(targetURLs_after_rank, on
= ['url', 'phase'], how = 'left')
targetURLs_after_agg = targetURLs_after_agg.rename(columns = {'count':
'reach'})
targetURLs_after_agg['visi'] = (targetURLs_after_agg['search_volume'] /
targetURLs_after_agg['wavg_rank']).round(2)
targetURLs_after_agg
```

This results in the following:

	url	phase	reach	search_volume	wavg_rank	visi
0	https://www.on24.com/	after	17	4692.235294	12.9	363.74
1	https://www.on24.com/act-on/	after	2	1.000000	6.0	0.17
2	https://www.on24.com/always-on-engagement-hub/	after	12	1.000000	3.2	0.31
3	https://www.on24.com/blog/12-amazing-tips-to-boost-webinar-registration-and-attendance/	after	2	5400.000000	9.0	600.00
4	https://www.on24.com/blog/4-quick-and-easy-webinar-formats-you-can-use-right-now/	after	2	40.000000	18.0	2.22
...	...	...	...	...	...	...
92	https://www.on24.com/sugarcrm/	after	4	5.500000	4.0	1.38
93	https://www.on24.com/tableau-3/	after	4	240.500000	10.0	24.05
94	https://www.on24.com/webinar-benchmarks/	after	3	1.000000	7.3	0.14
95	https://www.on24.com/wp-content/uploads/2020/06/on24_case-study-Ingram-Micro_20200521.pdf	after	2	1.000000	9.5	0.11
96	https://www.on24.com/zapier/	after	5	30.400000	10.7	2.84

97 rows × 6 columns

```
target_urls_deltas = targetURLs_before_agg.merge(targetURLs_after_agg, on =
['url'], how = 'left')
target_urls_deltas = target_urls_deltas.rename(columns = {'reach_x':
'before_reach', 'reach_y': 'after_reach',
                    'search_volume_x': 'before_sv', 'search_volume_y':
                    'after_sv',
```

```python
                            'wavg_rank_x': 'before_rank', 'wavg_rank_y':
                            'after_rank',
                            'visi_x': 'before_visi', 'visi_y': 'after_visi'})

target_urls_deltas = target_urls_deltas[['url', 'before_reach', 'before_
sv', 'before_rank', 'before_visi',
            'after_reach', 'after_sv', 'after_rank', 'after_visi']]

target_urls_deltas['after_reach'] = np.where(target_urls_deltas['after_
reach'].isnull(),
            0, target_urls_deltas['after_reach'])
target_urls_deltas['after_sv'] = np.where(target_urls_deltas['after_sv'].
isnull(),
            target_urls_deltas['before_sv'], target_urls_deltas['after_sv'])
target_urls_deltas['after_rank'] = np.where(target_urls_deltas['after_
rank'].isnull(),
            100, target_urls_deltas['after_rank'])
target_urls_deltas['after_visi'] = np.where(target_urls_deltas['after_
visi'].isnull(),
            0, target_urls_deltas['after_visi'])

target_urls_deltas['sv_delta'] = (target_urls_deltas['after_sv'] - target_
urls_deltas['before_sv'])
target_urls_deltas['rank_delta'] = (target_urls_deltas['before_rank'] -
target_urls_deltas['after_rank'])
target_urls_deltas['reach_delta'] = (target_urls_deltas['after_reach'] -
target_urls_deltas['before_reach'])
target_urls_deltas['visi_delta'] = (target_urls_deltas['after_visi'] -
target_urls_deltas['before_visi'])

target_urls_deltas = target_urls_deltas.sort_values(['visi_delta'],
ascending = False)
target_urls_deltas
```

This results in the following:

	url	before_reach	before_sv	before_rank	before_visi	after_
45	https://www.on24.com/live-webcast-elite/	5	16968.000000	15.1	1123.71	
44	https://www.on24.com/live-demo/	4	2700.000000	5.0	540.00	
3	https://www.on24.com/blog/12-amazing-tips-to-boost-webinar-registration-and-attendance/	7	772.285714	8.0	96.54	
0	https://www.on24.com/	12	125.750000	14.8	8.50	
17	https://www.on24.com/blog/how-webinars-work/#:~:text=Let's%20start%20with%20a%20simple,using%20other%20available%20interactive%20tools.	2	30.000000	1.0	30.00	
...	...	...	...	...	...	
39	https://www.on24.com/faqs/webinar-marketing-strategy-guide/	4	1450.500000	9.0	161.17	
21	https://www.on24.com/blog/on-demand-webinars-rules-everyone/	6	660.000000	6.1	108.20	
5	https://www.on24.com/blog/5-ways-to-drive-webinar-registrations-at-speed/	1	5400.000000	7.0	771.43	
79	https://www.on24.com/resources/upcoming-webinars/	2	1900.000000	1.0	1900.00	
15	https://www.on24.com/blog/how-webinars-work/	38	57057.631579	11.9	4794.76	

94 rows × 13 columns

```
winning_urls = target_urls_deltas['url'].head(10).tolist()

target_url_winners = pd.concat([targetURLs_before_agg, targetURLs_
after_agg])
target_url_winners = target_url_winners[target_url_winners['url'].
isin(winning_urls) ]

target_url_winners['phase'] = target_url_winners['phase'].
astype('category')
target_url_winners['phase'] = target_url_winners['phase'].cat.reorder_
categories(['after', 'before'])
target_url_winners
```

This results in the following:

	url	phase	reach	search_volume	wavg_rank	visi
0	https://www.on24.com/	before	12	125.750000	14.8	8.50
3	https://www.on24.com/blog/12-amazing-tips-to-boost-webinar-registration-and-attendance/	before	7	772.285714	8.0	96.54
17	https://www.on24.com/blog/how-webinars-work/#:~:text=Let's%20start%20with%20a%20simple,using%20other%20available%20interactive%20tools.	before	2	30.000000	1.0	30.00
29	https://www.on24.com/customer-stories/hubspot/	before	2	1000.000000	15.5	64.52
44	https://www.on24.com/live-demo/	before	4	2700.000000	5.0	540.00
45	https://www.on24.com/live-webcast-elite/	before	5	16968.000000	15.1	1123.71
57	https://www.on24.com/resources/asset/on24-webinar-benchmarks-report-special-edition-post-covid-trends/	before	2	10.000000	17.5	0.57
65	https://www.on24.com/resources/assets/cust-spotlight-salesforce/	before	2	1900.000000	15.0	126.67
82	https://www.on24.com/skype/	before	2	1.000000	11.0	0.09
91	https://www.on24.com/tableau-3/	before	4	240.500000	17.0	14.15
0	https://www.on24.com/	after	17	4692.235294	12.9	363.74
3	https://www.on24.com/blog/12-amazing-tips-to-boost-webinar-registration-and-attendance/	after	2	5400.000000	9.0	600.00

```
target_url_winners_plt = (
    ggplot(target_url_winners,
            aes(x = 'reorder(url, visi)', y = 'visi', fill = 'phase')) +
    geom_bar(stat = 'identity', position = 'dodge', alpha = 0.8) +
    labs(y = 'Visi', x = '') +
    coord_flip() +
    theme(legend_position = 'right',
            axis_text_y =element_text(rotation=0, hjust=1, size = 12),
            legend_title = element_blank()
            )
)

target_url_winners_plt.save(filename = 'images/8_target_url_winners_plt.
png', height=5, width=10, units = 'in', dpi=1000)
target_url_winners_plt
```

The Live Webcast Elite is the page that gained the most impressions, which is due to gaining positions on searches for "webcast" as seen earlier (Figure 10-14).

Figure 10-14. *URL visibility gainers for ON24 before and after*

If we had website analytics data such as Google, we could merge it with the URLs to get an idea of how much traffic the rankings were worth and how closely it correlates with search volumes.

Let's take a look at the losing URLs:

```
losing_urls = target_urls_deltas['url'].tail(10).tolist()
print(losing_urls)

target_url_losers = pd.concat([targetURLs_before_agg, targetURLs_
after_agg])
```

```
target_url_losers = target_url_losers[target_url_losers['url'].
isin(losing_urls) ]

target_url_losers['visi'] = (target_url_losers['search_volume'] / target_
url_losers['wavg_rank']).round(3)


target_url_losers['phase'] = target_url_losers['phase'].astype('category')
target_url_losers['phase'] = target_url_losers['phase'].cat.reorder_
categories(['after', 'before'])
target_url_losers
```

This results in the following:

	url	phase	reach	search_volume	wavg_rank	visi
0	https://www.on24.com/	before	12	125.750000	14.8	8.50
3	https://www.on24.com/blog/12-amazing-tips-to-boost-webinar-registration-and-attendance/	before	7	772.285714	8.0	96.54
17	https://www.on24.com/blog/how-webinars-work/#:~:text=Let's%20start%20with%20a%20simple,using%20other%20available%20interactive%20tools.	before	2	30.000000	1.0	30.00
29	https://www.on24.com/customer-stories/hubspot/	before	2	1000.000000	15.5	64.52
44	https://www.on24.com/live-demo/	before	4	2700.000000	5.0	540.00
45	https://www.on24.com/live-webcast-elite/	before	5	16968.000000	15.1	1123.71
57	https://www.on24.com/resources/asset/on24-webinar-benchmarks-report-special-edition-post-covid-trends/	before	2	10.000000	17.5	0.57
65	https://www.on24.com/resources/assets/cust-spotlight-salesforce/	before	2	1900.000000	15.0	126.67
82	https://www.on24.com/skype/	before	2	1.000000	11.0	0.09
91	https://www.on24.com/tableau-3/	before	4	240.500000	17.0	14.15
0	https://www.on24.com/	after	17	4692.235294	12.9	363.74
3	https://www.on24.com/blog/12-amazing-tips-to-boost-webinar-registration-and-attendance/	after	2	5400.000000	9.0	600.00
16	https://www.on24.com/blog/how-webinars-work/#:~:text=Let's%20start%20with%20a%20simple,using%20other%20available%20interactive%20tools.	after	4	140.000000	1.0	140.00
30	https://www.on24.com/customer-stories/hubspot/	after	2	1000.000000	9.0	111.11
44	https://www.on24.com/live-demo/	after	4	2700.000000	2.5	1080.00
45	https://www.on24.com/live-webcast-elite/	after	2	20270.000000	8.0	2533.75
60	https://www.on24.com/resources/asset/on24-webinar-benchmarks-report-special-edition-post-covid-trends/	after	2	50.000000	2.0	25.00
68	https://www.on24.com/resources/assets/cust-spotlight-salesforce/	after	2	1900.000000	9.0	211.11
84	https://www.on24.com/skype/	after	4	800.500000	14.5	55.21
93	https://www.on24.com/tableau-3/	after	4	240.500000	10.0	24.05

```
target_url_losers_plt = (
    ggplot(target_url_losers, aes(x = 'reorder(url, visi)', y = 'visi',
    fill = 'phase')) +
    geom_bar(stat = 'identity', position = 'dodge', alpha = 0.8) +
    position=position_stack(vjust=0.01)) +
    labs(y = 'Visi', x = '') +
    coord_flip() +
```

```
    theme(legend_position = 'right',
        axis_text_y =element_text(rotation=0, hjust=1, size = 12),
        legend_title = element_blank()
        )
)

target_url_losers_plt.save(filename = 'images/8_target_url_losers_plt.png',
                        height=5, width=10, units = 'in', dpi=1000)
target_url_losers_plt
```

"How webinars work" and "upcoming webinars" were the biggest losing URLs (Figure 10-15).

Figure 10-15. *URL visibility losers for ON24 before and after*

The `https://www.on24.com/blog/how-webinars-work/#:~:text=Let's%20` `start%20with%20a%20simple,using%20other%20available%20interactive%20tools` URL seems like it wasn't canonicalized (i.e., there was no defined rel="canonical" URL to consolidate any URL variant or duplicate to).

To all of this, one possible follow-up would be to use Google Search Console data to extract the search queries for each URL and see

- Whether the search intent is shared within the URL

- If the URLs have the content to satisfy the queries generating the impressions

Another possible follow-up would be to segment the URLs and keywords according to their content type. This could help determine if there were any general patterns that could explain and speed up the recovery process.

Segments

We return back to the SERPs to analyze how different site types fared in the Google update. The general approach will be to work out the most visible sites, before using the np.select() function to categorize and label these sites.

Top Competitors

To find the top competitor sites, we'll aggregate both before and after datasets to work out the visibility index derived from the reach and search volume weighted rank average:

```
players_before = getstat_bef_unique
print(players_before.columns)

players_before_rank = players_before.groupby('site').apply(wavg_rank_sv).
reset_index()
players_before_reach = players_before.groupby('site').agg({'count':
'sum'}).sort_values('count', ascending = False).reset_index()
players_before_agg = players_before_rank.merge(players_before_reach, on =
'site', how = 'left')
players_before_agg['visi'] = players_before_agg['count'] / players_before_
agg['wavg_rank']
players_before_agg = players_before_agg.sort_values('visi', ascending
= False)
players_before_agg
```

This results in the following:

	site	wavg_rank	count	visi
1707	google.com	1.8	927	515.000000
1864	hubspot.com	6.6	428	64.848485
1710	gotomeeting.com	6.4	254	39.687500
1488	facebook.com	1.8	60	33.333333
3388	scholar.google.com	1.0	33	33.000000
...	...	...	...	...
1374	enquiresolutions.com	20.0	1	0.050000
2924	paperpicks.com	20.0	1	0.050000
2948	pcisecuritystandards.org	20.0	1	0.050000
1344	emergencyreporting.com	20.0	1	0.050000
2083	isixsigma.com	20.0	1	0.050000

4432 rows × 4 columns

```
players_after = getstat_aft_unique
print(players_before.columns)

players_after_rank = players_after.groupby('site').apply(wavg_rank_sv).
reset_index()
players_after_reach = players_after.groupby('site').agg({'count': 'sum'}).
sort_values('count', ascending = False).reset_index()
players_after_agg = players_after_rank.merge(players_after_reach, on =
'site', how = 'left')
players_after_agg['visi'] = players_after_agg['count'] / players_after_
agg['wavg_rank']
players_after_agg = players_after_agg.sort_values('visi', ascending = False)
players_after_agg
```

This results in the following:

	site	wavg_rank	count	visi
1741	google.com	1.6	1027	641.875000
1912	hubspot.com	6.0	429	71.500000
1743	gotomeeting.com	5.1	238	46.666667
3425	scholar.google.com	1.0	31	31.000000
1520	facebook.com	2.0	57	28.500000
...	...	...	...	...
3519	shiftelearning.com	20.0	1	0.050000
2573	messinagroupinc.com	20.0	1	0.050000
1429	esker.com	20.0	1	0.050000
4397	wilsoncc.edu	20.0	1	0.050000
2072	integrations.clickmeeting.com	20.0	1	0.050000

4520 rows × 4 columns

To put the data aggregation together, we take the before dataset and exclude any sites appearing in the after dataset. The purpose is to perform an outer join with the after dataset, to capture every single site possible.

```
players_agg = players_before_agg[~players_before_agg['site'].isin(players_
after_agg['site'])]
players_agg = players_agg.merge(players_after_agg, how='outer',
indicator=True)
players_agg = players_agg.sort_values('visi', ascending = False)
players_agg.head(50)
```

This results in the following:

	site	wavg_rank	count	visi	_merge
875	google.com	1.6	1027	641.875000	right_only
876	hubspot.com	6.0	429	71.500000	right_only
877	gotomeeting.com	5.1	238	46.666667	right_only
878	scholar.google.com	1.0	31	31.000000	right_only
879	facebook.com	2.0	57	28.500000	right_only
880	en.m.wikipedia.org	2.8	79	28.214286	right_only
881	m.youtube.com	6.4	180	28.125000	right_only
882	capterra.com	8.2	199	24.268293	right_only
883	medium.com	9.4	226	24.042553	right_only
884	ventureharbour.com	5.4	103	19.074074	right_only

Now that we have all of the sites in descending order of priority, we can start categorizing the domains by site type. Using the hopefully now familiar np.select() function, we will categorize the sites manually, creating a list of our conditions that create lists of sites and then mapping these to a separate list of category names:

```
site_conds = [
    players_agg['site'].str.contains('|'.join(['google.com', 'youtube.com'])),
    players_agg['site'].str.contains('|'.join(['wikipedia.org'])),
    players_agg['site'].str.contains('|'.join(['medium.com', 'forbes.com',
    'hbr.org', 'smartinsights.com', 'mckinsey.com',
                            'techradar.com','searchenginejournal.com',
                            'cmswire.com', 'entrepreneur.com',
                            'pcmag.com', 'elearningindustry.com',
                            'businessnewsdaily.com'])),
    players_agg['site'].isin(['on24.com', 'gotomeeting.com', 'marketo.com',
    'zoom.us', 'livestorm.co', 'hubspot.com', 'drift.com',
    'salesforce.com', 'clickmeeting.com', 'liferay.com',
```

```
                            'qualtrics.com', 'workcast.com',
                            'livewebinar.com', 'getresponse.com',
                            'brightwork.com',
                            'superoffice.com', 'myownconference.com',
                            'info.workcast.com', 'tallyfy.com',
                            'readytalk.com', 'eventbrite.com', 'sitecore.
                            com', 'pgi.com', '3cx.com', 'walkme.com',
                            'venngage.com', 'tableau.com', 'netsuite.
                            com', 'zoominfo.com', 'sproutsocial.com']),
    players_agg['site'].isin([ 'neilpatel.com', 'ventureharbour.com',
    'wordstream.com', 'business.tutsplus.com',
                            'convinceandconvert.com',
                            'growthmarketingpro.com',
                            'marketinginsidergroup.com',
                            'adamenfroy.com', 'danielwaas.com',
                            'newbreedmarketing.com']),
    players_agg['site'].str.contains('|'.join(['trustradius.com', 'g2.com',
    'capterra.com', 'softwareadvice.com'])),
    players_agg['site'].str.contains('|'.join(['facebook.com', 'linkedin.
    com', 'business.linkedin.com'])),
    players_agg['site'].str.contains('|'.join(['.edu', '.ac.uk']))
]
```

Create a list of the values we want to assign for each condition. The categories in this case are based on their business model or site purpose:

```
segment_values = ['search', 'reference', 'publisher', 'martech',
'consulting', 'reviews', 'social_media', 'education']
```

Create a new column and use np.select to assign values to it using our lists as arguments:

```
players_agg['segment'] = np.select(site_conds, segment_values, default =
'other')
players_agg.head(5)
```

This results in the following:

	site	wavg_rank	count	visi	_merge	segment
875	google.com	1.6	1027	641.875000	right_only	search
876	hubspot.com	6.0	429	71.500000	right_only	martech
877	gotomeeting.com	5.1	238	46.666667	right_only	martech
878	scholar.google.com	1.0	31	31.000000	right_only	search
879	facebook.com	2.0	57	28.500000	right_only	social_media

The sites are categorized. We'll now look at the sites classed as other. This is useful because if we see any sites we think are important enough to be categorized as not "other," then we can update the conditions earlier.

```
players_agg[players_agg['segment'] == 'other'].head(20)
```

This results in the following:

	site	wavg_rank	count	visi	_merge	segment
901	abm.com	1.1	12	10.909091	right_only	other
909	toolshero.com	1.0	9	9.000000	right_only	other
923	eventmanagercom	5.9	45	7.627119	right_only	other
930	shrm.org	6.2	45	7.258065	right_only	other
931	productplan.com	2.5	18	7.200000	right_only	other
935	enterprisetraining.com	2.6	16	6.153846	right_only	other
936	indeed.com	3.1	19	6.129032	right_only	other
937	enonic.com	1.0	6	6.000000	right_only	other
944	en.rockcontent.com	12.0	69	5.750000	right_only	other
946	highspot.com	4.2	24	5.714286	right_only	other

```
players_agg_map = players_agg[['site', 'segment']]
players_agg_map
```

This results in the following:

	site	segment
875	google.com	search
876	hubspot.com	martech
877	gotomeeting.com	martech
878	scholar.google.com	search
879	facebook.com	social_media
...	...	...
778	360learning.com	other
779	scholarworks.umass.edu	education
780	cpe.kennesaw.edu	education
781	unisys.com	other
5394	integrations.clickmeeting.com	other

5395 rows × 2 columns

There you have a mapping dataframe which will be used to give segmented SERPs insights, starting with visibility.

Visibility

With the sites categorized, we can now compare performance by site type before and after the update.

As usual, we'll aggregate the before and after datasets. Only this time, we will also merge the site type labels.

Start with the before dataset:

```
before_sector_unq_reach = getstat_bef_unique.merge(players_agg_map, on =
'site', how = 'left')
```

We filter for the top 10 to calculate our reach statistics, which we'll need for our visibility calculations later on:

```
before_sector_unq_reach = before_sector_unq_reach[before_sector_unq_
reach['rank'] < 11 ]

before_sector_agg_reach = before_sector_unq_reach.groupby(['segment']).
agg({'count': 'sum'}).reset_index()
before_sector_agg_reach = before_sector_agg_reach.rename(columns =
{'count': 'reach'})
before_sector_agg_reach = before_sector_agg_reach[['segment', 'reach']]
before_sector_agg_reach['reach'] = np.where(before_sector_agg_
reach['reach'].isnull(),
       0, before_sector_agg_reach['reach'])
```

The same logic and operation is applied to the after dataset:

```
after_sector_unq_reach = getstat_aft_unique.merge(players_agg_map, on =
'site', how = 'left')
after_sector_unq_reach = after_sector_unq_reach[after_sector_unq_
reach['rank'] < 11 ]

after_sector_agg_reach = after_sector_unq_reach.groupby(['segment']).
agg({'count': 'sum'}).reset_index()
after_sector_agg_reach = after_sector_agg_reach.rename(columns = {'count':
'reach'})
after_sector_agg_reach['reach'] = np.where(after_sector_agg_reach['reach'].
isnull(), 0, after_sector_agg_reach['reach'])

after_sector_agg_reach = after_sector_agg_reach[['segment', 'reach']]
after_sector_agg_reach.sort_values('reach', ascending = False).head(10)
```

This results in the following:

	segment	reach
3	other	7395
2	martech	1666
7	search	1244
4	publisher	617
6	reviews	475
0	consulting	470
1	education	226
8	social_media	118
5	reference	100

"Other" as a site type segment dominates the statistics in terms of reach. We may want to filter this out later on. Now for the weighted average rankings by search volume, which will include the wavg_rank_sv() function defined earlier.

```
before_sector_unq_visi = before_sector_unq_svranks.merge(before_sector_agg_
reach, on = 'segment', how = 'left')
before_sector_unq_visi['reach'] = np.where(before_sector_unq_visi['reach'].
isnull(), 0, before_sector_unq_visi['reach'])

before_sector_unq_visi['wavg_rank'] = np.where(before_sector_unq_
visi['wavg_rank'].isnull(), 100, before_sector_unq_visi['wavg_rank'])

before_sector_unq_visi['visibility'] = before_sector_unq_visi['reach'] /
before_sector_unq_visi['wavg_rank']
before_sector_unq_visi = before_sector_unq_visi.sort_values('visibility',
ascending = False)

after_sector_unq_visi = after_sector_unq_svranks.merge(after_sector_agg_
reach, on = 'segment', how = 'left')
after_sector_unq_visi['reach'] = np.where(after_sector_unq_visi['reach'].
isnull(), 0, after_sector_unq_visi['reach'])
```

```
after_sector_unq_visi['wavg_rank'] = np.where(after_sector_unq_visi['wavg_
rank'].isnull(), 100, after_sector_unq_visi['wavg_rank'])
```

```
after_sectaor_unq_visi['visibility'] = after_sector_unq_visi['reach'] /
after_sector_unq_visi['wavg_rank']
after_sector_unq_visi = after_sector_unq_visi.sort_values('visibility',
ascending = False)
after_sector_unq_visi
```

This results in the following:

	segment	wavg_rank	reach	visibility
3	other	13.9	7395	532.014388
7	search	3.8	1244	327.368421
2	martech	8.8	1666	189.318182
4	publisher	10.2	617	60.490196
6	reviews	11.7	475	40.598291
0	consulting	11.8	470	39.830508
8	social_media	3.9	118	30.256410
1	education	7.5	226	30.133333
5	reference	4.8	100	20.833333

As well as reach, other performs well in the search volume weighted rank stakes and therefore in overall visibility. With the before and after segmented datasets aggregated, we can now join them:

```
compare_sector_visi_players = before_sector_unq_visi.merge(after_sector_
unq_visi, on = ['segment'], how = 'outer')
compare_sector_visi_players = compare_sector_visi_players.rename(columns =
{'wavg_rank_x': 'before_rank', 'wavg_rank_y': 'after_rank',
                    'reach_x': 'before_reach', 'reach_y': 'after_reach',
                    'visibility_x': 'before_visi', 'visibility_y':
                    'after_visi'
                })
```

```
compare_sector_visi_players['before_visi'] = np.where(compare_sector_visi_
players['before_visi'].isnull(),
                0, compare_sector_visi_players['before_visi'])
compare_sector_visi_players['after_visi'] = np.where(compare_sector_visi_
players['after_visi'].isnull(),
                0, compare_sector_visi_players['after_visi'])
compare_sector_visi_players['delta_visi'] = compare_sector_visi_
players['after_visi'] - compare_sector_visi_players['before_visi']

compare_sector_visi_players = compare_sector_visi_players.sort_
values('delta_visi')

compare_sector_visi_players.head(10)
```

This results in the following:

	segment	before_rank	before_reach	before_visi	after_rank	after_reach	after_visi	delta_visi
6	reference	3.2	91	28.437500	4.8	100	20.833333	-7.604167
4	consulting	12.2	468	38.360656	11.8	470	39.830508	1.469853
1	search	3.5	1140	325.714286	3.8	1244	327.368421	1.654135
5	reviews	12.7	444	34.960630	11.7	475	40.598291	5.637661
0	other	13.3	7000	526.315789	13.9	7395	532.014388	5.698599
7	education	9.6	215	22.395833	7.5	226	30.133333	7.737500
8	social_media	7.7	104	13.506494	3.9	118	30.256410	16.749917
2	martech	9.4	1595	169.680851	8.8	1666	189.318182	19.637331
3	publisher	13.4	538	40.149254	10.2	617	60.490196	20.340942

The only site group that lost were reference sites like Wikipedia, dictionaries, and so on. Their reach increased by 11%, but their rankings declined by almost two places on average. This could be that nonreference sites are churning out more value adding articles which are crowding out generic sites like Wikipedia that have no expertises in those areas.

Let's reshape the data for visualization:

```
compare_sector_visi_players_long = compare_sector_visi_players[['segment',
'before_visi','after_visi']]
compare_sector_visi_players_long = compare_sector_visi_players_long.
melt(id_vars = ['segment'], var_name='Phase',
                                  value_name='Visi')
compare_sector_visi_players_long['Phase'] = compare_sector_visi_players_
long['Phase'].str.replace('_visi', '')

compare_sector_visi_players_long['Phase'] = compare_sector_visi_players_
long['Phase'].astype('category')
compare_sector_visi_players_long['Phase'] = compare_sector_visi_players_
long['Phase'].cat.reorder_categories(['after',
                                  'before'])

compare_sector_visi_players_long.head(10)
```

This results in the following:

	segment	Phase	Visi
0	reference	before	28.437500
1	consulting	before	38.360656
2	search	before	325.714286
3	reviews	before	34.960630
4	other	before	526.315789
5	education	before	22.395833
6	social_media	before	13.506494
7	martech	before	169.680851
8	publisher	before	40.149254
9	reference	after	20.833333

```
compare_sector_visi_players_long_plt = (
    ggplot(compare_sector_visi_players_long, aes(x = 'reorder(segment,
    Visi)', y = 'Visi', fill = 'Phase')) +
    geom_bar(stat = 'identity', position = 'dodge', alpha = 0.8) +
    labs(y = 'Visibility', x = ' ') +
    coord_flip() +
    theme(legend_position = 'right', axis_text_x=element_text(rotation=0,
    hjust=1, size = 12))
)

compare_sector_visi_players_long_plt.save(filename = 'images/11_compare_
sector_visi_players_long_plt.png', height=5, width=10, units = 'in',
dpi=1000)
Compare_sector_visi_players_long_plt
```

So other than reference sites, every other category gained, including martech and publishers which gained the most (Figure 10-16).

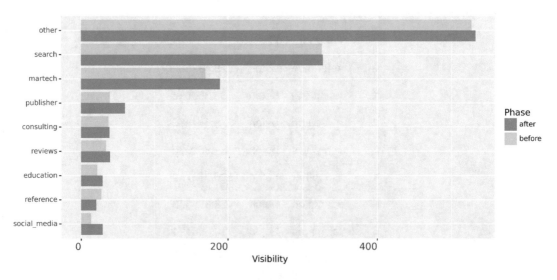

Figure 10-16. *Visibility before and after by site type*

Snippets

In addition to visibility, we can dissect result types by segment too. Although there are many visualizations that can be done by segment, we've chosen snippets so that we can introduce a heatmap visualization technique.

This time, we'll aggregate on snippets and segments, having performed the join for both before and after datasets:

```
before_sector_unq_snippets = getstat_bef_unique.merge(players_agg_map, on = 'site', how = 'left')
before_sector_agg_snippets = before_sector_unq_snippets.
groupby(['snippets', 'segment']).agg({'count': 'sum'}).reset_index()

before_sector_agg_snippets = before_sector_agg_snippets[['snippets', 'segment', 'count']]
before_sector_agg_snippets = before_sector_agg_snippets.rename(columns = {'count': 'reach'})

after_sector_unq_snippets = getstat_aft_unique.merge(players_agg_map, on = 'site', how = 'left')
after_sector_agg_snippets = after_sector_unq_snippets.groupby(['snippets', 'segment']).agg({'count': 'sum'}).reset_index()

after_sector_agg_snippets = after_sector_agg_snippets[['snippets', 'segment', 'count']]
after_sector_agg_snippets = after_sector_agg_snippets.rename(columns = {'count': 'reach'})
after_sector_agg_snippets.sort_values('reach', ascending = False).head(10)
```

This results in the following:

	snippets	segment	reach
90	organic	other	10805
61	image / organic	other	4603
89	organic	martech	1698
98	people also ask	search	862
17	amp / organic	other	779
60	image / organic	martech	699
91	organic	publisher	605
87	organic	consulting	470
78	interesting finds	other	366
93	organic	reviews	356

For post update, we can see that much of other's reach are organic, images, and AMP post update. How does that compare pre- and post update?

```
compare_sector_snippets = before_sector_agg_snippets.merge(after_sector_
agg_snippets, on = ['snippets', 'segment'], how = 'outer')
compare_sector_snippets = compare_sector_snippets.rename(columns =
{'reach_x': 'before_reach', 'reach_y': 'after_reach'})
compare_sector_snippets['before_reach'] = np.where(compare_sector_snippets
['before_reach'].isnull(), 0, compare_sector_snippets['before_reach'])
compare_sector_snippets['after_reach'] = np.where(compare_sector_snippets
['after_reach'].isnull(), 0, compare_sector_snippets['after_reach'])
compare_sector_snippets['delta_reach'] = compare_sector_snippets['after_
reach'] - compare_sector_snippets['before_reach']

compare_sector_snippets = compare_sector_snippets.sort_
values('delta_reach')
compare_sector_snippets.head(10)
```

This results in the following:

	snippets	segment	before_reach	after_reach	delta_reach
94	organic	reviews	407.0	356.0	-51.0
56	faq / organic	other	233.0	204.0	-29.0
65	image / organic	reviews	151.0	132.0	-19.0
89	organic	education	365.0	350.0	-15.0
16	amp / organic	martech	100.0	87.0	-13.0
95	organic	search	72.0	60.0	-12.0
88	organic	consulting	480.0	470.0	-10.0
38	answers / paragraph	other	112.0	103.0	-9.0
98	organic / sitelinks	other	18.0	10.0	-8.0
101	placesv3 / ratings	other	13.0	6.0	-7.0

Review sites lost the most reach in the organic listings and Google images. Martech lost out on AMP results.

```
compare_sector_snippets.tail(10)
```

This results in the following:

	snippets	segment	before_reach	after_reach	delta_reach
52	carousel / videos	search	258.0	280.0	22.0
63	image / organic	publisher	225.0	265.0	40.0
17	amp / organic	other	729.0	779.0	50.0
92	organic	publisher	553.0	605.0	52.0
79	interesting finds	other	306.0	366.0	60.0
61	image / organic	martech	639.0	699.0	60.0
57	faq / organic	reviews	41.0	129.0	88.0
99	people also ask	search	760.0	862.0	102.0
91	organic	other	10410.0	10805.0	395.0
62	image / organic	other	4194.0	4603.0	409.0

By contrast, publisher sites appear to have displaced review sites on images and organic results. Since we're more interested in result types other than the regular organic links, we'll strip these out and visualize. Otherwise, we'll end up with charts that show a massive weight for organic links while dwarfing out the rest of the result types.

```
compare_sector_snippets_graphdf = compare_sector_snippets[compare_sector_
snippets['snippets'] != 'organic']

compare_sector_snippets_plt = (
    ggplot(compare_sector_snippets_graphdf,
           aes(x = 'segment', y = 'snippets', fill = 'delta_reach')) +
    geom_tile(stat = 'identity', alpha = 0.6) +
    labs(y = '', x = '') +
    theme_classic() +
    theme(legend_position = 'right',
          axis_text_x=element_text(rotation=90, hjust=1)
          )
)

compare_sector_snippets_plt.save(filename = 'images/12_compare_sector_
snippets_plt.png', height=5, width=10, units = 'in', dpi=1000)
compare_sector_snippets_plt
```

The heatmap in Figure 10-17 uses color as the third dimension to display where the major changes in reach were for the different site segments (bottom) and result types (vertical).

This results in the following:

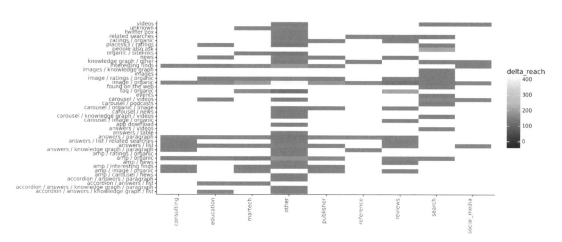

Figure 10-17. *Heatmap showing the difference in top 10 counts by site type and result type*

The major change that stands out is Google image results for the other segment. The rest appears inconsequential by contrast. The heatmap is an example of how three-dimensional categorical data can be visualized.

Summary

We focused on analyzing performance following algorithm updates with a view to explaining what happened and possible extraction of insights for areas of further research and recommendation generation.

Not only did we evaluate methods for establishing visibility changes, our general approach was to analyze the general SERP trends before segmenting by result types, cannibals. Then we looked at the target site level, seeing the changes by keyword, query length, and URLs.

We also tried evaluating general SERP trends by grouping sites into site category segments to give a richer analysis of the SERPs by visibility and snippets. While the visualization of data before and after the core update doesn't always reveal the causes of any algorithm update, some patterns can be learned to inform further areas of research for recommendation generation. The data can always be joined with other data sources and use the techniques outlined in competitor analysis to uncover potential ranking factor hypotheses for split testing.

The next chapter discusses the future of SEO.

CHAPTER 11

The Future of SEO

The exploration of applying data science methods to SEO ultimately leads to further questions on the evolution of SEO as an industry and a profession.

As paid search has increasingly become more automated and commoditized, could the same happen to SEO?

To answer that question, let's reflect on how data science has been applied to SEO in this book, the limitations and opportunities for automation.

Aggregation

Aggregation is a data analysis technique where data is summarized at a group level, for example, the average number of clicks by content group. In Microsoft Excel, this would be achieved using pivot tables.

Aggregation is something that can and should be achieved by automation. Aggregation has been applied to help us understand too many different areas of SEO to name (covered in this book). With the code supplied, this could be integrated into cloud-based data pipelines for SEO applications to load data warehouses and power dashboards.

Aggregation for most use cases is good enough. However, for scientific exploration of the data and hypothesis testing, we need to consider the distribution of the data, its variation, and other statistical properties.

Aggregation for reporting and many analytical aspects of SEO consulting can certainly be automated, which would be carried out in the cloud. This is pretty straightforward to do given the data pipeline technology by Apache that's already in place.

© Andreas Voniatis 2023
A. Voniatis, *Data-Driven SEO with Python*, https://doi.org/10.1007/978-1-4842-9175-7_11

Distributions

The statistical distribution has power because we're able to understand what is normal and therefore identify anything that is significantly above or below performance.

We used distributions to find growth keywords where keywords in Google Search Console (GSC) had impressions above the 95th percentile for their ranking position.

We also used distributions to identify content that lacked internal links in a website architecture and hierarchy.

This can also be automated in the cloud which could lead to applications being released for the SEO industry to automate the identification of internal link opportunities. There is a slight risk of course that all internal links are done on a pure distribution basis and not factoring in content not intended for search channels which would need to be part of the software design.

String Matching

String matching using the Sorensen-Dice algorithm has been applied to help match content to URLs for use cases such as keyword mapping, migration planning, and others.

The results are decent as it's relatively quick and scales well, but it relies on descriptive URLs and title tags in the first instance. It also relies on preprocessing such as removing the website domain portion of the URL before applying string matching, which is easily automated. Less easy to work around is the human judgment of what is similar enough for the title and URL slug to be a good match. Should the threshold be 70%, 83%, or 92%?

That is less easy and probably would require some self-learning in the form of an AI, more specifically a recurrent neural network (RNN). It's not impossible, of course, as you would need to determine what a good and less good outcome metric is to know how to train a model. Plus, you'd need at least a million data points to train the model.

A key question for keyword mapping will be "what is the metric that shows which URL is the best for users searching for X keyword." An RNN could be good here as it could learn from top ranking SERP content, reduce it to an object, and then compare site content against that object to map the keyword to.

For redirecting expired URLs (with backlinks) to live URLs with a 200 HTTP response, it could be more straightforward and not require an AI. You might use a decision tree–based algorithm using user behavior to inform what is the best replacement URL, that is, users on "URL A" would always go to URL X out of the other URL choices available.

A non-AI-based solution doesn't rely on millions of SERP or Internet data and would therefore be (relatively) inexpensive to construct in-house. The AI-based solutions on the other hand are likely to either be built as a SaaS or by a mega enterprise brand that relies on organic traffic like Amazon, Quora, or Tripadvisor.

Clustering

In this book, clustering has been applied to determine search intent by comparing search results at scale. The principles are based on comparing distances between data points, and wherever a distance is relatively small, a cluster exists. Word stemming hasn't been applied in this book as it lacks the required precision despite the speed.

Clustering is useful not only for understanding keywords but also for grouping keywords for reporting performance and grouping content to set the website hierarchy. Your imagination as an SEO expert is the limit.

Applications already exist in the SEO marketplace for clustering keywords according to search intent by comparing search results, so this can and already has been automated in the cloud by Artios, Keyword Insights, Keyword Cupid, SEO Scout and others.

Machine Learning (ML) Modeling

Trying to understand how things work in organic search is one of the many pleasures and trials of SEO, especially in the current era of Google updates. Modeling aided by machine learning helps in that regard because you're effectively feeding data into a machine learning algorithm, and it will show the most influential factors behind the algorithm update.

Machine learning models could most certainly be deployed into the cloud as part of automated applications to highlight most influential SERP factors and qualities of backlinks as part of a wider dashboard intelligence system or a stand-alone application.

Because no neural network is required, this is relatively cheap to build and deploy, leaving the SEO professionals to understand the model outputs and how to apply them.

Set Theory

Set theory is where we compare sets (think lists) of data like keywords and compare them to another set. This can be used to see the difference between two datasets. This was used for content gap analysis to find common keywords (i.e., where the keywords of two websites intersect) and to find the gap between the target site and the core keyword set.

This is pretty straightforward and can easily be automated using tools like SEMRush and AHREFs. So why do it in Python? Because it's free and it gives you more control over the degree of intersection required.

Knowing the perfect degree of intersection is less clear because it would require research and development to work out the degree of intersects required which for one will depend on the number of sites being intersected.

However, the skill is knowing which competitors to include in the set in the first place which may not be so easy for a machine to discern.

What Computers Can and Can't Do

From the preceding text, we see a common pattern, that is, when it comes to straightforward tasks such as responding to statistical properties or decisions based on numerical patterns, computers excel (albeit with some data cleanup and preparation).

When matters are somewhat fuzzy and subject to interpretation such as language, computers can still rise to the challenge, but that requires significant resources to get the data and train an AI model.

For the SEO Experts

We should learn Python for data analysis because that's how we generate insights and recommendations in response. It doesn't make sense not to make use of technology and data science thinking to solve SEO problems, especially when the SERPs are determined by a consistent (and therefore more predictable) algorithm.

Not only will learning Python and data science help future-proof your career as an SEO, it will give you a deeper appreciation for how search engines work (given they are mathematical) and enable you to devote much more time and energy toward defining hypotheses to test and create SEO solutions. Spending less time collecting and analyzing data and more time responding to the data is the order of the day.

You'll also be in a far better position to work with software engineers when it comes to specifying cloud apps, be it reporting, automation, or otherwise.

Of course, creativity comes from knowledge, so the more you know about SEO, the better the questions you will be able to ask of the data, producing better insights as a consequence and much more targeted recommendations.

Summary

In this chapter, we consolidated at a very high level the ideas and techniques used to make SEO data driven:

- Aggregation

- Distributions

- String matching

- Clustering

- Machine learning (ML) modeling

- Set theory

We also examined what computers can and can't do and provided a reminder why SEO experts should turn to data science.

Here's to the future of SEO.

Index

A

A/A testing
  aa_means dataframe, 314
  aa_model.summary(), 319
  aa_test_box_plt, 317
  dataframe, 313
  data structures, 312
  date range, 314
  groups' distribution, 317
  histogram plots, 316
  .merge() function, 315
  NegativeBinomial() model, 319
  optimization, 315
  pretest and test period groups, 318
  p-value, 319
  SearchPilot, 311, 312
  sigma, 315
  statistical model, 318
  statistical properties, 313
  .summary() attribute, 319
  test period, 313
Accelerated mobile pages (AMP), 505
Account-based marketing, 175
Additional features, 130, 475
Adobe Analytics, 343
Aggregation, 67, 81, 105, 131, 186, 205, 218, 253, 256, 276, 368, 449, 474, 480, 513, 521, 522, 563
AHREFs, 98, 201, 216, 244, 249, 266, 343, 566
Akaike information criterion (AIC), 32
Alternative methods, 118

Amazon Web Services (AWS), 5, 300
anchor_levels_issues_count_plt graphic, 116
anchor_rel_stats_site_agg_plt plot, 121
Anchor texts
  anchor_issues_count_plt, 113
  HREF, 113
  issues by site level, 114, 116
  nondescriptive, 113
  search engines and users, 111
  Sitebulb, 111
Anchor text words, 122–125
Andreas, 5, 245
Antispam algorithms, 200
API libraries, 345
API output, 128
API response, 128, 346, 347
Append() function, 168
apply_pcn function, 379
astype() function, 490
Augmented Dickey-Fuller method (ADF), 29
Authority, 199, 200, 236, 237, 241
  aggregations, 205
  backlinks, 201, 202
  data and cleaning, 203
  data features, 206
  dataframe, 204, 212
  descriptive statistics, 206
  distribution, 207
  domain rating, 207
  domains, 204

Authority (*cont.*)

    links, 200

    math approach, 210

    rankings, 201

    search engines, 199

    SEO harder, 200

    SEO industry, 200

    spreadsheet, 202

Authority preoptimization, 69

Authority scores, 74, 75

Automation, 374, 563, 567

averageSessionDuration, 365

B

Backlink domain, 209, 210

Bayesian information criterion
    (BIC), 32, 33

Beige trench coats, 44

Best practices for webinars, 151

BlackHatWorld forums, 303

Box plot distribution, 87, 88, 90

C

Cannibalization, 469, 477, 512–520

Cannibalized SERP

    generic and brand hybrid
        keywords, 520

    keyword, 518

Categorical analysis, 108

Change point analysis, 437–440

Child nodes, 379–381, 385, 386, 390

Child URL node folders, 405

Chrome Developer Tools, 148

Click-through rate (CTR), 8

Cloud computing services, 5

Cloud web crawlers, 343

CLS_cwv_landscape_plt, 137

Cluster headings, 191–197

Clustering, 38–39, 191, 565

Clusters, 39, 52, 54

Column reallocation, 71–74, 76

Combining site level and page authority

    orphaned URLs, 110

    underindexed URLs, 111

    underlinked URLs, 110

Comparative averages and variations, 89

Competitive market, 57

Competitor analysis, 245

    AHREFs files, 266

    cache age, 272

    competitiveness, 255

    concat() function, 267

    crawl_path, 261

    dataframe, 257, 258

    derive new features, 270

    domain-wide features, 248

    groupby() function, 260

    keywords, 252

    linear regression, 247

    machine learning, 245

    merge() function, 269

    rank and search result, 256

    rank checking tool, 246

    ranking, 246

    ranking factors, 245, 247

    ranking pages, 260

    robust analysis, 245

    search engines, 248

    SEO analysis, 245

    SERPs data, 245, 254, 268

    string format columns, 265

    tag branding proportion, 247

    tracking code, 272

    variable, 246

visibility metric, 256

zero errors, 253

Competitor_count_cumsum_plt plot, 233

Competitors, 4, 104, 141, 160, 207, 255, 259

Computational advertising, 2

Content

content consolidation, 151

content creation, 151

data sources, 152

keyword mapping, 152–159

user query, 152

Content creation (planning landing page content)

cleaning and selecting headings, 187–191

cluster headings, 191–197

crawling, 179–182

extracting the headings, 182–188

hostname, 178

reflections, 197

SERP data, 176–182

TLD, 178

URLs, 175

verticals, 175

Content gap analysis

combinations function, 168

content gaps, 160

content intersection, 169–171

core content set, 160

dataframe, 172–174

getting the data, 161–168

list and set functions, 172

mapping customer demand, 160

search engines, 160

SEMRush files, 161

SEMRush site, 171

Content intersection, 169–171

Content management system (CMS), 293

Core Web Vitals (CWV), 298

Google initiative, 125

initiative, 63, 125–141, 298, 362

landscape, 125–134, 136, 138–141

onsite CWV, 141–150

technical SEO, 125

web developments, 125

Crawl data, 58, 59, 65, 78, 111, 117, 142, 154, 243, 268, 270, 401, 403, 454, 456

Crawl depth, 82, 85, 86, 91, 94

Crawling software, 65, 419

Crawling tools, 64, 152

Creating effective webinars, 194

Cumulative average backlink traffic, 230

Cumulative Layout Shift (CLS), 130, 137

Cumulative sum, 212, 215, 231, 232

Custom function, 218

CWV metric values, 126

CWV scores, 128, 144, 146, 362, 365

D

Dashboard

data sources, 343

ETL (*see* Extract, transform and load (ETL))

SEO, 367, 370, 563

types, data sources, 343

Data-driven approach

CWV, 63, 125–150

internal link optimization (*see* Internal link optimization)

modeling page authority, 63–76

Data-driven keyword research, 62

Data-driven SEO, 2, 63, 64, 151, 238

DataForSEO SERPs API, 40, 248, 351–356

Dataframe, 15, 18, 20, 21, 23, 25, 42, 43, 45,
    61, 62, 66, 67, 78, 79, 82, 93, 98, 130
Data post migration, 446, 454
Data science, 151, 566
    automatable, 5
    cheap, 5
    data rich, 4
Data sources, 7–8, 19, 152, 248, 343, 344,
    365, 469
Data visualization, 462, 483
Data warehouse, 300, 344, 345, 365,
    370, 563
Decision tree-based algorithm, 248,
    290, 565
Dedupe, 477–479
Deduplicate lists, 170
Defining ABM, 175
depthauth_stats_plt, 110
Describe() function, 225, 281, 283
Destination URLs, 117–119, 243, 402, 422
df.info(), 348
diag_conds, 463, 464
Diagnosis, 457, 458, 461, 463–465
Distilled ODN, 301, 311
Distributions, 16, 17, 63, 64, 67–70, 75, 76,
    84–88, 90, 100, 101, 103, 107, 111,
    145–150, 202, 207, 208, 212, 225,
    226, 240, 308, 310, 311, 316, 564
DNA sequencing, 153
Documentation, 435, 449, 450, 453, 463
Domain authority, 206–208, 216
Domain rating (DR), 201, 207–210, 212,
    215, 216, 221, 244, 249
Domains
    create new columns, 482
    device search result types, 485
    HubSpot, 481, 482
    rankings, 493, 494

reach, 479, 480
reach stratified, 485–493
rename columns, 481
separate panels by phase as
    parameter, 502
visibility, 496–504
WAVG search volume, 495, 496
WorkCast, 482, 483
drop_col function, 165

E

Eliminate NAs, 288–289
Experiment
    ab_assign_box_plt, 336
    ab_assign_log_box_plt, 338
    ab_assign_plt, 335
    ab_group, 339
    A/B group, 332
    ab_model.summary(), 339
    A/B tests, 327
    analytics data, 331
    array, 339
    dataframe, 329
    dataset, 332
    distribution, test group, 335
    histogram, 334
    hypothesis, 328
    outcomes, 340
    pd.concat dataframe, 333
    p-value, 340
    simul_abgroup_trend.head(), 333
    simul_abgroup_trend_plt, 334
    test_analytics_expanded, 331, 332
    test and control, 328
    test and control groups, 333, 334
    test and control sessions, 337
    website analytics software, 329

Experiment design
    A/A testing (*see* A/A testing)
    actual split test, 305
    APIs, 304
    dataframe, 306
    data types, 305
    distribution of sessions, 307
    Pandas dataframe, 306
    sample size
        basic principles, 320
        dataframe, 322
        factor, 320
        level of statistical significance, 322
        levels of significance, 322
        minimum URLs, 323, 324
        parameters, 320
        python_rzip function, 321
        run_simulations, 321
        SEO experiment, 320
        split_ab_dev dataframe, 327
        test and control groups, 326
        testing_days, 322
        test landing pages, 325, 326
        urls_control dataframe, 326
    standard deviation (sd) value, 307
    to_datetime() function, 306
    website analytics data, 305
    website analytics package, 305
    zero inflation, 308–311
Extract, transform and load (ETL),
    344, 375
    extract process, 345
        DataForSEO SERPs API,
        351, 353–356
        Google Analytics (GA), 345–348, 350
        Google Search Console (GSC),
        356–360, 362
        PageSpeed API, 362–365

loading data, 370–372
transforming data, 365–367, 369, 370

F

facet_wrap() function, 502, 504
FCP_cwv_landscape_plt, 138
FID_cwv_landscape_plt, 136
Financial securities, 2
First Contentful Paint (FCP), 129, 138, 148
First Input Delay (FID), 136, 150
Forecasts
    client pitches and reporting, 24
    decomposing, 27-29
    exploring your data, 25–27
    future, 35–38
    model test, 33–37
    SARIMA, 30–33
The future of SEO
    aggregation, 563
    clustering, 565
    distribution, 564
    machine learning (ML) modeling, 565
    SEO experts, 566
    set theory, 566
    string matching, 564, 565

G

geom_bar() function, 492
Geom_histogram function, 69
get_api_result, 352
getSTAT data, 471
Google, 1–4, 7, 29, 39, 54, 125, 132, 175,
    176, 191, 199, 200, 469
Google algorithm update
    cannibalization, 512–520
    dataset, 475

Google algorithm update (*cont.*)
  dedupe, 477–479
  domains (*see* Domains)
  getstat_after, 477
  getSTAT data, 471
  import SERPs data, getSTAT, 470
  keywords
    token length, 520–525
    token length deep dive, 525–533
  np.select() function, 474
  ON24, 471
  result types, 504–512
  segments
    np.select() function, 544
    snippets, 557–561
    top competitors, 544–550
    visibility, 551–557
  strip_subdomains, 473
  target level
    keywords, 533–536
    pages, 537–543
  urisplit function, 473
  zero search volumes, 474
Google Analytics (GA), 3, 344–348, 350,
    365, 375, 413, 437
  and GSC URLs, 418
  tabular exports, 413
  URLs, 417
  version 4, 345
Google Cloud Platform (GCP), 5, 128, 300,
    358, 362
Google Data Studio (GDS), 300, 344
Google PageSpeed API, 126, 345, 362
Google rank, 132, 133, 135, 136, 138, 247,
    259, 298, 300, 453, 492, 493, 504
Google Search Console (GSC), 3, 344, 345,
    356–360, 362, 416, 437, 444, 448,
    452–454, 460, 461, 469, 564

  activation, 18, 19
  data, 8
  data explore, 15–18
  filter and export to CSV, 18
  import, clean, and arrange the
    data, 9, 10
  position data into whole numbers, 12
  search queries, 8
  segment average and variation, 13–15
  segment by query type, 10
Google's knowledge, 191
Google Trends, 25, 30
  multiple keywords, 20–23
  ps4 and ps5, 38
  Python, 19
  single keywords, 19
  time series data, 19
  visualizing, 23, 24
GoToMeeting, 497, 498, 531
Groupby aggregation function, 67, 81
groupby() function, 158, 231, 260, 275,
    465, 512
gsc_ba_diag, 454
GSC traffic data, 426

H

Heading, 154, 175, 182–194, 271, 304
Heatmap, 111, 117, 463, 557, 560, 561
Hindering search engines, 151
HTTP protocol, 273
HubSpot, 480–482, 486, 528, 533
Hypothesis generation
  competitor analysis, 302
  conference events, 303
  industry peers, 303
  past experiment failures, 304
  recent website updates, 303

SEO performance, 302
social media, 302, 303
team's ideas, 303
website articles, 302, 303

I, J

ia_current_mapping, 395
Inbound internal links, 89, 105, 108
Inbound links, 77, 79, 89, 97, 98
Indexable URLs, 68, 73, 75, 117, 142, 145
Individual CWV metrics, 132
Inexact (data) science of SEO
    channel's diminishing value, 2
    high costs, 4
    lacking sample data, 2, 3
    making ads look, 2
    noisy feedback loop, 1
    things can't be measured, 3
Internal link optimization, 63, 150
    anchor text relevance, 117–125
    Anchor texts, 111–116
    content type, 107–111
    crawl dataframe, 79
    external inbound link data, 79
    hyperlinked URL's, 77
    inbound links, 77
    link dataframe, 78
    by page authority, 97–106
    probability, 77
    Sitebulb, 78
    Sitebulb auditing software, 77
    by site level, 81–97
    URLs, 79
    URLs with backlinks, 80
    website optimization, 77
Internal links distribution, 99
intlink_dist_plt plot, 89

K

keysv_df, 48
Keyword mapping
    approaches, 152
    definition, 152
    string matching, 153–159
Keyword research
    data-driven methods, 7
    data sources, 7
    forecasts, 24–38
    Google Search Console (GSC), 8–19
    Google Trends, 19–24
    search intent, 38–57
    SERP competitors, 57–62
Keywords, 533–536
    token length, 520–525
    token length deep dive, 525–533
Keywords_dict, 167, 169

L

LCP_cwv_landscape_plt plot, 134
Levenshtein distance, 46
Life insurance, 39
Linear models, 277
Link acquisition program, 237
Link capital, 235, 237
Link capital velocity, 238
Link checkers, 343
Link quality, 202, 206, 208, 209, 212, 216, 221, 225–231
Link velocity, 234, 235
Link volumes, 212, 231–233

Irrel_anchors, 118
Irrelevant anchors, 120–122
Irrelevant anchor texts, 121

Listdir() function, 217
Live Webcast Elite, 541
Live webinar, 536
Logarithmic scale, 87, 228
Logarized internal links, 90
log_intlinks, 89
log_pa, 101–103
Log page authority, 103
Long short-term memory (LSTM), 26
Looker Studio bar chart, 373
Looker Studio graph, 373, 374

M

Machine learning (ML), 152, 243, 245, 248,
    270, 274, 284, 292, 293, 296, 299,
    300, 565
Management, 449, 450
Management content, 463
Many-to-many relationship, 119
Marketing channels, 3
The mean, 494
Median, 89, 212, 225, 227, 283, 284
Medium, 483, 500, 528
melt() function, 184, 489
Mens jeans, 4
Metrics, 129, 144, 202, 224, 225, 249, 267,
    269, 292, 293, 346, 348, 367, 520
Migration forensics
    analysis impact, 442–454
    diagnostics, 454–463
    segmented time trends, 440–442
    segmenting URLs, 423–436
    time trends and change point
        analysis, 437–440
    traffic trend, 426–436
Migration mapping, 377, 412, 467
Migration planning, 412, 564

Migration URLs, 377, 394–396, 403, 404,
    406–408, 410–412, 467
MinMaxScaler(), 278
ML algorithm, 260, 295
ML model, 292, 293
ML modeling, 565
ML processes, 270
ML software library, 260
Modeling page authority, 150
    approach, 64
    calculating new distribution, 70–74, 76
    dataframe, 66
    examining authority
        distribution, 67–69
    filters, 66, 67
    Sitebulb desktop crawler, 65
Modeling SERPs, 289
Multicollinearity, 282
Multiple audit measurements, 3

N

Natural language processing (NLP), 377,
    389, 394, 412, 467
Near identical code, 130
Near Zero Variance (NZVs), 279
    API, 279
    highvar_variables, 280
    scaled_images column, 281
    search query, 280
    title_relevance, 281
new_branch, 396, 397
Non-CWV factors, 141
Nonindexable URLs, 68, 84
Nonnumeric columns, 277
np.select() function, 11, 71, 402, 458, 463,
    464, 474, 544, 547, 548

O

old_branch, 397, 398, 402

ON24, 470, 471, 480, 481, 486, 488, 493, 498, 499, 533, 536, 537, 541, 543

One hot encoding (OHE), 286–288

Online webinars, 536

Onsite indexable URLs, 142

Open source data science tools, 4, 5

Organic results, 2, 560

Orphaned URLs, 64, 82, 93, 110

ove_intlink_dist_plt, 84

P

pageauth_newdist_plt, 75

page_authority_dist_plt, 100, 101

Page authority level, 67, 107, 111

page_authority_trans_dist_plt, 103

PageRank, 67, 97, 98, 100, 101, 103, 105, 106

PageSpeed API, 126–128, 362–365

PageSpeed data, 129

Paid search ads, 39

Pandas dataframe, 50, 66, 252, 306

parent_child_map dataframe, 380, 384

parent_child_nodes, 379

Parent URL node folders, 405

Pattern identification, 144

PCMag, 500

perf_crawl, 456

perf_diags, 463

perf_recs dataframe, 463, 465

Phase, 491

Plot impressions *vs.* rank_bracket, 16, 17

plot intlink_dist_plt, 87

Power network, 238–241

PS4, 26–29, 34, 35, 38

PS5, 24, 26–29, 31, 34, 35, 38

Q

Quantile, 14–17, 91, 93

Query data *vs.* expected average, 15

"Quick and dirty" analysis, 107, 221

R

Random forest, 248, 290

Rank checking tool, 4, 246, 248, 391

Ranking factors, 245, 247, 249, 254, 260, 275, 281–283, 286, 291, 294–300, 469

Ranking position, 2, 8, 12–18, 246, 259, 279, 292, 293, 446, 473, 493, 495, 564

Rankings, 3, 39, 60, 125, 493, 494

RankScience, 301

Rank tracking costs, 39

Reach, 485

Reallocation authority, 69

Recurrent neural network (RNN), 564

Referring domains, 98, 204, 207, 209, 214–216, 219, 223–228, 231, 233, 239, 240, 243

Referring URL, 78, 119

Repetitive work, 5

Root Mean Squared Error (RMSE), 35, 292, 293

r-squared, 222, 292, 293, 340

S

Salesforce webinars, 536

SARIMA, 26, 30–33

Screaming Frog, 58, 249

Python, 11, 19, 202, 203, 566

Python code, 391

Search engine, 1, 2, 7, 63, 64, 66, 77, 97, 111, 122, 125, 151, 156, 160, 199, 212, 214, 228, 244, 246, 255, 303, 477, 566

Search engine optimization (SEO), 1–5, 7, 8, 13, 19, 54, 57, 63, 64, 76, 77, 85, 118, 151, 152, 200, 221, 238, 245, 260, 281, 289, 291, 295, 299, 300, 302, 303, 320, 341, 343, 345, 373, 565

Search Engine Results Pages (SERPs), 4, 16, 39–46, 50, 57, 58, 62, 126, 127, 176, 185, 191, 192, 194, 245, 248–255, 257, 260, 268, 469, 505

Search intent, 53, 192
    convert SERPs URL into string, 41–43
    core updates, 39
    DataForSEO's SERP API, 40
    keyword content mapping, 39
    Ladies trench coats, 39
    Life insurance, 39
    paid search ads, 39
    queries, 38
    rank tracking costs, 39
    SERPs comparison, 43–57
    Split-Apply-Combine (SAC), 41
    Trench coats, 39

Search query, 3, 8, 9, 11, 39, 246, 249, 280, 520

Search volume, 3, 48–50, 56, 253, 255, 471, 494–496, 520, 541, 544

Segment, 4, 11–15, 17, 145, 433, 436, 443, 448, 453, 454, 544

SEMRush, 57, 160–162, 171, 173, 201, 223, 566

semrush_csvs, 161

SEMRush domain, 222

SEMRush files, 161

SEMRush visibility, 222, 224, 231

SEO benefits, 125, 141

SEO campaigns and operations, 4

SEO manager, 85

SEO rank checking tool, 391

SERP competitors
    extract keywords from page title, 60, 61
    filter and clean data, 58–60
    SEMRush, 57
    SERPs data, 61, 62

SERP dataframe, 192

SERP results, 16, 518, 520

SERPs comparison, 43–57

SERPs data, 61, 62, 126, 390, 391, 394

SERPs model, 4

Serps_raw dataframe, 252

set_post_data, 352

Set theory, 566, 567

Single-level factor (SLFs), 274
    dataset, 275
    parameterized URLs, 276
    ranking URL titles, 274

SIS_cwv_landscape_plt, 133

Site architecture, 39, 108, 564

Sitebulb crawl data, 78, 142

Site depth, 64, 82, 90, 119, 152

Site migration, 377, 412, 454, 467

Snippets, 504, 505, 512, 557–561

Sorensen-Dice, 46, 118, 153, 422, 564

speed_ex_plt chart, 141

Speed Index Score (SIS), 130, 132

Speed score, 133, 146

Split A/B test, 293, 299, 301, 312

Split heading, 190

Standard deviations, 3, 8, 13, 225, 307, 366–368

Statistical distribution, 564

Statistically robust, 14, 245

stop_doms list, 490

String matching, 564, 565
    cross-product merge, 156
    dataframe, 155–157
    DNA sequencing, 153
    groupby() function, 158
    libraries, 153
    np.where(), 158
    simi column, 157
    Sitebulb, 154
    Sorensen-Dice, 153
    sorensen_dice function, 157
    string distance, 159
    to URLs, 154
    values, 156

String matching, 117, 152–159, 564–565

String methods, 250, 251

String similarity, 118, 156, 157, 394, 411

Structured Query Language (SQL), 343, 344, 370–373

T

target_CLS_plt, 147

target_crawl_unmigrated, 401

target_FCP_plt, 148

target_FID_plt, 150

target_LCP_plt, 149

target_speedDist_plt plot, 146

Technical SEO
    data-driven approach (*see* Data-driven approach)
    search engines and websites interaction, 63

Tech SEO diagnosis, 412, 460, 461, 466

TF-IDF, 152

Think vlookup/index match, 15

Time series data, 19, 23, 27–29, 412, 437

TLD extract package, 177

Token length, 253, 520–525

Token size, 521, 523

Top-level domain (TLD), 177, 178

Touched Interiors, 227

Traffic post migration, 453

Traffic/ranking changes, 377
    parent and child nodes, 379–385
    separate migration documents, 385–389
    site levels, 378
    site taxonomy/hierarchy, 378

Travel nodes, 386

Two-step aggregation approach, 521, 522

U

Underindexed URLs, 111

Underlinked page authority URLs
    optimal threshold, 104
    pageauth_agged_plt, 106
    PageRank, 105
    site-level approach, 106

Underlinked site-level URLs
    average internal links, 90, 91
    code exports, 97
    depth_uidx_plt, 95
    depth_uidx_prop_plt, 96
    intlinks_agged table, 96
    list comprehension, 94
    lower levels, 95
    orphaned URLs, 93
    percentile number, 90
    place marking, 94
    quantiles, 91, 93

Upper quantile, 15, 16

Urisplit() function, 263

URL by site level, 87

URL Rating, 66, 67
452 URLs, 420
URLs by site level, 83, 96, 97
URL strings, 41, 377, 389, 396, 398,
        405, 467
URL structure, 389, 395–398, 404, 406
URL visibility, 541, 543
User experience (UX), 151, 469
User query, 151, 152

V

Variance inflation factor (VIF), 282, 283
Visibility, 496–504, 531, 551–557
Visualization, 300, 373–374, 441, 462, 500,
        555, 557, 561

W

wavg_rank, 445, 495
wavg_rank_imps, 445

wavg_rank_sv() function, 552
WAVG search volume, 495–496
Webinar, 535
Webinar best practices, 191, 197
Webinar events, 536
Webmaster tools, 343
Webmaster World, 303
Website analytics, 305, 329, 343, 345, 541
Website analytics data, 305, 541
Winning benchmark, 245, 247, 250, 299
Wordcloud function, 124
WorkCast, 482, 483, 488, 489, 492, 499

X, Y

xbox series x, 24

Z

Zero inflation, 308–311
Zero string similarity, 394

Printed in the United States
by Baker & Taylor Publisher Services